Digital Innovations in Architecture, Engineering and Construction

Series Editors

Diogo Ribeiro⊙, Department of Civil Engineering, Polytechnic Institute of Porto, Porto, Portugal

M. Z. Naser, Glenn Department of Civil Engineering, Clemson University, Clemson, USA

Rudi Stouffs, Department of Architecture, National University of Singapore, Singapore, Singapore

Marzia Bolpagni, Mace Group, London, UK

The Architecture, Engineering and Construction (AEC) industry is experiencing an unprecedented transformation from conventional labor-intensive activities to automation using innovative digital technologies and processes. This new paradigm also requires systemic changes focused on social, economic and sustainability aspects. Within the scope of Industry 4.0, digital technologies are a key factor in interconnecting information between the physical built environment and the digital virtual ecosystem. The most advanced virtual ecosystems allow to simulate the built to enable a real-time data-driven decision-making. This Book Series promotes and expedites the dissemination of recent research, advances, and applications in the field of digital innovations in the AEC industry. Topics of interest include but are not limited to:

- Industrialization: digital fabrication, modularization, cobotics, lean.
- Material innovations: bio-inspired, nano and recycled materials.
- Reality capture: computer vision, photogrammetry, laser scanning, drones.
- Extended reality: augmented, virtual and mixed reality.
- Sustainability and circular building economy.
- Interoperability: building/city information modeling.
- Interactive and adaptive architecture.
- Computational design: data-driven, generative and performance-based design.
- Simulation and analysis: digital twins, virtual cities.
- Data analytics: artificial intelligence, machine/deep learning.
- Health and safety: mobile and wearable devices, QR codes, RFID.
- Big data: GIS, IoT, sensors, cloud computing.
- Smart transactions, cybersecurity, gamification, blockchain.
- Quality and project management, business models, legal prospective.
- Risk and disaster management.

Alberto Paoluzzi · Giorgio Scorzelli

BIM Geometry with Julia Plasm—Functional Language for CAD Programming

Volume 1: Mathematics and Software Engineering

 Springer

Alberto Paoluzzi
Roma Tre University
Rome, Italy

Giorgio Scorzelli
Scientific Computing and Imaging Institute
University of Utah
Salt Lake City, UT, USA

ISSN 2731-7269   ISSN 2731-7277 (electronic)
Digital Innovations in Architecture, Engineering and Construction
ISBN 978-3-031-90243-7   ISBN 978-3-031-90244-4 (eBook)
https://doi.org/10.1007/978-3-031-90244-4

This Springer imprint is published by the registered company Springer Nature Switzerland AG
The registered company address is: Gewerbestrasse 11, 6330 Cham, Switzerland

If disposing of this product, please recycle the paper.

Son of man, ... show them
the design and plan of the Temple, its exits and
entrances, its shape, how all of it is arranged, the
entire design and all its principles.

Give them all this in writing so that they can see
and take note of its design and the way it is all
arranged and carry it out.

(Ezekiel 43, 10–11)

Foreword

When I started my journey in the world of solid modeling, I was immediately fascinated by the ability of CAD software to describe and invent 3D shapes, whether of buildings, mechanical artifacts or natural features. The emphasis at the time was mostly on interactive techniques, made popular by the well liked "point and click" interface. However, as a young disciple of Professor Alberto Paoluzzi in his University of Rome laboratory, the questions that filled my mind were more about "what is the essence of computing with shapes": How do we best describe 3D objects? How do we express and compute the results of complex operations between them? Are the algorithms robust for any configuration of input? This book represents the culmination and synthesis of several decades of work by Prof. Paoluzzi and his group on their distinctive and original quest to answer those questions: a metalanguage to define and manipulate geometric shapes in a robust and expressive manner. This research has been pursued along two intertwined directions.

On one hand the book describes a strong, consistent theoretical framework that constitutes an algebra of shapes, leading to the ability to compute with shapes, combining them using powerful operators, and reasoning about them. This framework builds on a particular class of cell decomposition, the chain complexes. The result of this approach is the geometric modeling language Plasm. On the other hand, the authors have developed a continuous refinement of programming language-based approaches to solid modeling, exploring the expressivity of subsequent generations of programming paradigms such as symbolic, functional, and finally landing on a multi-paradigm, efficient language such as Julia. The advantages of a language-based approach to modeling are manifold. Expressions can be manipulated, combined and reused. They evolve to constitute a library of parameterized shapes, just like libraries of functions and APIs are used and reused in programming to build ever more complex software systems. The functional programming style adopted by the authors, pioneered by John Backus, adds an element of formal elegance, allowing concise, powerful expression of geometric concepts. This book comprehensively covers both the theory and practice of language-based geometric modeling. Researchers and practitioners will be guided through the underlying theoretical framework, both of the topology and geometry concepts as well as the functional language formalism, in an incremental, accessible manner. The many examples sprinkled throughout the book will make the expressive qualities of the approach self-evident. I feel that this work is particularly compelling at this time: advances in Artificial Intelligence are opening new possibilities in the exploration of a machine's ability to reason about the real world. To learn and interact with our three-dimensional physical words, most probably, the machine will need a model, which the machine will build via exploration and sensor data. While several paradigms for such representation are being proposed, an approach with built-in geometric consistency and language-based expressivity seems well poised to help advance the state of the art. I wholeheartedly recommend this book to any researchers or practitioners who, like me, are asking themselves: "what is the essence of computing with shapes"?

New York, january 2025   *Fausto Bernardini*

Preface

Imagine designing complex 3D geometries —from architectural marvels to intricate engineering models —with just a few lines of code. This is the power of Plasm, a multidimensional geometry engine that revolutionizes the way we think about programming for solid modeling. Developed under Italy's 'Edilizia' Finalized Project, Plasm emerged as a pioneering metalanguage for Building Information Modeling (BIM). Spearheaded by A.P. and G.S., it evolved across multiple programming languages while retaining its core focus on multidimensional geometry and functional programming.

The Programming Language for Solid Modeling began development at the end of the last century, created under the Italian Research Council (CNR) as a subproject of the "Edilizia" Finalized Project for Building Design and Construction. Plasm was the first and likely only metalanguage capable of generating data, code, and models for solid modeling design, particularly in Building Information Modeling, a specialized subdomain of Computer-Aided Design for the architecture, building, and construction sector. An outstanding CAD research group and lab emerged around this project, involving the University of Rome La Sapienza and several U.S. universities between the 20th and 21st centuries. A.P. served as the principal investigator in this effort, and G.S. graduated with honors, receiving a Matra OpenCascade award.

From the very start, Plasm was a uniquely different CAD system. Instead of the usual point-and-click interface, which was popular then and familiar to most users, it introduced a geometric metalanguage. Developers and professionals could experiment with a few lines of code using innovative primitive-shape generators and new prototyping methods for current designs, which could be stored in code libraries for future use. While A.P. used it in Mechanical, Civil, and Computer Engineering courses, G.S. migrated it from the original Common Lisp implementation to a mix of Scheme and C++ (for the geometry kernel), then to Python and C++, and finally to the current Julia development. This migration consistently maintained the original multidimensional and functional nature, a legacy of functional programming

(FP) and functional language (FL) pioneered by Backus and his team at IBM Research.

In the last ten years, the scientific programming field has witnessed the emergence of the Julia language, which is as fast as C and C++ but as user-friendly as Python and Matlab—porting PLaSM to Julia and giving the community a multidimensional geometry engine in the form of the Plasm.jl package was a natural step and much easier than we had anticipated. We hope it will provide the geometric tools that Julia currently lacks and more. While writing this book, we simultaneously developed the package and enjoyed Julia's powerful computing capabilities with sparse matrices and multi-framework environments (local, cloud, web).

Over many years of teaching, implementing, and applying graphics and geometric methods, algorithms, tools, and languages, we learned that young engineers and architects (as well as advanced high school students) are highly receptive to learning programming methods with languages where they can creatively construct and experiment 3D objects that originate in their minds. Spatial concepts can be rapidly tested and refined using a fast language that offers an interactive environment for checking and implementing ideas and exercises, even directly on a notebook. Therefore, our experience motivated us to write this book, in which we aimed to distinguish theoretical concepts from their practical, immediate application and visualization.

This book is primarily for developers and advanced professionals, such as software engineers and architects, who are creating interactive tools for BIM and CAD. It aims to help them provide platforms for parametric shapes using a highly effective yet straightforward tool. However, some sections are also beneficial for students eager to learn how to create shape prototypes (for example, in school projects associated with Architectural Composition, Design Methods, Built Environment, and Urban Design) that can subsequently be imported into a ray tracer to produce stunning visual effects.

Some familiarity with Julia and/or programming allows immediate text comprehension. Still, it is not needed because the Plasm language provides a simple, working introduction to higher-level functional programming, consistently enforcing the idea that any small program is a transformer between some data input and the corresponding program output. Many examples also aim to show the extreme simplicity of Plasm generative methods, which serve as short templates for readers' projects on creating shapes with symbolic language. Indeed, some mathematical knowledge will aid in the ongoing abstraction required to design and implement the geometric concepts the book aims to present to students and professionals for optimal results in creating highly complex shapes and environments. Specifically, given the brevity of each Plasm code compared to other methods for shape creation, the authors and the publisher should promote Plasm as a metalanguage for the metaverse.

One distinctive feature of the Plasm approach to geometric algorithms is its rejection of specialized and complex data structures. In contrast, previous solid modeling libraries use incompatible, intricately linked data structures that require highly skilled software developers. Instead, Plasm depends solely on cellular descriptions via chains of cells, represented as lists of indices to columns of coordinate matrices (for points) or binary arrays (for incidence relations). It has even been utilized with children to engage them in basic programming through geometric constructions.

Additionally, we would like to highlight that the Plasm package's non-monolithic structure significantly supported its development and upcoming maintenance. At the time of this writing, it leverages the collaboration of over 100 Julia helper packages, which aligns with the collaborative design of the extensive Julia ecosystem.

Chapter 1 is specifically designed for software developers unfamiliar with Julia. The second chapter introduces Plasm, showcasing some primitives with the highest compactness to engage younger audiences. Chapter three explores computational topology for software developers looking for a deeper understanding of Plasm's internal mechanics. Chapter four covers the fundamental concepts of geometry programming, focusing on the application of coordinate transformations necessary for mapping various project cells and chains. Chapter five explores more advanced patterns of shape generation, highlighting some unconventional shape parameterization permitted by the Plasm language. Chapter six outlines and illustrates the homological process that calculates the spatial partition generated by any set of geometric shapes. In contrast, chapter seven uses the concepts of generators and atoms within this spatial arrangement to present a new approach to Boolean solid algebras. In particular, it explores how Plasm simplifies any solid algebraic expression into an equivalent finite Boolean algebra form, extracting a binary chain of atoms, which can be easily reduced to a boundary representation model.

Plasm would not have been born without a chance encounter with two IBM researchers at a SIAM conference on geometric design at Phoenix. This was followed by the mail-in receipt of Backus' group's FL manual, which described the novel research language for programming at the function level. A.P. was captivated and immediately began theses with his top students to extend it with a geometric primitive type in Common Lisp. Two years later, the Plasm design was presented to his group in a seminar held in the Backus just-retired room at IBM Almaden. Numerous theses on geometric design were subsequently completed by computer engineering students at La Sapienza and Roma Tre Universities, culminating in the publication of the book Geometric Programming for Computer-Aided Design (GP4CAD) by Wiley and a SUR Award from IBM, which included a substantial donation of hardware and software for a new CADLAB at Roma Tre (2003). After 2000, other students ported Plasm from Common Lisp to Scheme, increasing its portability and didactic impact. In the same period, G.S. wrote a multidimensional geometric kernel in C++. He then ported it to Python a few years later and maintained it until the Julia embedding emerged.

The decades-long Plasm project involved around one hundred master's and Ph.D. theses at La Sapienza and Roma Tre universities. It would be im-

possible to acknowledge all the individuals who contributed and were trained
to conduct research alongside these scholars. Some mentors, friends, and col-
leagues warrant special mention: Gianfranco Carrara, Antonio Ruberti, Antonio Bottaro, John E. Hopcroft, Chandrajit Bajaj, Fausto Bernardini, Valerio
Pascucci, Michele Vicentino, Franco Milicchio, Antonio DiCarlo, and Vadim
Shapiro. Without any of them, this project would not have endured. The
Italian CNR, the Italian National Research Program, and IBM Corporation
provided significant funding for the project and the CADLAB, where many
young Italian software engineers were introduced to geometric computing.
We hope that the Julia Plasm software and this accompanying volume will
persist and offer the Julia community the best geometric development environment, particularly as a metalanguage for applying generative intelligence
in architectural design and construction engineering[1]. We wholeheartedly encourage readers to experiment with Plasm and contribute to its ecosystem.
We believe it can inspire architects and software engineers of all ages to explore creative applications of geometry programming.

You are invited to explore the Plasm.jl package through this book, experiment with its powerful generative methods, and contribute to its expanding
ecosystem. Together, they can shape the future of geometric intelligence.

Rome,   *Alberto Paoluzzi*
january 2025   *Giorgio Scorzelli*

---

[1] This book was partially supported by PR FESR Lazio 2021-2027 Program through
Project: "Digital Twin City" under Grant CUP: B83C23003970007.

Contents

1   Introduction to Julia Programming 1
  1.1 Basic syntax and type system 2
  1.2 Functions and collections 6
      1.2.1 Julia functions 7
      1.2.2 Collections 11
  1.3 Matrix computations.................................... 14
  1.4 Linear algebra and sparse arrays 17
  1.5 Parallel and distributed computing 21
      1.5.1 Parallel Programming............................. 21
      1.5.2 Multiprocessing and Distributed Computing 27
      1.5.3 Programming the GPU 28
  1.6 Modules and packages 32
  References 34

2   The Julia Package Plasm.jl 35
  2.1 Backus' functional programming........................ 36
  2.2 FL-based PLaSM in Julia syntax 37
  2.3 Geometric Programming at Function Level 40
  2.4 Plasm.jl modules and geometric types 47
  2.5 Geometric Programming Examples 53
  References 57

3   Topology primer 61
  3.1 Preliminaries.. 62
      3.1.1 Motivation 62
      3.1.2 Definitions 63
      3.1.3 Basic geometry.................................. 65
  3.2 Cellular models 66
      3.2.1 Cubical complex 69
      3.2.2 Simplicial complex............................... 77
      3.2.3 Product of complexes 90

3.3   From cells to chains and cochains 91
3.4   Chain complex 92
       3.4.1   Linear chain spaces 94
       3.4.2   Linear chain operators 96
3.5   Cochain complex 101
       3.5.1   Cochain integration with Plasm 103
       3.5.2   Mechanical properties with structure products 107
References 113

4 Geometric models 115
   4.1   Geometric Transformations 116
          4.1.1   Vector space 116
          4.1.2   Affine space 122
          4.1.3   Convex space 124
          4.1.4   Space Transformations 126
   4.2   Representation Schemes 137
          4.2.1   Boundary 138
          4.2.2   Decompositive 141
          4.2.3   Enumerative 141
          4.2.4   Constructive Solid Geometry (CSG) 144
          4.2.5   Primitive Instancing 146
   4.3   Assembly of geometric objects 147
          4.3.1   Hierarchical graphs 147
          4.3.2   Hierarchical structures in Plasm 150
          4.3.3   Structure graphs in Hpc and STRUCT objects 153
   4.4   Attach properties to geometry 156
   References 157

5 Symbolic modeling with Julia Plasm.jl 159
   5.1   Parametric primitives 160
          5.1.1   Generators of solid classes 163
          5.1.2   Primitive design devices 174
   5.2   Topological operators 183
          5.2.1   Incidence operators 184
          5.2.2   Adiacency operators 186
          5.2.3   Atomic decomposition 187
   5.3   Parametric manifold mapping 189
          5.3.1   Curve generation methods 195
          5.3.2   Surface generation methods 201
          5.3.3   Solid generation methods 209
   References 214

6   Space arrangement pipeline 215
  6.1   Cellular and boundary models 216
          6.1.1   Cellular models 216
          6.1.2   Boundary models.............................. 217
          6.1.3   Julia Plasm representation scheme 218
  6.2   Arrangement of space 221
          6.2.1   Homology of spaces 221
          6.2.2   Introduction to arrangement pipeline 224
  6.3   Homological pipeline 227
          6.3.1   Starting from a comic tale 228
          6.3.2   Generators assembly 228
          6.3.3   Two-dimensional splitting tasks 229
          6.3.4   From two to three dimensions 230
          6.3.5   Topological Gift Wrapping....................... 231
  6.4   Commented examples 233
          6.4.1   Boundary 237
          6.4.2   Decompositive 237
          6.4.3   Enumerative 237
          6.4.4   Constructive Solid Geometry (CSG) 239
          6.4.5   Primitive Instancing 240
  References 241

7   Boolean solid algebras 243
  7.1   Plasm and solid representations........................ 244
  7.2   Finite Boolean Algebras 250
          7.2.1   Solid algebra expression resolution 253
  7.3   Subspaces of cycles and boundaries 256
          7.3.1   Plasm Boolean pipeline......................... 256
          7.3.2   Chain, cycle, and boundary subspaces.............. 257
          7.3.3   Compendium of homological method................ 260
  7.4   Solid and Boolean Algebras 264
          7.4.1   Solid algebraic expressions...................... 264
          7.4.2   Solid operations in Plasm expressions 270
          7.4.3   Plasm examples 271
  7.5   Atoms × generators = truth table 277
          7.5.1   Julia and Boolean operations 277
          7.5.2   Stepwise Boolean example computation 279
  References 286

Index 291

Chapter 1
Introduction to Julia Programming

This chapter introduces Julia's multiparadigm approach to programming, which uniquely blends high-level expressiveness with remarkable computational speed. Although not intended as a programming tutorial, the chapter assumes familiarity with basic coding concepts from other languages.

Programming languages, broadly speaking, can be categorized into subsets based on their programming paradigms—distinct styles that dictate how computations are structured, how data is managed, and how problems are solved. Some languages strictly adhere to a single paradigm, while others embrace flexibility by supporting multiple paradigms. For example, Smalltalk epitomizes object-oriented programming, Haskell and ML champion functional programming, and Prolog exemplifies declarative programming. By contrast, languages like C++, Java, Python, and Julia adopt a multiparadigm philosophy, enabling developers to choose or combine styles as needed.

Julia stands out as a language crafted with scientific computing in mind. It achieves a rare synergy between performance, simplicity, and expressiveness, addressing the long-standing two-language problem [3]. By unifying rapid prototyping and high-performance optimization within a single framework, Julia empowers users to write code that is both elegant and efficient.

A. Paoluzzi and G. Scorzelli, *BIM Geometry with Julia Plasm—Functional Language for CAD Programming*, Digital Innovations in Architecture, Engineering and Construction, https://doi.org/10.1007/978-3-031-90244-4_1

1.1 Basic syntax and type system

Julia combines the user-friendliness of productivity languages like Python and MATLAB with the performance features of languages like C++ and Fortran. Syntactically, Julia is straightforward to write and fast to debug, and it also benefits from a robust ecosystem of packages.

Basic syntax

Julia closely resembles Python in general-purpose programming and MAT-LAB in performing algebraic operations with matrices and vectors. Readers only need a few basic syntax concepts to understand the scripts presented in this book. For a detailed comparison of Julia, Python, and R, including various scripting examples, readers can consult the web at the URL string https://cheatsheets.quantecon.org pages.

Comment syntax and behavior: Comments in Julia begin with the character #. Any text that follows this symbol on the same line is considered a comment and is ignored by the compiler. Multiline comments, or block comments, begin with #= and end with the matching =#.

Numeric types and syntax: Julia supports several basic numeric types, including `Int64`, `Float64`, `Complex{Float64}`, and `Rational{Int64}`. The `String` values are enclosed in double quotes, for example, "string", while `Char` literals are written with single quotes, such as 'a' and 'b'. Julia also offers a full range of arithmetic infix operators, bitwise operators, Boolean values and operations, and primitive comparison operators.

Julia as a functional language: Although Julia is a multiparadigm language, it can accurately be described as a functional language because everything in Julia is an expression that evaluates to a value.

Variables

Variables are initialized when a value is assigned to them. Accessing an undeclared variable results in an error. Variable names must begin with a letter or underscore and can consist of any combination of alphanumeric characters, including underscores and exclamation marks. Additionally, many Unicode characters, such as Greek letters like π or mathematical symbols like $\in$, can be used by typing their LaTeX names followed by the Tab key.

Naming Conventions

The following naming conventions are recommended:

1. For long names, word separation can be done using underscores, but it's generally discouraged unless the name would otherwise be difficult to read.
2. `Type` names should begin with a capital letter and employ *UpperCamel-Case* for word separation, as in `AbstractFloat`.
3. `Function` and macro names should be in lowercase and should not use underscores.
4. Mutating functions or in-place functions that modify their input should have names that end with !. However, this practice is rarely suitable for large data structures.

Control flow

Julia provides a variety of control flow constructs to manage the flow of your program:

Compound expressions: Use the `begin-end` block or the `;` separator to group multiple expressions together.

Conditional Evaluation: Use the construct `if-elseif-else` or the ternary operator `<>?<>:<>` to make decisions based on conditions.

Short-circuit evaluation: Logical operators `&&` (and) and `||` (or) can be chained to evaluate conditions efficiently.

Repeated evaluation: Loops are available in Julia, including `while` and `for` constructs, which allow you to iterate over values or execute code blocks repeatedly.

for. . .end: loops iterate over iterables;

while. . .end: lops while a condition predicate is `true`, and terminates when the predicate becomes `false`.

Iterable types: include `Range`, `Array`, `Set`, `Dict`, and `AbstractString`. In loops, you can use `in` instead of range generators;

Exception handling: Use `try-catch` blocks to handle unexpected conditions, and `error` and `throw` functions to raise exceptions. Exceptions are thrown when an unexpected condition occurs.

Tasks (coroutines): Julia's tasks are a type of generalized subroutine that can be suspended and resumed using the `yield` to function. This allows for efficient and flexible execution of tasks.

Type system

Julia has a powerful type system where every value has a type, but the variables do not. To determine the type of a value, you can use the `typeof` function:

```
abcd = "abcd"
typeof(abcd)   # => String
```

In Julia, types are first-class values, meaning they can be passed as arguments to functions or returned from them. The type `DataType` represents all types, including itself.

Julia's type system is dynamic, yet it offers some of the benefits of static type systems. Julia can generate highly efficient code by allowing values to be annotated with specific types and enabling type reasoning. More importantly, it deeply integrates method dispatch based on the types of function arguments in the language.

In Julia, a type is concrete if it can be instantiated; that is, some type `T` is concrete if there exists at least one value `v` such that `typeof(v)` # => `T`. Abstract types cannot be instantiated; they are used to create hierarchies of types, useful to generate fast code. Other types can parameterize both abstract and concrete types. They can also be parameterized by symbols and values of types such that `isbits` returns true.

Hierarchical relationships between types are explicitly declared. One distinctive feature of Julia's type system is that concrete types may not subtype each other. All concrete types are final: may only have abstract types as supertypes.

Users can define new types in Julia, similar to records or structs in other languages. These types are created using the `struct` keyword. Julia automatically provides a default constructor, a function named after the type. The arguments to this constructor correspond to the properties of the type, listed in the same order as in the type definition [13]. These struct-based types are concrete types. They can be instantiated but cannot have subtypes.

The other category of types is abstract types, defined as:

```
abstract <name> end
```

Abstract types cannot be instantiated but can have subtypes. For example, as its name suggests, `AbstractString` is an abstract type.

Type relations and hierarchy

The method `supertype(T::DataType)` returns the `DataType` father of `T`. The subtype operator is a predicate `T1 <: T2` which returns `true` when values of type `T1` are also of type `T2`.

The supertype operator `T1 >: T2` is equivalent to `T2 <: T1`. Two examples follow (see also Figure 1.1):

```
subtypes # => (generic function with two methods)
subtypes(Number) # => [Complex, Real]|
```

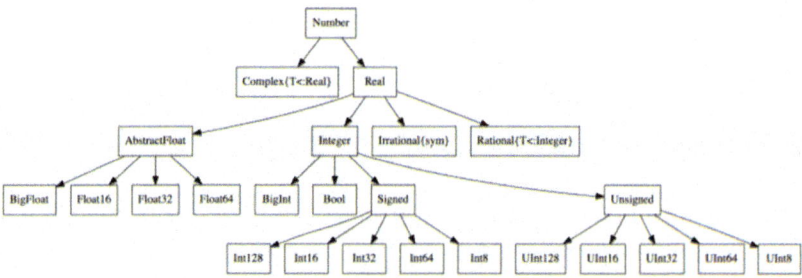

Fig. 1.1 The hierarchy tree rooted by Julia **Number** type.

Types in Julia serve several purposes: they are utilized for documentation, optimization, and method dispatch. Although types are not statically checked, they assist the JIT compiler in generating faster code.

The :: operator can attach type annotations to expressions and variables. However, as a general rule, avoiding type annotations and allowing the compiler to infer types is better. Writing generic code improves flexibility and makes it easier to interact with other packages.

The **abstract type** keyword declares a type that cannot be instantiated. Instead, it serves as a node in the type graph, representing sets of related concrete types, specifically its descendant types. Abstract types define the conceptual hierarchy that makes Julia's type system more than just a collection of object implementations.

For example, **Number** has the supertype **Any**, while **Real** is an abstract subtype of **Number** (see Figure 1.1).

A **primitive** type declares a concrete type whose data consists solely of a series of bits. Classic examples of primitive types include integers and floating-point values. Below are examples of built-in primitive-type declarations from Julia's implementation:

```
primitive type Int32 <: Signed32 end
primitive type Float64 <: AbstractFloat64 end
primitive type Char32 end
primitive type Bool <: Integer8 end
```

Type declarations can be applied in the **global** scope by adding type annotations to global variables. This practice helps ensure type stability, which is crucial for optimizing performance, as it allows the compiler to generate more efficient code by avoiding unnecessary runtime type checks (for a detailed explanation of type stability, see [17]).

```
julia> x :: Int = 10
10
julia> x = 3.5
```

```
ERROR: InexactError: Int64(3.5)
```

1.2 Functions and collections

The core of Julia's programming paradigm is built around its collection types, which are the fundamental data structures that hold and manipulate data. Such collection types include:

Arrays: A collection of elements of the same type, stored in a contiguous memory block. Arrays are the most common and widely used collection type in Julia.

Tuples: An immutable collection of elements of different types stored in a single memory block. Tuples are similar to arrays but immutable, meaning their contents cannot be changed after creation.

Dictionaries (or Hashes):   A collection of key-value pairs, where each key is unique and maps to a specific value. Dictionaries are used to store and retrieve data efficiently.

Sets: An unordered collection of unique elements, which can be used to perform set operations such as union, intersection, and difference.

These collection types are the building blocks of Julia's programming model, and provide a flexible and efficient way to work with data.

Programming model

In addition to collection types, Julia's `Function` type plays a vital role in the language. Functions are first-class citizens in Julia, meaning they can be passed as arguments to other functions, returned as values from functions, and stored in data structures. These functions define computations that take input data and produce output results. They are used to implement algorithms, conduct data processing, and construct complex logic.

The significance of `Function` type in Julia lies in its ability to represent any computable transformation from input data to output results. This means that functions can be used to model various problems, from simple arithmetic operations to complex scientific simulations. The Function type is also used to implement Julia's macro system, allowing developers to extend the language.

Overall, combining collection types and `Function` type provides a powerful and flexible foundation for Julia's programming model, making it an ideal language for a wide range of applications, from data analysis and machine learning to scientific computing and web development.

1.2.1 Julia functions

In Julia, a function is an object that maps a tuple of argument values to a
return value. Without the keyword, it returns the last computed value to
the calling function. All arguments to functions are passed by reference (i.e.,
by memory address). Julia functions are not pure mathematical functions
because they can change and be influenced by the global state of the program.

Definition 1.1 (Basic syntax for defining functions)

```julia
julia> function f(x,y)
           x + y
end
f (generic function with 1 method)
```

A generic function can be used for different formal arguments. In simple
words, whenever a function is defined with arguments of a new type, the Julia
compiler will generate a different version, called method for that function.
The same happens when a function name is invoked with various numbers
and types of arguments. See Julia Multiple Dispatch [16].

The keyword 'function' is used to create new functions. Functions return
the value of their final expression.

Definition 1.2 (Statement functions) There is a concise definition of
functions, similar to statement functions in older Fortran. Here is the definition with formal arguments and its application to actual argument values:

```julia
julia> f_add(x,y) = x + y   #=
f_add (generic function with 1 method)    =#

julia> f_add(2,3) # => 5
```

Definition 1.3 (Multiple return values) Functions can also return multiple values, as a tuple.

```julia
julia> f(x,y) = x+y, x-y, x*y    #=
f (generic function with 1 method)    =#

julia> a,b,c = f(2,4)
(6, -2, 8)
```

Definition 1.4 (Variable number of arguments) You may define functions whose < head > takes a variable number of positional arguments:

```
function varargs(args...) <body> end
```

of course followed by a body block of instructions and by the end token.

The token "..." is called a splat. We just used it to define a function head. The splat can also be used in a function call, where it will splat an array or tuple contents into the argument list:
add(list...) is equivalent to add(5,6,7,8,9) when list = [5,6,7,8,9].

Definition 1.5 (Optional positional arguments) You may define functions in Julia with positional arguments optional. With an assigned default value and hence not necessarily with an assigned value at runtime.

```
julia> function defaults(a, b, x=5, y=6)
 return "$a $b and $x $y"
 end
defaults (generic function with three methods)
\index{Functions!Optional positional arguments!default value} \
    index{Functions!default!}

julia> defaults(0, 0)
"0 0 and 5 10"
```

The $ is used to interpolate the values of variables or expressions into strings.

Definition 1.6 (keyword-optional arguments) You may also define functions that take keyword-optional arguments.

```
julia> function keyword_args(;k1=4, name2="hello")
           return Dict("k1" => k1, "name2" => name2)
       end
keyword_args (generic function with 1 method)
```

Note the ; character before the optional arguments

Coding 1.2.1 (Function call) A corresponding function call follows, showing the parameter names **k1** and **name2** together with the assigned values, used to increase the code readability:

```
julia> keyword_args(name2="ciao")
Dict{String, Any} with 2 entries:
  "name2" => "ciao"
  "k1"    => 4
```

You can combine all types of arguments in the same function, with keyword arguments defined in the last positions, following the ; character. They can be invoked in any quantity and order, and may be replaced by the default value.   □

Functional programming style

Julia allows the programmer to efficiently use several functional programming style traits, as shown in the following section.

Coding 1.2.2 In other words, Julia has first class functions. In the following scriptlet, the function `create_adder` returns an `adder` function:

```julia
function create_adder(x)
    adder = function(y)
        return x + y
    end
    return adder
end
create_adder(3)(4)   # => 7
```

Above, the `adder` variable contains an unnamed value of the `DataType` type and returns it, which will be used to retrieve the second actual parameter value in function applications. It is also possible to name the internal function if desired:

```julia
function create_adder(x)
    function adder(y)
        x + y
    end
    adder
end
```

To be called, e.g., `add_10 = create_adder(10); add_10(3)` #=> 13.

Definition 1.7 (lambda syntax) The lambda syntax or "stabby lambda syntax" is used to create anonymous functions: `(x -> x + 2)(3)` # => 5, where the lambda expression is `x -> x + 2`. The arguments are before the characters "`->`", and after the stab we have the value-generating expression.

```julia
function create_adder(x)
    y -> x + y
end
```

Of course, it is possible to use a tuple of arguments in the lambda form of functions: `((x,y) -> x + y)(2, 3)` # => 5. This function is identical to the `create_adder` implementation above.

Coding 1.2.3 Even more, like a proper functional language, with curried parameters, we may have:

```julia
add = (x -> y -> x + y);
add(2)(3) # => 5.
```

A curried function is a function that takes multiple parameters one at a time by taking the first argument and returning a series of functions, each of which takes the following argument until all the parameters have been fixed. The function application will be completed when the resulting value is returned. Note that the number of arrows is equal to the number of applications.   □

Julia higher-order functions

The map(), filter(), and reduce() functions are three fundamental higher-order functions that are found in almost every programming language today. In Julia, we have this syntax for built-in higher-order functions:

```
map(add_10, [1,2,3])              # => [11, 12, 13]
filter(x -> x > 5, [3, 4, 5, 6, 7]) # => [6, 7]
reduce(*, [2; 3; 4]; init=-1)     # => -24
```

Remark 1.1 (Sintax) Note two different uses of semicolon: within the Array value stands for column elements; after it provides named default values.

reduce takes two arguments: a function f and a collection A. The function f must accept two parameters, and then reduce iterates through the collection, updating result = f(result, elt) with one element at a time. The keyword argument init specifies the initial value for the reductions. For +, *, max, and min, the init argument is optional.

```
julia> a = reshape(Vector(1:16), (4,4)) #=
  4×4 Matrix{Int64}:
   1   5    9   13
   2   6   10   14
   3   7   11   15
   4   8   12   16    =#
julia> reduce(max, a, dims=2)        #=
  4×1 Matrix{Int64}:
   13
   14
   15
   16     =#
julia> reduce(max, a, dims=1)        #=
  1×4 Matrix{Int64}:
   4   8   12   16       =#
```

You may also consider using foldl or foldr, with fixed associativity direction: foldr(op, it; [init]) is like reduce, but with guaranteed right associativity. Or foldl(op, itr; [init]) is like reduce, but with guaranteed left associativity, foldr(*, 1:5; init=1) # => 120 in this case also `foldr(*, 1:5)` `# => 120`. Such recursive functionals transform any binary operator into a *n*-ary operator and might implement many geometric programming patterns. The last may be used to implement factorial for small positive integer numbers.

Remark 1.2 When redefining a method or adding new methods, it is essential to realize that these changes do not take effect immediately.

This is key to Julia's ability to statically infer and compile code to run fast without the usual Just In Time (JIT) tricks and overhead. Indeed, any new method definition will not be immediately visible to the current runtime environment, including `Tasks` and `Threads` (and any previously defined `@generated` functions, an excellent metaprogramming tool [7]).

1.2.2 Collections

The Julia `Base` package includes a variety of functions and macros that are suitable for scientific and numerical computing, yet it is as extensive as many general-purpose programming languages. Additional functionality is accessible from a growing collection of available packages.

Arrays, tuples, dictionaries, and sets are the proper Julia collections. All of them (except for sets) can be accessed using integer indices. User-defined collections must adhere to the Iterable Collections protocol.

Beware, Julia indexes everything starting from `1` (like MATLAB and Fortran), not `0` (like most languages, including C, C++, Rust, and Java). Additionally, iterating over collections is recommended (`map`, `for` loops, etc.). The notation `$(expr)` can be used to interpolate a value inside a string, making complex printing quite straightforward. Note that parentheses `()` are mandatory when `expr` is not a single token.

Arrays

Julia offers a complete set of basic arithmetic and bitwise operators for all its numeric primitive types, along with portable, efficient implementations of a wide range of standard mathematical functions.

AbstractArray{T, N}   serves as the supertype for N-dimensional arrays (or array-like types) that contain elements of the parametric type T and have a dimension of the parametric type N. `Array` and other types are subtypes of this.

AbstractVector{T} and AbstractMatrix{T}   are the supertype for one- or two-dimensional arrays (or array-like types) with elements of type T. They are aliases for `AbstractArray{T,1}` and `AbstractArray{T,2}`, respectively.

A broadcast dot operator (`.`) maps any Julia operator or function to all elements of a collection: `f.[a,b,c]` => `[f(a),f(b),f(c)]`.

Tuples

In Julia, `typeof(Tuple)` # => `DataType` is an immutable collection of distinct values that can be of the same or different data types, separated by commas.

Tuples are a heterogeneous collection of values. They resemble arrays in Julia, except that arrays can only contain values of the same `Datatype`. The values within a tuple cannot be changed because tuples are immutable. The entire tuple assigned to a variable can only be replaced with a new tuple value.

The sequence of values stored in a tuple can be of any type, with integers indexing them. Although it's not necessary, defining a tuple with parentheses around the sequence of values is encouraged. This makes it easier to understand Julia's tuples.

Coding 1.2.4 The `tuple` function returns a tuple from given objects:

```
tuple(1, 'b', pi) # => (1, 'b', π).
```

Coding 1.2.5 The function `ntuple(f :: Function, n :: Integer)` creates a tuple of length n, computing each element as `f(i)`, where `i` is the index of the element:

```
ntuple(i -> 2i, 4) # => (2, 4, 6, 8).
```

Dictionary

A lookup table is a simple yet effective method for organizing various types of data. Given a single piece of information—referred to as a key—such as a number, string, or symbol, the goal is to retrieve its corresponding value. This functionality is provided by the `Dictionary` object, referred to as `Dict` in Julia. It is described as an associative collection because it maps keys to values.

Definition 1.8 (Standard dictionary type Dict) The Julia implementation uses `hash` as the hashing function for the key and `isequal` to determine the equality.

The Julia developer can redefine these two functions for custom types to override how they are stored in a hash table. Any hash function must compute an integer hash code so that isequal(x,y) implies hash(x) == hash(y).

Dictionaries can be created by passing pair objects constructed with => to a Dict constructor: Dict("A"=>1, "B"=>2). This call will attempt to infer type information from the keys and values (specifically, this example creates a Dict{String, Int64}). To explicitly specify types, use the syntax Dict{KeyType,ValueType}().

Coding 1.2.6 Construction syntax directly using pairs key => value:

```
julia> Dict{String, Int32}("A"=>1, "B"=>2)    #=
Dict{String, Int32} with 2 entries:
  "B" => 2
  "A" => 1              =#
```

Coding 1.2.7 Dictionaries can also be created using generators:

```
julia> f = i->2i; h = Dict(i => f(i) for i=1:10)    #=
Dict{Int64, Int64} with 10 entries:
  5 => 10
  4 => 8
  ⋮ => ⋮              =#
```

Remark 1.3 (Generators) In Julia, a generator is a special function that creates a sequence of values on the fly without storing them all in memory at once. Generators are often used to create large datasets, iterate over complex data structures, or perform computations requiring large data processing.

Remark 1.4 (Unordered dataset) Note that the pairs in a Dict are not ordered according to the generation sequence. OrderedDict in DataStructures package is used for this purpose.

Given a dictionary D, the syntax D[x] returns the value associated with the key x (if it exists) or throws an error, and D[x] = y stores the key-value pair x => y in D, replacing any existing value for the key x.

Multiple arguments to D[...] are converted to tuples; for example, the syntax D[x,y] is equivalent to D[(x,y)], i.e., it refers to the value keyed by the tuple (x,y).

Set data type

In Julia, a Set is an unordered collection of unique elements, similar to a Python set or a Ruby set. Sets in Julia are implemented as hash tables, making them efficient for membership testing and insertion. Julia's `Set` type is immutable by default but can be converted to a mutable `Set` using the `mutable` function.

Definition 1.9 Sets are mutable containers that provide fast membership and insertion testing. The type `Set` is derived from `AbstractSet`type: `Set{T } <: AbstractSet{T}` .

`Set` datatype enjoys efficient implementations of set operations such as `in`, `union`, and `intersect`. Elements in a `Set` are unique, as determined by the element definition of `isequal` . The order of elements in a `Set` is an implementation detail and cannot be relied on.

In Julia, `Set`s supports various operators such as union ($\cup$), intersection ($\cap$), and difference ($\backslash$) using the `union`, `intersect`, and `setdiff` functions, respectively, allowing for efficient manipulation of sets.

Coding 1.2.8 Some `Set` examples follow:

```
julia> s = Set("aaBca")      #=
Set{Char} with 3 elements:
  'a'
  'c'
  'B'     =#
julia> push!(s, 'b')      #=
Set{Char} with 4 elements:
  'a'
  'c'
  'b'
  'B'     =#
julia> union([4 2 3 4 4], 1:3, 3.0)      #=
4-element Vector{Float64}:
  4.0
  2.0
  3.0
  1.0     =#
```

Let's note the type promotion of all `Set` elements to the same type `Float64`.□

1.3 Matrix computations

A primitive type is a concrete type whose data consist of old bits. Classic examples of primitive types are integers and floating-point values. Julia's primitive numeric types called bits types (integers, both signed and unsigned, Booleans, and floats) are `Int8`, `UInt8`, `Int16`, `UInt16`, `Int32`, `UInt32`, `Int64`, `UInt64`, `Int128`, `UInt128`, `Bool`, `Float16`, `Float32`, `Float64`, depending on the number of bits. Additionally, full support for `Complex` and `Complex` numbers is built on top of these primitive numeric types.

Vector and Matrix Types

Array *type*

In Julia, the `Array` type holds both "bits" values as well as "boxed" values. Boxed variables are heap-allocated and tracked by the garbage collector. The distinction lies in whether the value is stored inline (in the directly allocated memory of the array, meaning without following pointers) or if the array's memory is merely a collection of pointers to objects allocated elsewhere. In terms of performance, accessing values inline has a significant advantage over following a pointer to the actual values

Julia provides a first-class array implementation without treating arrays in any special way. The array library is implemented almost entirely in Julia itself and derives its performance from the compiler, just like any other code written in Julia.

Vector *and* Matrix

The `Vector{T}` type is a 1-dimensional dense array with elements of type T, often used to represent a mathematical vector. It is an alias for `Array{T,1}`. The method `Vector{T}(undef, n)` constructs an uninitialized object `Array{T,1}` of length n. Similarly, the `Base.Matrix{T}` type is an alias for `Array{T,2}` in the `Base` package.

1. The parameterized type `Matrix{T}` `<:` `AbstractMatrix{T}` is a two-dimensional dense array with elements of type T, frequently used to represent a mathematical matrix.
2. The `Matrix{T}(undef,m,n)` method constructs an uninitialized `Array{T,2}` of size $m \times n$.

In Julia, `Matrix{Number}` is a type that represents a matrix of numbers, where `Number` is a type parameter that can be any numeric type, such as `Int`, `Float64`, `Complex`, or `Rational`. It provides various methods for matrix operations, such as:

1. Basic arithmetics such as addition, subtraction, multiplication, and division.
2. Matrix multiplication and matrix-vector multiplication.
3. Transpose, conjugate transpose, and other matrix transformations.
4. Solving systems of linear equations.
5. Eigenvalue decomposition and other matrix factorizations.

This type is a core component of Julia's numerical computing capabilities, offering a flexible and efficient method for working with matrices in scientific computing, linear algebra, and machine learning applications.

Vector *operations*

We see here how to create and manipulate vectors in Julia and how Julia notation differs from mathematical notation.

To create the 3-vector $x = (8, -4, 3.5) = \begin{pmatrix} 8 \\ -4 \\ 3.5 \end{pmatrix}$, use: x = [8,-4,3.5], but

x = [8;-4;3.5] also works.

Be careful for similar-looking expressions, because (8,-4,3.5), [8,-4,3.5] and [8 -4 3.5] are not equivalent in Julia. They are a tuple, a column vector, and a row matrix, respectively.

To obtain an integer range from i to j (for $i \leq j$), we can use a colon iterator i:j. The assignment x = collect(1:10) returns the array x. To specify an increment size, include an increment argument in the middle:

```
x = collect(1:0.1:10)' #= >
[1.0 1.1 1.2 1.3 1.4 … 9.4 9.5 9.6 9.7 9.8 9.9 10.0].      =#
```

The range from 1 to 10 with a step size of 0.1 is shown above. Take note of the apex for transposition, as it is necessary only for typographical reasons here.

1. Indexes run from 1 to n = length(x), and x_2 is x[2];
2. you can also set an element, e.g., x[3] = 10.5;
3. use a range to select more than one element; x[2:3] selects the second and third elements;
4. x[end] selects the last element;
5. to select the even significant elements of x use

```
x[1:2:end]' # => [1.0 1.2 1.4 … 9.6 9.8 10.0].
```

To form a **stacked** Vector made by vectors a=[1;2] and b=[3;4;5] use w=[a...,b...] or w=[a;b].

The expression w = [a,b] would return a Vector{Vector{Int64}} value with 2 elements of type Vector{Int64}. Stacked vectors can be used as a list of lists. To access an element in w = [a,b] use w[2][2] # => 4

Many more vector operations are defined in the Base package. In Julia, a scalar and a vector can be added using the dot (broadcast) operator. The scalar is added to each entry of the vector: [2,4,8] .+ 3 # => [5,7,11].

Scalar-vector multiplication uses * because the operator is linear:

[2,4,8] * 3 # => [6,12,24]. Both expressions are commutative.

In Julia syntax, like in MATLAB and differently from Python: + and * operate on congruent arrays, otherwise promote the arguments to same type.

Matrix *operations*

Julia provides a concise and intuitive syntax for creating matrices. A matrix can be defined using the following notation: `A = [1 2 3; 4 5 6]`. In this syntax, spaces separate the elements within a row, and semicolons separate the rows. Additionally, the `size` function can be utilized to obtain the dimensions of a matrix, such as the number of rows (`m`) and columns (`n`), as illustrated in

```
size(A) # => (2, 3).
```

The matrix-building syntax, which uses brackets and row separators (`;`), can also be employed to create block matrices, provided the component submatrices or vectors are consistent. Julia's support for multidimensional arrays is enhanced by its native implementation of various algebra operations, which can be accessed by loading the **LinearAlgebra** package by using **LinearAlgebra**. This package offers a variety of basic operations, including matrix `tr` (trace), `det` (determinant), and `inv` (inverse), along with more advanced functions for determining eigenvalues and eigenvectors. These operations are crucial for numerous scientific, engineering, and data analysis applications, and they are readily accessible in Julia.

```
A = [-4. -17.; 2. 2.]    #=
2×2 Matrix{Float64}:
 -4.0   -17.0
  2.0     2.0     =#
eigvals(A)   #=
2-element Vector{ComplexF64}:
 -1.0 - 5.0im
 -1.0 + 5.0im   =#
eigvecs(A)   #=
2×2 Matrix{ComplexF64}:
  0.945905-0.0im         0.945905+0.0im
 -0.166924+0.278207im   -0.166924-0.278207im   =#
```

1.4 Linear algebra and sparse arrays

In addition to supporting multi-dimensional arrays, Julia offers native implementations of various common and valuable linear algebra operations, which can be accessed by using the package LinearAlgebra. A simple example is provided below. Note the promotion of values in matrix A that significantly enhances both writing and execution performance:

```
using LinearAlgebra
A = [3. 9 8; 5 9 1; 4 8 2]    #=
```

```
3x3 Matrix{Int64}:
 3.0  9.0  8.0
 5.0  9.0  1.0
 4.0  8.0  2.0 =#
```

```
A'     #=
3x3 adjoint(::Matrix{Int64}) with eltype Int64:
 3.0  5.0  4.0
 9.0  9.0  8.0
 8.0  1.0  2.0 =#
```

As we already know, several basic operations, including `tr`, `det`, `rank`, and `inv` are supported by the `LinearAlgebra` package:

```
tr(A)    # => 14.0
det(A)   # => 7.999999999999996
rank(A) # => 3
inv(A)             #=
3x3 Matrix{Float64}:
  1.25    5.75   -7.875
 -0.75   -3.25    4.625
  0.5     1.5    -2.25      =#
```

The inner product $a^\top b$ is written as `LinearAlgebra.dot(a,b)`. of course, `Vector`s a and b must have the same `length`.

The norm of the x vector, i.e., $\sqrt{(x_1^2 + x_2^2 + \cdots + x_n^2)}/n$ is given by `LinearAlgebra.norm(x)`. The distance of two vectors $dist(x,y) = \|y - x\|$ is `LinearAlgebra.norm(y-x)`. Of course, the qualification `LinearAlgebra.` is not explicitly required after declaration for exported package symbols.

The function Root Mean Square: $rms(x) = \sqrt{((x_1^2 + \cdots + x_n^2)/n)} = \|x\|/\sqrt{n}$ can be expressed as

```
x = [0.543101 0.335506]  #=
1x2 Matrix{Float64}:
 0.543101  0.335506       =#
rms(x) = LinearAlgebra.norm(x)/sqrt(length(x))   #=   function
    definition
rms (generic function with 1 method)    =#
rms(x)                        #=    function application
0.45139931240367087           =#
```

From the Rosetta Code [18], we present several methods for implementing or using the `rms(x)` function. In fact, there are various ways to achieve this utilizing the built-in functions in Julia, given an array `A = rand(10)` of values

The formula can be implemented directly using the comma broadcast:

```
sqrt(sum(A.^2)/length(A))
```

or shorter by `using Statistics` package: `sqrt(mean(A.^2))`. The implicit allocation of a new array by `(A.^2)` can be avoided by using `sum` as a higher-order function: `sqrt(sum(x -> x*x, A)/length(A))`. Of course, one can also use an explicit loop for near–C performance:

```julia
function rms(A)
    s = 0.0
    for a in A
        s += a*a
    end
    return sqrt(s/length(A))
end
```

It may be even better to utilize the built-in `norm` function, which calculates the square root of the sum of the squares of the entries of `A` in a manner that prevents the potential for spurious floating-point overflow (particularly if the entries of `A` are so large that they could overflow when squared): `norm(A)/sqrt(length(A))`.

Remark 1.5 (Matrix left division) Solving a linear system `Ax = b` by explicit inverse `x = LinearAlgebra.inv(A) * b` is not recommended for big `A` matrices. A better approach would be to compute `x = A \ b`. In addition, Julia provides many factorizations that can be used to speed up problems such as linear solve or matrix exponentiation by pre-factorizing a matrix into a form more amenable (for performance or memory reasons) to the problem.

Sparse Arrays

In numerical analysis and scientific computing, a sparse matrix is a matrix that has a large number of zero elements, allowing for optimization of storage and computation by only storing and processing the non-zero values. This proves particularly beneficial when working with large datasets, as it significantly reduces memory usage and enhances computational efficiency. Various storage schemes exist for sparse vectors and matrices, each offering distinct advantages and disadvantages, including compressed sparse row (CSR), compressed sparse column (CSC)—the default in Julia—and coordinate (COO) format.

One of the most straightforward storage schemes for sparse matrices is coordinate format, also known as storage by triples, which represents each non-zero element as a triple `(i, j, v)`, where `i` is the row index, `j` is the column index and `v` is the corresponding value. In Julia, this format is implemented in the `SparseArray` package using the `sparse()` function, which takes three input arrays: `I` for the row indices, `J` for the column indices, and `V` for the values, all of which are of type `Int` or `Float64`. This enables efficient storage and retrieval of sparse matrix elements.

Remark 1.6 (Julia implementation) Typical implementation schemes for sparse matrices are the CSR (compressed sparse row) and CSC (compressed sparse column). Julia uses, by now, only the CSC scheme and supports sparse vectors and sparse matrices in the `SparseArrays` `stdlib` module. Sparse arrays contain enough zeros that storing them in a special data structure leads to savings in space and execution time compared to dense arrays.

Julia sparse matrices

As mentioned above, in Julia, sparse matrices are stored in the Compressed Sparse Column (CSC) format. Julia sparse matrices have the type `SparseMatrixCSC{Tv, Ti}`, where `Tv` is the type of stored values, and `Ti` is the integer type to store column pointers and row indices.

The internal representation of type `SparseMatrixCSC`, which is widely used in the implementation of `Plasm` topological computations, is as follows:

```julia
struct SparseMatrixCSC{Tv,Ti<:Integer} <:
    AbstractSparseMatrixCSC{Tv,Ti}
    m :: Int                  # Number of rows
    n :: Int                  # Number of columns
    colptr :: Vector{Ti}      # Column j in colptr[j]:(colptr[j+1]-1)
    rowval :: Vector{Ti}      # Row indices of stored values
    nzval :: Vector{Tv}       # Stored values, typically nonzeros
end
```

In some applications, storing explicit zeros in a `SparseMatrixCSC` is convenient. The `nnz` function returns the number of elements explicitly stored in the sparse data structure, including non-structural zeros. To count the exact number of numerical nonzeros, use `zcount(!iszero, x)`, which inspects every stored element of a sparse matrix. `dropzeros`, and the in-place `dropzeros!`, can be used to remove stored zeros from the sparse matrix.

Julia sparse vectors

Sparse vectors are stored in a close analog to a compressed sparse column format for sparse matrices. In Julia, sparse vectors have the type `SparseVector{Tv, Ti}` where `Tv` is the type of the stored values and `Ti` is the integer type for the indices. The internal representation is as follows:

```julia
struct SparseVector{Tv,Ti<:Integer} <: AbstractSparseVector{Tv,Ti}
    n :: Int                  # Length of the sparse vector
    nzind :: Vector{Ti}       # Indices of stored values
    nzval :: Vector{Tv}       # Stored values, typically nonzeros
end
```

The `sparse()` function is a convenient way to create sparse arrays. For example, to construct a sparse matrix, one can provide a vector `I` of row indices, a vector `J` of column indices, and a vector `V` of stored values, which is also known as the COO (coordinates) format. The `sparse` function returns a sparse matrix `S` such that `S[I[k], J[k]] = V[k]`.

Similarly, the `sparsevec()` function constructs a sparse vector `R` from a vector `I` of row indices and a vector `V` of stored values, where `R[I[k]] = V[k]`. This allows for efficient creation of sparse arrays and vectors in Julia.

The inverse of the `sparse` and `sparsevec` functions is `findnz`, which retrieves the input used to create the sparse array. `findall(!iszero, x)` returns the Cartesian indices of nonzero entries in `x` (including stored entries equal to zero). Details and examples can be found in the Sparse Vectors and Matrices section of the standard library reference [12].

1.5 Parallel and distributed computing

Julia simplifies the implementation of complex parallel algorithms by providing built-in support for concurrency and optimized computation. In Julia, implementing concurrency and multithreading is straightforward, requiring only adherence to specific coding guidelines and the use of a few intuitive macros.

Julia's parallelism is pervasive, spanning from low-level instruction execution to distributed computing, and is often implicit. The language natively supports task-based control flows for parallel execution, including cooperative multitasking and preemptive thread-based multitasking.

Julia's multithreading model enables tasks to be executed simultaneously on multiple threads or CPU cores, while sharing memory, providing a flexible and efficient way to leverage parallel processing.

1.5.1 Parallel Programming

To successfully implement parallelization, you need to follow a three-step process. First, you need to (1) identify a decomposition of your problem into independent tasks that can be computed concurrently.

However, if the number of tasks exceeds the number of threads, the overhead of scheduling can outweigh the benefits of parallelization, emphasizing the importance of (2) balancing task granularity with thread count.

The third step is to determine how to efficiently assign tasks to threads, which is essential for achieving optimal parallel performance. [2].

Julia's manual outlines four primary models of parallel computing [10], which can be leveraged for concurrent and parallel programming:

1. **Asynchronous tasks or coroutines**: In this model, tasks run concurrently using coroutines or asynchronous functions. Coroutines are functions that can be paused and resumed at specific points, allowing other tasks to run in between. This model is useful for I/O-bound tasks, such as network requests or file operations, where waiting for external resources is common. Julia's `Task` type and `@async` macro offer support for asynchronous tasks.
2. **Multi-threading**: In this model, multiple threads are created to execute tasks simultaneously. Each thread runs a different part of the program, and the operating system schedules the threads to run on available CPU cores. Julia's `@spawn` macro and `Threads` module support this multithreading.
3. **Distributed computing**: Tasks are run on multiple machines or nodes, often using a network protocol for communication. This approach is useful for large-scale calculations that require significant processing power or memory. Julia's `Distributed` module and `@distributed` macro support distributed computing.
4. **GPU computing**: In this model, tasks run on a Graphics Processing Unit (GPU), which is built for parallel processing. GPUs have thousands of cores, making them perfect for tasks needing extensive parallelism, such as scientific simulations, machine learning, and data processing. Julia's `CuArrays` package and `@cu` macro support GPU computing.

These models can be used individually or in combination to achieve efficient parallelism in Julia.

Asynchronous tasks (coroutines)

Coroutines are program components that enable execution to be suspended and resumed, generalizing subroutines (computation with effects rather than functions that return a value) for cooperative multitasking. This multitasking style means the operating system does not perform a context switch at the system level from one running process to another. Instead, tasks voluntarily yield control periodically or when idle or logically blocked. Tasks serve as a control flow feature that allows computations to be flexibly suspended and resumed. Julia's tasks facilitate the suspension and resumption of computations for I/O, event handling, producer-consumer processes, and similar patterns to run multiple applications concurrently.

Macros for concurrency

Objects of type `Task` can be created by the macro `@task x`, where x is any expression, usually `begin`; ...; `end`. The `Task` object can be assigned to a variable:

`t = @task x`. After creation, the task must be started; it is started by calling `schedule(t)`, where it is added to a queue of tasks waiting for resources for execution. It is common to create a task and `schedule` it immediately. The macro `@async` is provided for that purpose: `@async x` is equivalent to `schedule(@task x)`.

The macro `@async` wraps an expression in a `Task` and adds it to the local machine's scheduler queue. The values can be interpolated into `@async` via `$`, which copies the value directly into the underlying constructed closure. This allows you to insert the value of a variable, isolating the asynchronous code from changes to the variable's value in the current task.

It is strongly encouraged always to favor `Threads.@spawn` over `@async`. This is because the use of `@async` turns off the migration of the parent task across worker threads in the current implementation of Julia. Thus, seemingly innocent use of `@async` in a library function can significantly impact the performance of very different parts of user applications [11].

The macro `@sync` waits until all lexically enclosed uses of `@async`, `@spawn`, `@spawnat` and `@distributed` are complete. In practice, it works as a parallel barrier in other languages. One of the simplest examples is to use `@sync/` `@async` for nonblocking I/O. For instance, you might want to download ten web pages. If you do it in a simple blocking loop, most of the time Julia does nothing but wait for the network:

```julia
URLs = ["https://discourse.julialang.org/" for i=1:10]
results = Vector(10)
@time for (i, URL) in enumerate(URLs)
  results[i] = Requests.get(URL)
end
```

But network I/O in Julia is nonblocking, so you can use task machinery to accelerate:

```julia
@sync for (i, url) in enumerate(urls)
    @async results[i] = Requests.get(url)
end
```

In the following example, each `println` request is run in a separate task using `Threads.@spawn` and then waits for the result of all of them because of `@sync`. When one of these tasks encounters the I/O operation, it gives away control so that other tasks can use the CPU. When I/O operations are finished, the task is resumed via `@sync`.

```julia
julia> Threads.nthreads() # => 4
```

You can use `--threads auto` (or `-t auto`) to start Julia with the number of threads available on the system, both physical (cores number) or logical.

```julia
julia> @sync begin
```

```
    Threads.@spawn
        println("Thread-id $(Threads.threadid()), task 1")
    Threads.@spawn
        println("Thread-id $(Threads.threadid()), task 2")
end;
# =>
Thread-id 3, task 1
Thread-id 1, task 2
```

Tasks communicate through Channels. While not strictly parallel computing, Julia allows you to schedule Tasks across multiple threads. When a piece
of computing work (essentially executing a specific function) is designated as
a Task, it becomes possible to interrupt it by switching to another Task. The
original Task can later be resumed, at which point it will continue right where
it left off.

Initially, this may seem similar to a function call. However, there are two
key differences. First, switching tasks does not consume space, allowing for an
unlimited number of task switches without affecting the call stack. Second,
switching between tasks can happen in any order, unlike function calls, where
the called function must complete execution before control returns to the
calling function [1].

In summary, Julia's tasks are not threads; they are coroutines that can
be scheduled asynchronously on a single thread or multiplexed onto a thread
pool. All I/O in Julia is non-blocking and yields to the scheduler, creating a
more cooperative environment

Data parallelism

For data parallelism, a more abstract description is appropriate. It also helps
in writing more reusable code; for example, using the same code for single-
threaded, multithreaded, and distributed computing.

In particular, it is important to use libraries that help you describe what to
compute rather than how to compute. Basically, it means using the generalized forms of map and reduce and learning how to express your computation
in terms of them. Luckily, if you already know how to write iterator comprehensions, there is not much more to learn to access a wide range of data
parallel computations. [1].

Local and distributed workers

Just like how multi-threading is configured, you need to set up multiple
worker processes to achieve a speedup. You can start Julia with -p auto and
-t auto to maximize the number of processes and tasks sharing memory.

Mapping is the most frequently used function in data parallelism. In sequential code, we have: map(f, vect) which is evaluated as [f(vect[i])

for i in vect]; but Julia's standard library `Distributed.jl` contains a
function `pmap` as a distributed version of `map`, so that you can write:

```julia
using Distributed
pmap(f,vect)
```

In addition, the `Folds.jl` package provides a unified interface for sequential, threaded, and distributed folds. Most of the functions can be used with
iterator comprehensions. You can use the `Folds` library and write: `using
Folds; Folds.map(f,vect)` with good speedups on large or distributed col-
lections.

```julia
using Folds
pmap(f, [::AbstractWorkerPool], c...; distributed=true, [...])
```

Remark 1.7 (Fold meaning) The term "fold" in computer science and functional programming refers to a higher-order function that recursively processes a data structure (typically a list or tree) by applying an operation to
combine its elements. In imperative languages, this is often referred to as a
"reduce" operation.

The above transforms the collection `c` by applying `f` to each element using available workers and tasks. For multiple collection arguments, apply `f`
elementwise. Note that `f` must be made available to all worker processes;
see [5] for details. If a worker pool is not specified, all available workers are
used; that is, the default worker pool is used. By default, `pmap` distributes
the computation over all specified workers. To use only the local process and
distribute over tasks, specify `distributed=false`.

Julia's iterator comprehension syntax is a powerful tool for composing
mapping, filtering, and flattening. Recall that sequential mapping can be
written as an array or as an iterator comprehension:

```julia
b1 = map(x -> x + 1, 1:3);          # => [2,3,4]
b2 = [x + 1 for x in 1:3];          # array comprehension
b3 = collect(x + 1 for x in 1:3);   # iterator comprehension
@assert b1 == b2 == b3              # => true
```

The iterator comprehension can be executed with threads via `Folds.
collect` [1]:

```julia
b4 = Folds.collect(x + 1 for x in 1:3)
@assert b1 == b4                    # => true
```

Functions such as `sum`, `prod`, `maximum`, and `all` are examples of reduction
(aka `fold`) that can be parallelized. Using `Folds.jl`, a sum of an iterator
created by the comprehension syntax can be easily parallelized by

```
d = Folds.sum(x + 1 for x in 1:3).
```

Multi-threading

Multithreading enables programmers to speed up their programs by leveraging concurrent execution of multiple threads, with each thread assigned to a different CPU core. On the surface, this type of programming may seem easy, but in practice, it can be difficult to ensure correctness and to obtain a significant speedup.

Julia supports two different models for multithreaded programming: loop parallelism with the `@threads` macro and task parallelism with the `Threads.@spawn` macro, which is a low-level basic construct. Julia's multithreading provides the ability to schedule tasks simultaneously on more than one thread or CPU core, sharing memory.

Remark 1.8 (Loop parallelism) This is the simplest way to achieve parallelism on a PC or a single large multi-core server. Julia's multi-threading is composable. When one multi-threaded function calls another, Julia will automatically schedule all the threads on available resources.

Although Julia threads can communicate through shared memory, it is notoriously difficult to write correct and data-race-free multithreaded code. Julia's Channels are thread-safe and can be used for safe communication. The best way to ensure data-race freedom is to acquire a lock around any data access that might be observed from multiple threads.

By default, Julia starts with a single thread of execution. `Threads.nthreads()` returns the number of threads, set by the parameter -t when starting Julia, for example, `-t auto` or `--threads` n. The function `Threads.threadid()` returns the integer `id` of the current thread.

Julia supports parallel loops using the `Threads.@threads` macro. This macro is affixed in front of a `for` loop to indicate to Julia that the loop is a multi-threaded region. Julia supports accessing and modifying values atomically, that is, in a thread-safe way to avoid race conditions.

When a program's threads are busy with many tasks, tasks may experience delays that can negatively impact the program's responsiveness and interactivity. To address this, you can specify that a task is `interactive`.

External libraries, such as those called via `ccall`, pose a problem for Julia's task-based I/O mechanism. If a C library performs a blocking operation, it prevents the Julia scheduler from executing any other tasks until the call returns.

In conclusion, let's note that there are a few specific limitations and warnings to be aware of when using threads in Julia [9].

1.5.2 Multiprocessing and Distributed Computing

Most modern computers have more than one CPU, and multiple computers can be linked together in a cluster. Using the power of these multiple CPUs enables many computations to be done faster.

Two major factors influence performance: the speed of the CPUs themselves and the speed of their memory access. In a cluster, it is clear that a given CPU will have the fastest access to RAM within the same computer (or node). Perhaps more surprisingly, similar problems are relevant on a typical multicore laptop, due to differences in the speed of the main memory and the cache [6].

Distributed *module*

Module `Distributed` is part of the standard library shipped with Julia. It implements distributed-memory parallel computing.

The `Distributed` standard library (in `stdlib`) runs multiple Julia processes with separate memory spaces. These can be on the same computer or multiple computers. The `Distributed` standard library provides the capability for remote execution of a Julia function. This basic building block makes it possible to build many different kinds of distributed computing abstractions. Packages like `DistributedArrays.jl` are an example of such an abstraction. On the other hand, packages like `MPI.jl` and `Elemental.jl` provide access to the existing `MPI` ecosystem of libraries.

Consequently, a good multiprocessing environment should allow for control over a particular CPU's "ownership" of a chunk of memory. Julia provides a multiprocessing environment based on message passing to allow programs to run on multiple processes in separate memory domains at once [6].

Distributed programming in Julia is built on two primitives: remote references and remote calls. A remote reference is an object that can be used in any process to refer to an object stored in a particular process. Remote references come in two flavors: `Future` and `RemoteChannel`. A remote call is a request by one process to call a specific function on certain arguments on another (possibly the same) process.

Distributed *arrays*

Much easier is parallel programming in Julia by making use of `DistributedArrays.jl`. This computational abstraction uses the stdlib `Distributed` to implement a `GlobalArray` interface. Large computations are often organized around large arrays of data. In these cases, distributing arrays among several processes is a particularly natural way to obtain parallelism.

`DistributedArrays.jl` provides a global array interface `DArray` using `stdlib Distributed`, a collection of tools for distributed parallel processing.

The DArray is distributed across a set of workers. Each worker can read and write from its local portion of the array, and each worker has read-only access to the portions of the array held by other workers. This provides a ready answer to the question of how a program should be divided among machines and combines the memory resources of multiple machines, allowing the use of arrays too large to fit on one machine.

A DArray has an element type and dimensions similar to an Array. A DArray can also use arbitrary array-like types to represent the local chunks that store actual data. The data in a DArray is distributed by dividing the index space into some number of blocks in each dimension.

Using DistributedArrays, common arrays can be constructed with distributed data structures. E.g., d = DistributedArrays; d.zeros, d.ones, d.rand, d.randn, d.fill. The constructor that builds a distributed array is:

```
DArray(init, dims, [procs, dist])
```

1. The parameter init is a function that accepts a tuple of index ranges. This function should allocate a local chunk of the distributed array and initialize it for the specified indices.
2. dims is the overall size of the distributed array.
3. procs optionally specifies the vector of process IDs to use. If not specified, the array is distributed on all worker processes only. Typically, when running in distributed mode, i.e., nprocs() > 1, this would mean that there is no chunk of the distributed array on the process hosting the interactive julia prompt.
4. dist is an integer vector that specifies how many chunks the distributed array should be divided into in each dimension.

For example, the dfill function that creates a distributed array and fills it with a value v is implemented as:

```
dfill(v, args...) = DArray(I->fill(v, map(length,I)), args...)
```

1.5.3 Programming the GPU

While the kernel functions for GPUs are usually written in a C/C++ dialect, the Julia GPU compiler provides the ability to run Julia code natively on GPUs. There is also a rich ecosystem of Julia packages that target GPUs. The JuliaGPU.org website lists capabilities, supported GPUs, related packages, and documentation. Julia has several GPU packages, each supporting different programming models.

JuliaGPU is a GitHub organization established to consolidate the various packages for programming GPUs in Julia, facilitating high-performance GPU programming in a high-level language. JuliaGPU is vendor-neutral; some content is accessible for all supported GPU backends. Additionally, JuliaGPU includes vendor-specific tools and APIs. At the 2021 JuliaCon workshop, they demonstrated three major GPU programming packages: `CUDA.jl` for Nvidia GPUs, `AMDGPU.jl` for AMD GPUs, and `oneAPI.jl` for Intel GPUs. The different methods for programming GPUs with these packages range from generic array operations that emphasize ease of use to hardware-specific kernels used when performance is critical.

GPU functions (kernels) are inherently parallel, so writing GPU kernels is at least as tricky as writing low-level parallel CPU code, but the difference in hardware adds quite a bit of complexity. Most algorithms require arrays to manage all their data, which calls for a robust GPU array library.

GPUArrays.jl

`GPUArrays` is an abstract interface for GPU computations. Think of it as the `AbstractArray` interface in Julia `Base` but for GPUs. It is not intended for the end user, who should only use one of the packages built on `GPUArrays.jl`, such as `CUDA.jl` (Nvidia), `oneAPI.jl` (Intel), `AMDGPU.jl` (AMD), or `Metal.jl` (Apple), for different hardware support.

It allows you to write generic Julia code for all GPU platforms and implement common GPU algorithms. Like Julia `Base`, this includes `BLAS` wrapper[1], `maps`, `broadcast`s and `mapreduce`s. So when you inherit `GPUArrays` and overload the interface correctly, you will get a lot of functionality for free.

It is important to note that Julia allows you to write both GPU kernels and surrounding code in Julia itself, while running on most GPU hardware.

In complex analysis, the Julia set [19] of a holomorphic function consists of all the points whose behavior after repeated iterations of the function is chaotic in the sense that it can change drastically following a small initial perturbation. As one can see in [4], the computational example of Julia set with Julia (!) strongly motivates why one should move big-array computations to the GPU. For large arrays, one gets a solid 60-80x speed-up by moving the calculation to the GPU. Getting this speed-up was as simple as converting the Julia `array` to a `GPUArray`.

CUDA and OpenCL

A significant difference exists between CUDA and OpenCL, two prominent parallel programming frameworks used to write low-level GPU code. OpenCL

---

[1] The **BLAS** (Basic Linear Algebra Subprograms) are routines that provide standard building blocks for performing basic vector and matrix operations efficiently.

is a heterogeneous programming platform that enables applications to run across multiple platforms, including CPUs, GPUs, and other specialized hardware, providing a high degree of portability. In contrast, CUDA is a software framework specifically designed for Nvidia's GPU computations, offering optimized performance and functionality for Nvidia hardware. This distinction emphasizes each framework's unique strengths and limitations, making it crucial to choose the right tool for a specific project or application.

While `CUDA.jl` only supports Nvidia hardware, `OpenCL.jl` supports all types of hardware, although it may be a "bit rough around the edges". You must decide which one to use and stick with that decision [4].

One might think that the GPU performance suffers from being written in a dynamic language like Julia. Still, Julia's GPU performance should be pretty much on par with the raw performance of CUDA or OpenCL. Tim Besard, the creator of the Julia GPU compiler, did a great job integrating the LLVM Nvidia compilation pipeline to achieve the same or sometimes even better performance as the CUDA C code [4].

CUDA programming in Julia

In https://cuda.juliagpu.org/stable/tutorials/introduction/ you can find 'A gentle introduction to parallelization and GPU programming in Julia:

> Julia has first-class support for GPU programming: you can use high-level abstractions or obtain fine-grained control, all without ever leaving your favorite programming language. The purpose of this tutorial is to help Julia users take their first step in GPU computing. In this tutorial, you will compare CPU and GPU implementations of a simple calculation and learn about a few of the factors that influence the performance you obtain.

CUDA *package*

The `CUDA.jl` package is the main entry point for programming NVIDIA GPUs in Julia. The package makes it possible to do so at various abstraction levels, from easy-to-use arrays to handwritten kernels using low-level CUDA APIs. The following is synthesized from Reference [15] that you are invited to read.

```julia
using CUDA
x_d = CUDA.fill(1.0f0, N) # vector on GPU filled of 1.0 (Float32
    )
y_d = CUDA.fill(2.0f0, N) # vector stored on GPU filled of 2.0
```

Here, the symbol d means `device`, in contrast to `host`. Now, let us do the increment:

```julia
y_d .+= x_d
```

```
@test all(Array(y_d) .== 3.0f0) # => Test Passed
```

The statement `Array(y_d)` moves the data in `y_d` back to the host for testing. If we want to benchmark this, let us put it in a function:

```
function add_broadcast!(y, x)
    CUDA.@sync y .+= x
    return
end

add_broadcast! # => (generic function with 1 method)
@btime add_broadcast!($y_d, $x_d) # => 67.047 µs
  (84 allocations: 2.66 KiB)
```

The most interesting part of this is the call to `CUDA.@sync`. The CPU can assign jobs to the GPU and then do other stuff (such as assigning more jobs to the GPU) while the GPU completes its tasks. Wrapping the execution in a `CUDA.@sync` block will make the CPU block until the queued GPU tasks are done, similar to how `Base.@sync` waits for distributed CPU tasks.

Without such synchronization, the time it takes to launch the computation would be measured, not the time it takes to perform the calculation. But most of the time, you don't need to synchronize explicitly: many operations, like copying memory from the GPU to the CPU, implicitly synchronize execution.

CuArrays *package*

We can perform GPU computations at a high level using the `CuArray` type, without explicitly writing a kernel function. The `CuArrays` package in Julia provides several macros to facilitate GPU computing. Here are some of the most commonly used macros:

1. `@cu`: This macro is used to define a CUDA kernel, which is a function that runs on the GPU.

   ```
   @cu function my_kernel(x, y)
   # kernel code here
   end
   ```

   It's similar to the `@kernel` macro in the `CUDA.jl` package.
2. `@cuarray`: This macro is used to create a CuArray, which is a GPU-based array.

   ```
   A = @cuarray rand(100, 100)
   ```

3. `@cuindex`: This macro is used to create a CuIndex, which is a GPU-based index array.

   ```
   idx = @cuindex rand(100)
   ```

4. `@cuprintf`: This macro is used to print output to the console from a CUDA kernel.

```
@cuprintf "Hello from GPU!\n"
```

5. `@cufetch`: This macro is used to fetch data from the GPU to the CPU.

```
A = @cufetch A
```

6. `@cudestroy`: This macro is used to destroy a CuArray or CuIndex.

```
@cudestroy A
```

These macros provide a convenient way to work with GPU arrays and kernels in Julia. However, it is worth noting that the `CuArrays` package is still evolving and that the macros may change or be deprecated in future versions.

Remark 1.9 (Multiple cores) For the particular hardware used, the GPU computation was significantly faster than the single-threaded CPU computation, but the use of multiple CPU threads in multiple cores makes the CPU implementation competitive.

Remark 1.10 (About kernels) The high-level functionality of CUDA often means that you don't need to worry about writing kernels at low level. However, there are many cases where computations can be optimized using low-level manipulations [15].

1.6 Modules and packages

A Julia `module` is a named sequence of `Julia` code, typically contained in a file with the same name:

```
module NameOfModule; <some code>; end # => Main.NameOfModule
```

A module starts with the reserved word `module` followed by `NameOfModule` and is terminated by the `end` word. The modules in Julia help organize the code into coherent units and have the following features.

1. The modules are separate namespaces, each introducing a new global scope. This is helpful because it lets the same name be used for different functions or global variables without conflicts, as long as they are in separate modules.

2. Modules have facilities for detailed **namespace** management: each defines a set of names it **export**s, and can **import** names from other modules with **using** and **import**. The **import** keyword only operates on one name at a time. It does not add modules to the search path like **using** does. Additionally, import differs from using in that functions must be imported with import to extend them with new methods.
3. Modules can be precompiled for faster loading and may contain code for runtime initialization. Typically, in larger Julia packages, you will see module code organized into program files [8]. One can have multiple files per module and multiple modules per file. The reserved word **include** behaves as if the contents of the source file were evaluated in the global scope of the including module.

The recommended style is not to indent the module body, since that would typically lead to indenting entire files. Also, it is common to use **UpperCamelCase** for module names, just like for types.

A software library is a suite of data and programming code used to develop software programs and applications. In Julia, a library is called package.

A Julia package contains modules, tests, and documentation. It extends the core Julia functionality. You can share your code with the community by developing a package. You can create a Julia package using as a support the built-in package manager **PkgDev.jl** or the package **PkgTemplates.jl**. The second is easier for a novice.

A good tutorial on how a user can develop a package is [14]. The reader is warmly encouraged to view and try the suggested step-by-step development.

The Julia ecosystem contains over 9,000 packages registered in General registry, so finding the right package can be challenging. Fortunately, some services can help navigate the ecosystem, including:

1. JuliaHub: Service that includes a search for all documentation from registered open-source packages, code search, and navigation by tags/keywords.
2. Julia Packages: to browse Julia packages, filter by categories, and sort them by popularity, creation date, or last update date. Also supports browsing package
3. Julia.jl: a manually curated taxonomy of Julia packages (category information for JuliaPackages is derived from this as well).
4. JuliaRegistries/General: General is the default Julia package registry. Package registries are used by Julia's package manager Pkg.jl and include information about packages such as versions, dependencies, and compatibility constraints..

References

1. Arakaki., T.: Julia Data Parallel Computing. URL `https://juliafolds.github.io/data-parallelism/tutorials/quick-introduction/`. [retrieved june 27, 2023] pages 24, 25
2. Aubanel, E.: Elements of Parallel Computing. Chapman Hall/CRC Press (2016) pages 21
3. Bezanson, J., Edelman, A., Karpinski, S., Shah, V.B.: Julia: A fresh approach to numerical computing. SIAM Review **59**(1), 65–98 (2017). URL `https://doi.org/10.1137/141000671` pages 1
4. Danisc, S., Kavalar, M., Dombrowski, M., Markovics, P.: An Introduction to GPU Programming in Julia. URL `https://nextjournal.com/sdanisch/julia-gpu-programming`. [retrieved june 29, 2023] pages 29, 30
5. Julia: Manual: Code Availability and Loading Packages. URL `https://docs.julialang.org/en/v1/manual/distributed-computing/#code-availability`. [retrieved January 31, 2025] pages 25
6. Julia: Manual: Distributed-Computing.   URL   `https://docs.julialang.org/en/v1/manual/distributed-computing/#Multi-processing-and-Distributed-Computing`.   [retrieved june 29, 2023] pages 27
7. Julia: Manual: Metaprogramming. URL `https://docs.julialang.org/en/v1/manual/metaprogramming/`. [retrieved june 21, 2023] pages 11
8. Julia: Manual: Modules. URL `https://docs.julialang.org/en/v1/manual/modules/#modules`. [retrieved june 29, 2023] pages 33
9. Julia: Manual: Multi-Threading. URL `https://docs.julialang.org/en/v1/manual/multi-threading/#man-multithreading`. [retrieved june 29, 2023] pages 26
10. Julia: Manual: Parallel Computing. URL `https://docs.julialang.org/en/v1/manual/parallel-computing/#Parallel-Computing`. [retrieved june 26, 2023] pages 22
11. Julia: Manual: Parallel Computing. URL `https://docs.julialang.org/en/v1/base/parallel/#Tasks`. [retrieved june 27, 2023] pages 23
12. Julia: Manual: stdlib/SparseArrays. URL `https://docs.julialang.org/en/v1/stdlib/SparseArrays/#man-csc/`. [retrieved june 25, 2023] pages 21
13. Julia: Manual: Types. URL `https://docs.julialang.org/en/v1/manual/types/#man-types`. [retrieved june 30, 2023] pages 4
14. Julia Community: How to develop a Julia package. URL `https://julialang.org/contribute/developing_package/`. [retrieved june 29, 2023] pages 33
15. JuliaGPU: A gentle introduction to parallelization and GPU programming in Julia. URL `https://cuda.juliagpu.org/stable/tutorials/introduction/#Introduction`. [retrieved june 29, 2023] pages 30, 32
16. Kamiski, B.: Julia for Data Analysis. Manning (2023) pages 7
17. Mainon, P.: Writing type-stable julia code (2021). URL `https://blog.sintef.com/industry-en/writing-type-stable-julia-code/` pages 5
18. Rosetta Code: Averages/Root mean square. URL `https://rosettacode.org/wiki/Averages/Root_mean_square#Julia`. [retrieved june 25, 2023] pages 18
19. Weisstein, E.W.: Julia set. From MathWorld–A Wolfram Web Resource URL `https://mathworld.wolfram.com/JuliaSet.html` pages 29

Chapter 2
The Julia Package **Plasm.jl**

`Plasm.jl` is a Julia package for advanced geometric programming specifically designed to generate complex engineering and architectural models represented as maps of topological polyhedra. This includes various shapes, such as piecewise linear approximations of curved objects and cellular complexes composed of simplicial, cubical, and polyhedral cells.

`Plasm.jl` is an open-source implementation of the original `PLaSM` (Programming Language for Solid Modeling) project, funded by the Italian National Research Council (CNR) to advance building design and support the growth of industrialized construction. The project introduced a geometric type within the FL language at the Function Level, developed by Backus and his team at IBM Almaden in the 1990s. Plasm was originally written in Common Lisp and later ported to Scheme, C++, and Python before being adapted to Julia. In this final environment, the language has found an ideal platform to leverage its mathematical foundation rooted in algebraic topology and Boolean algebra.

This chapter introduces the reader to developing geometric models of solid objects with Julia and Plasm, as well as the primary primitive operators that Plasm imports from the original FL into Julia.

© The Author(s), under exclusive license to Springer Nature Switzerland AG 2026   35
A. Paoluzzi and G. Scorzelli, *BIM Geometry with Julia Plasm—Functional Language for CAD Programming*, Digital Innovations in Architecture, Engineering and Construction, https://doi.org/10.1007/978-3-031-90244-4_2

2.1 Backus' functional programming

John W. Backus led the IBM team that invented and implemented the FOR-
TRAN (Formula Translation) language. Developed by IBM from 1954 to 1956
for scientific and engineering applications, FORTRAN was the first high-level
scientific and technical programming language and later dominated the field
of scientific computing. In the last decades, it has primarily been used on
supercomputers. Julia
citeBEKS14 is now recognized as the leading heir to the long tradition of
scientific computation.

FP (Functional Programming) and FL (Function Level)

The 1977 ACM Turing Award, in honor and recognition of Turing's contribution to the field of computing, was presented to John Backus at the ACM
Annual Conference in Seattle. In introducing the recipient, Jean E. Sammet,
Chairman of the Awards Committee, stated that he was receiving that year's
Turing Award for Fortran and the BNF (Backus Normal Form). Therefore,
everyone expected Backus to describe the work of implementing FORTRAN
at IBM in the 1950s.

Conversely, the Backus Turing Lecture was entitled "Can Programming Be
Liberated from the von Neumann Style? A Functional Style and Its Algebra of
Programs" [1]. This lecture was enormously influential and opened a decade
and a half of renewed research on functional languages.

A simple Functional Programming (FP) system is presented in [1], which
is based on combining forms for constructing programs from simpler programs. This system features a algebra of programs, where variables represent
functional programs and "operations" are functional forms that combine FP
programs. This paradigm supports function-level programming, allowing programs to be built from a set of generally useful primitives without needing
named variables. Only programs or functions can be assigned names. In Figure 2.1, we illustrate the algebraic rules of FP, which provide a foundation
for this functional programming system.

FL, a programming language developed by IBM Research at Almaden, is
a practical implementation of Backus' functional programming (FP) concept.
As described in [2], FL aims to provide a language in which it is easy to write
clear, concise, and efficient programs. The language is centered on a rich set
of functionals, which are forms for combining existing programs to construct
new ones. This focus on programming at the function level yields software
systems with a rich mathematical structure, making it easier to reason about
and check for validity.

Fig. 2.1 The primitive operations (combining forms) and notations [19] of FP programs. The semantics of FP is embodied in an underlying algebra of programs, a set of function-level equalities that may be used to transform programs and to reason about them.

Figure 1: Primitive Operations and Notation.

identity	$id{:}x = x$
error	$err{:}x = \perp$
selectors si	$s3{:} <x_1, ..., x_n> = x_3$
empty test	$isnull{:}x = \textbf{true}$ iff
	$x = <\ >$
tail	$tl{:} <x_1, ..., x_n>$
	$= <x_2, ..., x_n>$
append to left	$al{:} <x, <y_1, ..., y_n>>$
	$= <x, y_1, ..., y_n>$
constant	$\tilde{}x{:}y = x\ (y \neq \perp$ in FP.$)$
construction	$[f_1, ..., f_n]{:}x$
	$= <f_1{:}x, ..., f_n{:}x>$
composition	$(f \circ g){:}x = f{:}(g{:}x)$
conditional	$(p \to f;\ g){:}x =$
	$f{:}x$ if $p{:}x = \textbf{true}$
	$g{:}x$ if $p{:}x = \textbf{false}$
	$\perp$ otherwise
apply to all	$\alpha{:}f{:} <x_1, ..., x_n>$
	$= <f{:}x_1, ..., f{:}x_n>$
catenate	$cat{:} \ll x_1...>, <...x_n \gg$
	$= <x_1, ..., x_n>$
infix cat	$f ++ g = cat \circ [f, g]$

2.2 FL-based PLaSM in Julia syntax

This section provides an informal introduction to the Plasm language, along with a few examples [10, 9]. The goal is to familiarize the readers with the syntax and style of Plasm programs without delving into the technical details.

The interpreter and interactive graphical user interface (GUI) for PLaSM were developed at the Sapienza University of Rome in the 1990s. Building upon the FL syntax and semantics, the PLaSM implementation introduced a new geometric type called HPC (hierarchical polyhedral complex). It extended the FL programming style to support geometric computing within the broader domain of Computer-Aided Design.

The Wiley book Geometric Programming for Computer-Aided Design (GP4CAD) [9] offered a comprehensive collection of hundreds of small, simple programs written in the FL syntax. In the past decade, the command-line user interface (CLI), geometric computational kernel, and interactive visualizer of the original "classic" PLaSM have been successfully ported to Python [16] and C++ [12], respectively, leveraging the functional features of these languages [17]. Building on this momentum, a native and extended port to Julia is currently in progress [11], which is utilized in this book and is briefly described below in a few key points.

Here are the key characteristics of the functional programming paradigm:

1. The syntax rules are minimal and straightforward.
2. Each rule is simple and easy to understand.
3. The code is concise and clear, making it easy to read and maintain.
4. The meaning of a program is well-defined, as there is no concept of state.
5. Programs can be composed by concatenating and by nesting functions.
6. Functions can be used as both programs and data, allowing for greater flexibility.

In addition to the characteristics of functional programming languages, `Plasm` programming has some unique features that are worth noting. Here are the key features of the Plasm implementation in Julia:

1. Julia's native support for the function application and the binary function composition allows for concise and expressive code, enabling the use of `f(x)` and `g ∘ f`, respectively.
2. The FL sequence `<x`$_1$`, x`$_2$`, ..., x`$_n$`>` is implemented as a Julia array `[x`$_1$`, x`$_2$`, ..., x`$_n$`]`.
3. All primitive functions (programs) are pure, meaning they have no side effects, and are written in uppercase. Named expressions utilize capitalized names.
4. Each `Plasm` function is unary, meaning it accepts a single argument, which can be an array.
5. The geometric types `Hpc` and `Lar` have been extended to include an optional Julia dictionary of properties.

In the Julia `Plasm` implementation, the requirement for unary functions is occasionally relaxed to reduce the visual clutter. As noted in [9], where many codes discussed in this book are presented in their FL version, this approach has demonstrated significant advantages regarding program development style and efficiency.

Programs are functions

Generally speaking, a `Plasm` program is a *function*. When applied to some input *argument*, a program produces some output *value*. Two programs are usually connected by using functional composition so that the output of the first program is used as input to the second program. Starting from here, we may use the Julia syntax directly.

Program composition and application

The composition of `Plasm` functions, i.e., of `FL` programs with Julia syntax, works exactly as the composition of mathematical functions. For instance, the

application of the composite mathematical function $f \circ g$ to the argument x
is defined as:

$$(f \circ g)(x) \equiv f(g(x)).$$

This indicates that the function g is applied first to the x object, followed by
the application of the function f to the resulting value $g(x)$. The Julia Plasm
notation for this expression is identical:

```
(f∘g)(x) ≡ f(g(x))
```

In Julia, the binary function composition operator ∘ represents a composite function. The binary application of a function to an argument uses jux-
taposition, with the second argument enclosed in parentheses. For example,
(f ∘ g)(x) means that "f" is applied to the result of "g(x)". This notation
is native to Julia.

Naming objects

In the Plasm language, each value generated by the language can be assigned
a name using a definition construct. This construct can function with or
without explicit parameters. The definition consists of a head and a body.
The head specifies the name being defined, while the body is the subsequent
expression that describes the computational process generating the produced
value. The body may depend on parameters, either explicitly or implicitly.

For example, we may have the following.

```
object = (Fun3 ∘ Fun2 ∘ Fun1)(params);
```

The computational process that produces the object value can be thought
as the computational pipeline shown in Figure 2.2.

Fig. 2.2 Computational pipeline: object = (fun3 ∘ fun2 ∘ fun1)(params)

This passage describes a model generation that relies on implicit parameters. This means that the model's behavior is tied to the specific parameters
used to generate the object, and changing those parameters requires modifying the source code, recompiling, and re-evaluating the object identifier.

Parametrized objects

Here, we describe a parametric geometric model, which uses a generating function with formal parameters. These formal parameters are placeholders that can be replaced with actual arguments to produce different values.

For example, we may have the following.

```
object(params) = (Fun3 ∘ Fun2 ∘ Fun1)(params);

obj1 = object([p₁, p₂, ... , pₙ]);
obj2 = object([q₁, q₂, ... , qₙ]);
```

It is interesting to note that the generating function of a geometric `Plasm` model may accept parameters of type Any, including other geometric objects.

This statement highlights the flexibility and descriptive power of parametric geometric models. By allowing the generating function to accept parameters of any type, including other geometric objects, enables the creation of complex and hierarchical models that can capture a wide range of relationships and dependencies between different geometric elements.

2.3 Geometric Programming at Function Level

As we already know, the original **PLaSM** was a geometry-oriented extension of a subset of the **FL** language [2, 3], which is a pure functional language based on combinatory logic [Wikipedia]. In particular, the **FL** language makes use of both pre-defined and user-defined combinators, i.e., higher-order functions which are applied to functions to produce new functions. The small but very significant **FL** subset, which is used as the base environment of **Plasm** is summarized in this section.

Notice that here and in the remainder of this book, the infix symbol $\equiv$ is normally used to tell the reader that the expression on its left side evaluates to the value on its right side. Sometimes, this symbol is also used to denote an equivalence between syntactical forms.

Elements of FL syntax in Julia

Primitives **FL** objects are characters, numbers, and truth values. Primitive user-defined Julia types to create geometric models, added by **Plasm** to **FL**, are the **Hpc** and the **Lar** geometric types, which are explored in detail in the following chapters. Primitive objects, functions, applications, and sequences are expressions; sequences are expressions separated by commas and contained within a pair of square brackets, using the Julia's **Vector** type:

```
[5, fun]
```

An application expression exp1(exp2) applies the function obtained from evaluating exp1 to the argument obtained from evaluating exp2. Julia allows certain binary operators to be used in both infix and prefix forms:

```
1 + 3 ≡ +(1,3) ≡ 4
```

Application associates to the left, i.e., a sequence of repeated applications are evaluated from left to right. Note that this is only possible if all the applications, but possibly the last one, generate a new function to be applied to the following argument:

```
f(g)(h) ≡ (f(g))(h)
```

Application binds stronger than composition, i.e., applications are evaluated first before compositions, as is shown in the following example. Of course, the application of f must generate a function value:

```
f(g) ∘ h ≡ (f(g) ∘ h)
```

Combining forms and functions

In FL, a function-level approach focuses on defining new functions by composition, combining existing tasks in various ways to create novel functions. This approach yields a programming style centered on function-valued expressions, which differs significantly from the typical applicative programming style found in most Julia code. Notably, many geometric operators in Plasm that utilize the FL approach are defined without explicit arguments, departing from the standard applicative style. Some important FL combining forms and functions follow.

Construction

The combining form CONS, for construction, allows the application of a sequence of functions to an argument, so generating a sequence of applications:

```
CONS([f₁,...,fₙ])(x) ≡ [f₁(x),...,fₙ(x)]
CONS(f₁,...,fₙ)(x) ≡ [f₁(x),...,fₙ(x)]
```

A CONSed sequence of functions is a sort of vector function that can be composed with other functions or applied directly to the data, resulting in all the components being applied to them. For example, CONS([sin,cos,tan,atan]) when applied to the argument π returns the sequence of applications:

```
CONS(sin,cos,tan,atan)(π)    #=
4-element Vector{Float64}:
  0.0
 -1.0
  0.0
  1.2626272556789115    =#
```

Unfortunately, we could not use the original FL syntax for vector functions that used square brackets, which Julia uses for arrays. Therefore, we introduced the operator CONS: Array{Function} → Function. Julia's great functional flexibility allowed us to add the above better syntax as a new method, obtaining the same behavior with a variable number of function arguments and without the square brackets.

Apply-to-all

The combining form AA has a symmetric effect, i.e., it applies a function to a sequence of arguments, giving a sequence of applications. It is equivalent to the functional map of other languages, Julia included:

```
AA(f)([x₁,...,xₙ]) ≡ [f(x₁),...,f(xₙ)]
  ≡ map(f, [x₁,...,xₙ])
```

For example, we may apply the trigonometric SIN function to all the elements of a list of numeric expressions:

```
AA(SIN)([0, π/3, π/6, π/2])
≡ [SIN(0), SIN(π/3), SIN(π/6), SIN(π/2)]
≡ [0, 0.8660254037844382, 0.49999999999999956, 1.0];
```

The reader should note that numerical computations often introduce round-off and approximation errors. Recall that π is irrational and cannot be precisely represented using finite precision arithmetic. Additionally, functions like SIN are calculated using truncated series expansion.

Identity

The ID function returns its argument x :: Any unchanged:

```
ID(x) ≡ x
```

In other words, the application of the identity function to any argument, gives back the same argument:

```
ID(0.5) ≡ 0.5
ID(SIN) ≡ SIN
ID(SIN)(0) ≡ SIN(0) ≡ 0
```

Trigonometric `Plasm` functions, commonly employed in geometric programming and models, are implemented using Julia's functions directly.

Constant

The constant combining form `K` is evaluated as follows for whatever values x_1 :: Any and x_2 :: Any:

$$K(x_1)(x_2) \equiv x_1$$

In other words, the first application returns a constant function of value x_1, i.e. such that when applied to any argument x_2, always returns x_1. Some concrete examples follow:

```
K(0.5) ≡ Anonymous-Function
K(0.5)(10) ≡ 0.5
K(0.5)(100) ≡ 0.5
K(0.5)(SIN) ≡ 0.5
```

Composition

The binary composition of functions, denoted in Julia by the infix operator "∘", is defined in a standard mathematical way, as we already know:

$$(f \circ g)(x) \equiv f(g(x))$$

where ∘ is obtained via the LaTeX expression \circ followed by TAB character. This important typing mechanism is standard in Julia and allows the program code to use Greek letters and many mathematical symbols. The *n-ary* composition of functions is also allowed:

$$COMP([f, g, h])(x) \equiv (f \circ g \circ h)(x) \equiv f(g(h(x)))$$

In the following we have, using Julia's π, cos and acos [1], respectively:

```
(acos ∘ cos)(π) ≡ acos(cos(π)) ≡ acos(-1) ≡ 3.141592653589793
(cos ∘ acos)(-1) ≡ cos(acos(-1)) ≡ cos(π) ≡ -1.
COMP(acos, cos, acos)(-1) ≡ acos(cos(acos(-1))) ≡
    3.141592653589793
```

where `COMP` is the variadic[2] version of the Julia's ∘ operator.

---

[1] Which can be directly used in **Plasm** code, of course.

[2] A variadic function has an indefinite arity, meaning it can accept a variable number of arguments. In **Plasm**, the Boolean operations of UNION, INTERSECTION, DIFFERENCE, and XOR of solid objects are variadic. See Chapter 7.

Conditional combinator

This combinator IF has the following semantic: "IF the predicate p applied to object x is true, THEN apply f to x; ELSE apply g to x". This construct is very useful when it is necessary to apply different actions to input data depending on the value of some predicate evaluated on them, and is possibly more "natural" than the conditional statements available in other languages.

Formally, the conditional form IF([p, f, g]) is evaluated as follows:

```
IF([ p, f, g ])(x)
   ≡ f(x) if p(x) ≡ TRUE
   ≡ g(x) if p(x) ≡ FALSE
```

From a syntax viewpoint, we remark that the IF operator is a higher-order function that must be applied to a triplet of functions to return a function which is in turn applied to the input data.

A predicate is a function p: x::T → {true, false} where T is a Type. true and false are called truth values, and in Julia are Bool values. The predicate p is a function, as well f and g, to be alternatively executed depending on the truth value of the logical expression p(x). E.g., we have:

```
IF([ISINTPOS, K(true), K(false)])(1000) ≡ true
IF([ISINTPOS, K(true), K(false)])(-1000) ≡ false
```

where ISINTPOS is a predefined predicate that returns true when applied to some positive integer. The equivalent predicate in Julia:

```
x = 100; x isa Int && x>0 # => true
```

Insert Right/Left

The combining forms INSR and INSL allow the user to apply a binary function f, with signature[3] f: D × D → D, on a sequence of arguments of any length n. In other terms, we can write: INSR(f): D^n → D. Note that in the right-hand expressions below, f is always applied explicitly to a pair of arguments:

```
INSR(f)([x₁, x₂,..., xₙ]) ≡ f([x₁, INSR(f)([x₂,..., xₙ])])
INSL(f)([x₁,..., xₙ₋₁, xₙ]) ≡ f([INSL(f)([x₁,..., xₙ₋₁]), xₙ])
```

An interesting example of using the INSL combinator is given below, where the function BIGGER: Num × Num → Num is defined. The BIGGER function returns the maximum of two arguments; the BIGGEST: Num^n → Num does the same from a list of arguments of arbitrary length:

---

[3] The signature of a function f from a domain A to a codomain B is the ordered pair of sets (A, B). It is normally associated to f by writing $f : A → B$.

```
BIGGER    # predefined function
BIGGEST  = INSL(BIGGER)
SMALLER   # predefined function
SMALLEST = INSL(SMALLER)

BIGGER([-10, 0]) ≡ 0  # => true
BIGGEST([-10, 0, -100, 4, 22, -3, 88, 11]) ≡ 88#   # => true
SMALLEST([-10, 0, -100, 4, 22, -3, 88, 11]) ≡ -100#  # => true
```

Catenate

The CAT function concatenates any number of input sequences, thus creating a single output sequence:

```
CAT([[10,30,20],[11],[-7,8,12]]) == [10,30,20,11,-7,8,12]
```

A pair of concrete examples of how the CAT function is used follows. The second one is quite interesting: it gives a filter function used to select the non-negative elements of a number sequence:

```
CAT([[10,30,20],[11],[-7,8,12]]) # => [10,30,20,11,-7,8,12]
(CAT ∘ AA(IF([ LT(0), K([]), ID ])))([-101,23,-37.02,0.1,84])
≡ CAT([ [], [23], [], [0.1], [84] ])
≡ [23, 0.1, 84]
filter(x->x≥0, [-101,23,-37.02,0.1,84]) # Better !!
```

It is more useful to abstract a filter function with respect to a predicate and to an argument sequence, by showing this function semantics where curried LE stands for less or equal to its first argument. LT, GE, GT are similar.

```
FILTER(predicate)(sequence)
FILTER(LE(0))([-1,0,1,2,3,4]) == [-1, 0] # => true
```

A.P. recalls first encountering the comparison predicates LE (less than or equal), LT (less than), GE (greater than or equal), and GT (greater than) in FORTRAN 66, the first scientific programming language, also led by Backus.

Distribute Right/Left

The functions DISTR and DISTL are defined as:

```
DISTR([[a,b,c], x]) ≡ [[a,x], [b,x], [c,x]]
DISTL([x, [a,b,c]]) ≡ [[x,a], [x,b], [x,c]]
```

They accordingly transform a pair, constituted by an arbitrary expression and by an arbitrary sequence, into a sequence of pairs.

Two numeric examples

Example 2.3.1 (Euler number) Let us give an example of Plasm numeric use. The Euler number e is defined as the sum of the reciprocal of factorials. In particular:

$$e = \frac{1}{0!} + \frac{1}{1!} + \frac{1}{2!} + \cdots + \frac{1}{n!} + \cdots$$

We compute an approximation of e, named euler:

```
euler(n) = (ADD ∘ AA(DIV) ∘ DISTL)([1, AA(factorial)(0:n)])
euler(10) # => 2.7182818011463845
euler(20) # => 2.7182818284590455
```

of course, the result precision grows with the number of terms of the series.□

Example 2.3.2 (Factorial function) The factorial function is native in Julia: The number 20 is the highest positive integer for which the expression factorial(20) does not overflow out of memory assigned to an Int64 number. In Julia, you can use a type BigInt with the big function, converting a number to a maximum precision representation. Hence, redefine factorial as the FACT function below:

```
FACT(n) = factorial(big(n))
FACT(49)
608281864034267560872252163321295376887552831379210240000000000
typeof(FACT(49)) # => BigInt
```

Example 2.3.3 (Different implementation) A simpler and more elegant implementation, perfectly corresponding to the defining formula of the Euler number, is given below, where C is the currying combinator[4]:

```
EULER(n) = (sum ∘ AA(C(DIV)(1) ∘ FACT))(0:n)
```

The best Julia approximation of the Euler number is with n = 57 terms of the defining series, since all digits (80) of a BigFloat value are exact:

---

[4] In mathematics and computer science, currying, from logician Haskell Curry (1900, 1982) is the technique of translating the evaluation of a function that takes multiple arguments into evaluating a sequence of functions, each with a single argument.

```
EULER(56)   #=
2.7182818284590452353602874713526624977549695416224229154734
483565013216246202514    =#
EULER(57)   #=
2.7182818284590452353602874713526624977549695416224229154734
483565013216246202549    =#
EULER(58)   #= The 58th element of the Euler series
2.7182818284590452353602874713526624977549695416224229154734
483565013216246202549    =#
```

It could be interesting to investigate how the Julia compiler implements this pipelined formula and the execution times of different approaches.

2.4 Plasm.jl modules and geometric types

We spent several years researching the LAR (Linear Algebraic Representation) representation scheme and the algebraic operations [5, 6, 13, 14] with sparse matrices and solid models. While writing this book, this research developed a new Julia package for geometric programming called Plasm.jl. We aim to finish the software version 1.0 before the book is published.

In the first stage of this endeavor, the work has mainly consisted of porting to Julia the Pyplasm library [16] written in Python years ago, in turn ported from Scheme (initially Common Lisp), with geometric kernel in C++. Of course, both are open-source and downloadable from the web[5].

The second stage of this work was concentrated on a novel implementation of the interactive visualization interfaces in various computational platforms, the flexible implementation of the Lar data structure and its interfacing with the pre-existing Hpc data structure, and principally in a rediscussion and deep testing of the algorithmic pipeline used to compute the space arrangement induced by a given collection of geometric models.

Finally, we have implemented our novel approach to Boolean operations within algebras of rigid solid objects and developed a simple textual user interface for any possible formula of this solid algebra, reduced to an equivalent logical formula between the atoms of the space arrangement. Our software plan was to implement in version 1.0 a significant extension of the language related to computation of space arrangements [13] and Boolean solid algebras [14, 11]. See Chapters 6 and 7 for a complete discussion of this approach.

---

[5] Citation of pyplasm and plasm.jl URLs.

Plasm.jl modules

We warmly encourage readers to install the latest version of Julia and download our package, `Plasm.jl`, into their computational environment. Although this installation is not strictly necessary, web access will be available, and having a local environment to experiment freely is often beneficial, especially when learning a new programming language.

Plasm.jl

The main file of the package is conventionally named `Plasm.jl`, matching the package name itself. Its primary role is to create the run-time executable for producing `Plasm` models by importing external references (i.e., exported functionalities) from other packages and including Julia code from other modules in the package—specifically `fenvs.jl`, `hpc.jl`, `lar.jl`, `boolean.jl`, and `viever.jl`. Following Julia's naming convention, the package file name begins with an uppercase letter.

fenvs.jl

The `fenvs.jl` module, which stands for "functional environments," implements most of the small, exciting programs developed for the GP4CAD book [9]. It includes a rich set of primitives and tools for surface design, leveraging various methods such as parametric, algebraic, and topological approaches. These primitives allow the reader to experiment and explore different surface modeling techniques, providing a solid foundation for understanding geometric programming concepts.

In addition to the built-in functionalities, the `fenvs.jl` file is designed to be extensible. Readers with a computer graphics or CAD background are encouraged to develop and customize their functions to address specific needs or challenges. This flexibility empowers advanced users to adapt the module to various applications, particularly for curve solid modeling, fostering creativity and deeper engagement with the material for programmer readers.

hpc.jl

The primary data structure of the language is the `Hpc` (Hierarchical Polyhedral Complex) [15]. It is based on convex cells but is extended to support general polyhedral cells—polyhedra of any (low) dimension that are connected, potentially nonconvex, and may include interior holes. The `hpc.jl` file includes the design and implementation of the `Hpc` structure, a highly general and multidimensional geometric data structure. This structure facilitates the geometric and solid algorithms and tools presented in this book. As we show, the `Hpc` structure is simple, versatile, and widely applicable. In contrast to the many specialized and complex data structures created for solid modeling, our method relies exclusively on the linear algebra of sparse matrices and vectors, highlighting a topological perspective for geometric computing.

lar.jl

The `lar` subfolder contains a collection of submodules, including the latest procedures in `Plasm`, which utilize chain complexes and yet necessitate extensive testing on novel complex geometries. The subfolder relies on the `arrange2d.jl` and `arrange3d.jl` files, which partition 2D and 3D spaces, respectively, based on a given collection of solid geometric objects. This partitioning is employed to generate an algebra of space atoms. Additionally, other software components were developed years ago to assist designers, particularly for tasks such as domain integration of structure products and polynomials and the generation of multidimensional simplicial and cuboidal grids to model curved manifolds. The subfolder also includes files dedicated to low-level geometric operations and software functionalities.

boolean.jl

The `boolean.jl` module performs a specialized traversal of `Plasm` expressions to generate Boolean formulas for solid algebra. It supports the primitive operators `UNION`, `INTERSECTION`, `DIFFERENCE`, and `XOR`. The module converts the resulting hierarchical `Hpc` assembly into an equivalent flat `Hpc` structure, where all geometric components are contained within a single Cartesian coordinate frame. After flattening, the assembly is rearranged, and the resulting atoms are classified into a canonical sum of minterms. This process yields the logical solution for every Boolean set algebra formula, including the initial solid form. The procedure is inherently complex, and the related software is undergoing further optimization and low-level testing to ensure numerical robustness. In summary, it produces the Brep (boundary representation) solution for any `Plasm` solid algebra formulas.

viever.jl

The `viewer.jl` module provides an interactive geometry viewer for visualizing and interacting with the shapes generated by the `Plasm` codes. It supports rendering on a laptop terminal screen and within the HTML interface of a web browser. The web-based viewer was developed to display Plasm models online and facilitate writing rich-text examples and exercises in a web notebook, eliminating the need to install any software. Two submodules extend this functionality by integrating external libraries. `Viewer.glfw.jl` leverages GLFW, an open-source, multi-platform library for developing OpenGL, OpenGL ES, and Vulkan applications. Meanwhile, `viewer.meshcat.jl` utilizes `MeshCat.jl`, a remotely-controllable 3D viewer built on `three.js` library for web applications. `MeshCat.jl` allows local visualization and interaction with Plasm models directly in a web browser, such as within a `Jupyter` notebook controlled by a Julia server, local or remote.

Plasm.jl geometric types and Objects

Julia `Plasm` is based on three geometric types, with constructors and conversion methods:

Hpc Hierarchical Polyhedral Complex, for hierarchical assemblies and interactive visualization, even on notebooks.

Lar Linear Algebraic Representation, for instancing and computing with cellular complexes and chain operators.

Geo Geometry format, for enumeration of 0- and d-chains in hierarchical space domains, like octrees and potrees, and in nD triangulations.

The following chapters will explore the definitions, motivations, and examples of these data types.

An interactive language shell is an interactive computer programming environment running in a single terminal that accepts user input from a keyboard or a file, executes it, and prints the result to the display. A program developed in the Read-Eval-Print Loop (`REPL`) terminal environment is written and executed piecewise. This approach requires that the language's executable code can serve as an interpreter, i.e., that it combines the translation of source lines into machine code with immediate execution.

Let's begin by explaining the concept of a `REPL` environment, a key feature in many prog languages. At a computer terminal, users enter commands or code in real time, and the system instantly evaluates and displays the results. This interactive environment is available in languages such as Python, Ruby, and JavaScript. Julia provides a first-class `REPL`, which will be described in more detail later.

Here is a breakdown of how `REPL` works:

Read The `REPL` reads the input, usually from the user or a file.

Eval It evaluates the input, which means that it translates the input code into an executable format (such as bytecode or machine code).

Print The evaluation result is printed on the display, showing the user the output or any errors.

Loop] The process repeats, allowing for continuous interaction.

This environment is valuable for developers because it allows quick experimentation, small code snippet testing, and debugging. It is beneficial for learning and rapid development cycles. In the `REPL`, each line of code is typically processed one at a time, offering immediate feedback.

REPL editor

Julia has a full-featured interactive command line REPL (read-eval-print loop) built into the Julia executable [18, 7]. Once Julia is installed on your computer, to enter it, open a terminal and digit "julia" after the system prompt "$". The acknowledgment in Figure 2.3 will appear on the screen. Once the REPL starts, you will be at the Julia prompt [7].

The Julia REPL can operate in different prompt modes:

- Julia mode (the default),
- Help mode,
- Pkg mode, and
- Shell mode.

To enter Help, Pkg, or Shell mode, place the cursor at the beginning of the Julia mode prompt and type a question mark (?), a closing bracket (]), or a semicolon (;), respectively.

The help mode provides information about a function's meaning or use (arguments, examples, etc.), as the developer wrote in the function "docstring." The Pkg (package) mode covers many things, including managing package installations, developing packages, working with package registries, and more. The shell mode allows the terminal user to run shell commands from the Julia REPL.

To return to Julia mode, place the cursor at the beginning of the prompt and press Backspace [18]. A simple and valuable interactive blog to start using REPL is https://blog.glcs.io/julia-repl.

Fig. 2.3 To run Julia by invoking its name on the terminal's command line on a computer screen. Note the "using Plasm" after the REPL prompt "julia> ".

OhMyREPL package

The Julia package OhMyREPL hooks into the Julia REPL and gives it many new features [4]. It allows for several enhancements, including: syntax highlighting; bracket highlighting; bracket completion; rainbow brackets; markdown syntax highlighting; fuzzy REPL history search.

From the online Manual we read:

> "The Syntax highlighting pass transforms the input text in the REPL to highlighted text, highlighting keywords, operators, symbols, strings, etc., in different colors. Bracket highlighting makes matching brackets highlighted when the cursor is between an opening and closing bracket. Rainbow brackets are a feature that colors matching brackets in the same color (with nonmatching closing brackets shown in bold red). OhMyREPL will by default make code blocks written in markdown syntax (for example, in docstrings) highlighted with the color scheme used by the syntax highlighter."

Julia REPL workflow

The most basic Julia workflows involve using a text editor with the Julia command line. A common pattern includes the following elements [8]:

1. Put code under development in a temporary module. Create a file, say Tmp.jl, and include within it

   ```julia
   module Tmp
   <your definitions here>
   end
   ```

2. Put your test code in another file. Create another file, say tst.jl, which begins with

   ```julia
   import Tmp
   ```

   and includes tests for the contents of Tmp. The advantage of using import over using is that you can call reload("Tmp") instead of restarting the REPL when your definitions change. Of course, the downside is needing to prepend Tmp. to access the names defined in your module. (You can reduce this overhead by keeping your module name short.) Alternatively, you can wrap the contents of your test file in a module as

   ```julia
   module Test
   using Tmp
   <scratch work>
   end
   ```

3. The advantage is that you can now do `using Tmp` in your test code and avoid prepending `Tmp.` everywhere. The disadvantage is that the code can no longer be copied selectively to the `REPL` without some adjustments. Lather. Rinse. Repeat. Explore ideas on the Julia command prompt. Save good ideas in `tst.jl`. Occasionally, restart the `REPL`.

```
include("Tmp.jl")
include("tst.jl")
```

Julia's `REPL`, particularly when enhanced by `using OhMyREPL`, provides rich functionality that facilitates an efficient interactive workflow, which one of the authors of this book prefers [7].

2.5 Geometric Programming Examples

Here we discuss a couple of simple, unusual examples to show the compactness and expressive power of `Plasm` geometric package.

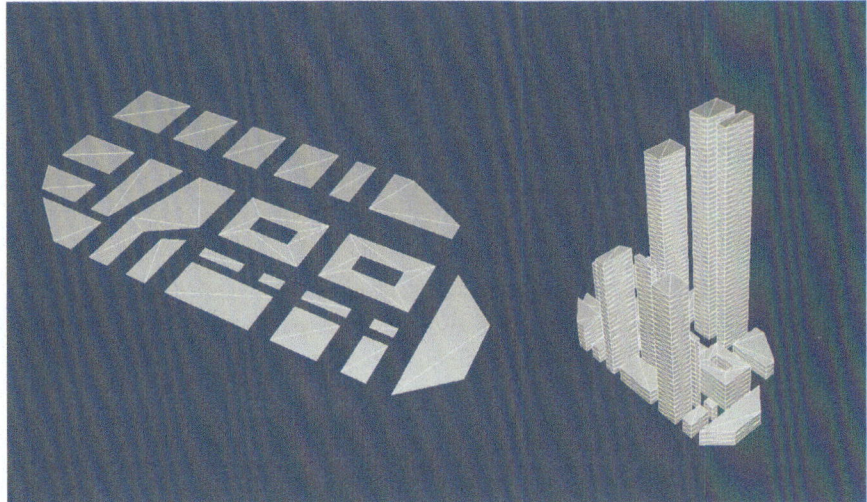

Fig. 2.4 (a) Hierarchical polyhedral complex (Hpc) 2D value embedded in 3D. A perspective projection generates the image. Each connected shape is a polyhedral cell, even nonconvex and with inner holes. Each triangle is a convex cell; (b) 3D Plasm view of the model value of Hpc type of Example 2.5.2.

Coding 2.5.1 (2D virtual Manhattan) Of course, we start our interactive geometric modeling session by telling the Julia compiler[6] to use the `Plasm` package. Then, we give the compiler the following code and data, possibly contained in a file `Manhattan2D.jl`. See the left image on Figure 2.5.

```
julia> using Plasm
```

We begin by defining the 2D coordinates of the vertices of our planar model within a Cartesian coordinate system in $\mathbb{E}^2$—that is, with a specified origin and orientation of the coordinate axes. These coordinates are derived from a pen drawing on a paper graph. When entered on the console, `Plasm` generates a dataset of type `Vector{Vector{Float64}}` containing 100 elements.

```
julia> verts =
    [[0.,0],[3,0],[5,0],[7,0],[8,0],[9.5,1],[10,1.5],[0,3],
[3,3],[5,3],[7,3],[8,3],[9.5,3],[0,4],[3,4],[5,4],[9.5,4],
[12,4],[9.5,5],[10,5],[12,5],[0,6],[3,6],[5,6],[0,7],[3,7],
[5,7],[9.5,7],[12,7],[9.5,8],[12,8],[0,9],[3,9],[5,9],[8,
9],[9,9],[12,9],[0,10],[3,10],[5,10],[8,10],[9,10],[9.5,10],
[10,10],[12,10],[6,11],[7,11],[0,12],[3,12],[9,12],[9.5,12],
[0,13],[3,13],[6,13],[7,13],[9,13],[9.5,13],[0,14],[3,14],[5,
14],[8,14],[9,14],[9.5,14],[10,14],[12,14],[0,15],[3,15],[5,
15],[8,15],[0,16],[6,16],[7,16],[9,17],[9.5,17],[10,17],[12,
17],[6,18],[7,18],[9,18],[9.5,18],[10,18],[12,18],[2,19],[3,
19],[5,19],[8,19],[9,19],[9.5,19],[10,19],[12,19],[5,20],[12,
20],[7,22],[10,22],[9,6],[12,6],[9,15],[9.5,15],[10,15],[12,
15]]
100-element Vector{Vector{Float64}}:      #=
    [0.0, 0.0]
    [3.0, 0.0]
       ⋮
    [12.0, 15.0]           =#
```

Next, we describe both convex and nonconvex cells using lists of vertex indices, represented by the user-defined Julia type `Cells ≡ Vector{Vector{Int64}}`. For simplicity, we manually decompose any nonconvex polygons into their convex components without introducing additional `Plasm` constructions. This is because the `Plasm` type `Hpc` can generate a polyhedral complex directly from a list of convex cells, provided with vertex coordinates and a `Vector{Vector{Int}}` of vertex indices.

```
julia> cells =
    [[1,2,9,8],[3,4,11,10],[5,6,13,12],[14,15,23,22],[16,
17,19,24],[7,18,21,20],[25,26,33,32],[27,95,28,35,34],[95,
96,29,28],[30,31,37,36],[38,39,49,48],[40,41,47,46],[41,61,
55,47],[55,61,60,54],[54,60,40,46],[42,43,51,50],[44,45,65,
```

---

[6] To load Plasm in Julia, open a terminal, start your Julia application, and after the prompt **julia>** write using Pkg; Pkg.add("Plasm ")

```
64],[52,53,59,58],[56,57,63,62],[66,67,84,83,70],[68,69,72,
71],[69,86,78,72],[78,86,85,77],[71,77,85,68],[97,98,74,
73],[99,100,76,75],[79,80,88,87],[81,82,90,89],[91,92,94,93]]
29-element Vector{Vector{Int64}}:     #=
[1, 2, 9, 8]
[3, 4, 11, 10]
  ⋮
[91, 92, 94, 93]        =#
```

Finally, the `verts` and `cells` are transformed into a geometric object of type `Hpc` using the `MKPOL` function ("Make Polyhedron"), and its memory address is stored in the Julia variable `model`.

```
julia> manhattan2d = MKPOL(verts,cells)
# Hpc ... ...
```

A `model` image can be generated by the `Plasm` viewer component within a system window named "Manhattan2D" for graphical interaction with input devices like touchscreen, mouse, joystick, and arrow buttons. Let's note the exact " character for `String` objects.

```
julia> VIEW( manhattan2d, title = "Manhattan2D" )
```

Fig. 2.5 Perspective projections of the 3D model generated by using the 2D cells defined in **Manhattan2D** example. (a) **VIEW** visualization of **Hpc** value of model ; (b) **VIEWCOMPLEX** color visualization of the **Lar** value.

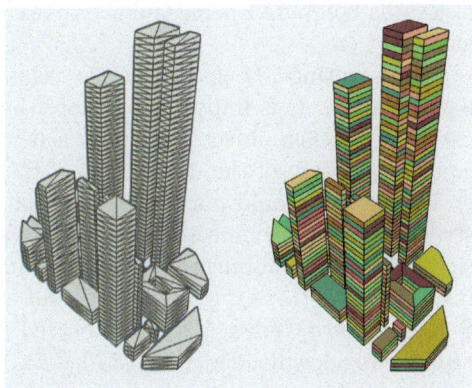

Coding 2.5.2 (3D virtual Manhattan) The model of Figure 2.5 is produced by (1) a vector `ManhattanH` of floors' numbers for ground cell; (2) the generation of corresponding 1D and 2D vectors of `Hpc` objects stored in `pols1D` and `pols2D`, respectively; (3) by Cartesian product of corresponding `Hpc` pairs (in 2D and 1D). See below:

```
ManhattanH = [1,3,1,11,1,2,1,1,1,8,15,1,1,1,1,8,1,15,8,
    1,2,2,2,2,5,9,1,1,1].*3
# 29-element Vector{Int64}:
storeys = CONS(AA(DIESIS)(ManhattanH))(.5)
# 29-element Vector{Vector{Float64}}:
pols1D = AA(QUOTE)(storeys)
# 29-element Vector{Hpc}:
pols2D = [MKPOL(verts,[cell]) for cell in cells]
# 29-element Vector{Hpc}:
```

The 3D model, stored in the `pols3D` variable, is produced by a nice mixture of Julia, `FL`, and `Plasm` primitives:

```
pols3D = AA(splat(*))(TRANS([pols2D, pols1D]))
# 29-element Vector{Hpc}:
VIEW(STRUCT(pols3D), title = "Manhattan3D")
VIEWCOMPLEX(LAR(STRUCT(pols3D)), title = "Manhattan3D")
```

Specifically, we define an array of virtual heights for each polygon in `Manhattan2D`. These values are converted into repeated story heights using `AA(DIESIS)`. Here, `DIESIS` refers to the # operator of FL-based `Plasm`, which cannot be reused in Julia because # denotes comments.

The heights are encoded as 1D `Hpc` polyhedra and stored in the `pols1D` array using the `QUOTE` operator. Similarly, a collection of 2D polyhedra is stored in the `pols2D` array. The Cartesian product * of corresponding `Hpc` objects is computed using Julia's `splat` operator and saved in the `pols3D` array.

The resulting 3D `Hpc` object is visualized in a system window named `Manhattan3D`. It is important to note that every polygon segment multiplication of two `Hpc` objects results in a new `Hpc` polyhedron with a dimension equal to the sum of the dimensions of the operands.

The reader should not forget that the `Plasm` language and the `Hpc` data structure are both multidimensional. We also note that a large number of significant programming examples with Julia and `Plasm` in Julia can be found inside the file `Plasm/src/fenvs.jl` and executed in the terminal by writing $ `julia ./src/fenvs.jl`, being located into the `Plasm.jl directory`, and after having downloaded and installed the package `Plasm.jl` in a recent `julia` environment.

Coding 2.5.3 (Solid 2D graph of a scalar function) A simple, unusual geometric example shows canonical geometric design constructs with `Plasm`. First, we build and show in Figure 2.6a the `Domain2D` of the parametric 2D solid `model` given in Figure 2.6b.

The `MAP` operator is applied to the function `Mapping2D` to be then applied to all vertices of a cell decomposition of `Domain2D`, in turn, generated as the Cartesian product * of two 1D cellular complexes, so producing a 2D cell complex of the mapped interval `Domain2D` with 36×6 squared cells.

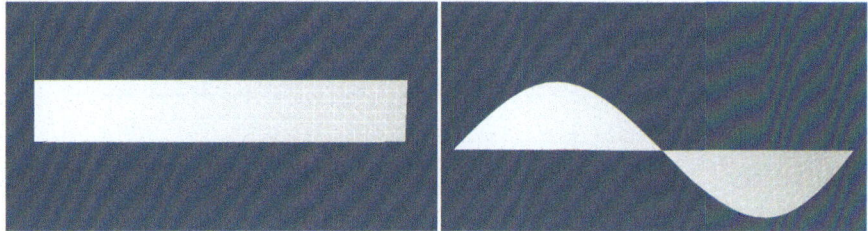

Fig. 2.6 (a) The cellular complex decomposing the function domain $[2\pi, 1]$; (b) the function range transformed after the function $p=(u, v) \mapsto [u, v\sin(u)]$ was mapped on the vertices of the 2D domain.

In Figure 2.6 we show the "solid" mapping (right) of a function of two arguments on its 2D parameter domain (left). Each element (points and cells) of the cellular domain are paired with a corresponding component of the function range. The figures show how the cells are mapped.

```
Domain2D = INTERVALS(2π)(36) * INTERVALS(1)(6);
Mapping2D = p->begin (u,v)=p; [u,sin(u)*v] end;
model = MAP(Mapping2D)(Domain2D);    # Domain2D curved by
    Mapping2D
```

Remark 2.1 (Introduction of MAP *operator)* The `model` domain's two parameters, `u,v`, are calculated by the $p \mapsto (...)$, anonymous `Julia` function, and applied to each vertex of `Domain2D` using the `MAP` operator. ☐

Geometric objects `Domain2D` and `model`—the input and curved output of the second-order `MAP` operator—are displayed in Figure 2.6:

```
VIEW(Domain2D)
VIEWCOMPLEX(LAR(model))
```

The 2D images of `Domain2D` and `model`, visualized in Figures 2.6 and 2.7, are available for display through the `Plasm` operators `VIEW` and `VIEWCOMPLEX`, respectively, and can be manipulated interactively by the user. ☐

References

1. Backus, J.: Can Programming Be Liberated from the von Neumann Style? A Functional Style and Its Algebra of Programs. Commun. ACM **21**(8), 613,641 (1978). DOI 10.1145/359576.359579. URL https://doi.org/10.1145/359576.359579 pages 36
2. Backus, J., Williams, J., Wimmers, E.: An introduction to the programming language FL. In: D. Turner (ed.) Research Topics in Functional Programming. Addison-Wesley, Reading, MA (1990) pages 36, 40

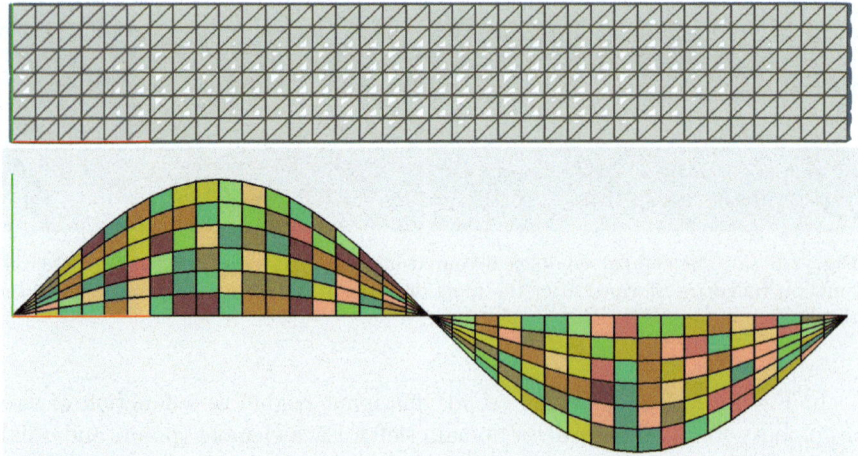

Fig. 2.7 The solid map of function Mapping2D : $\mathbb{E}^2 \to \mathbb{E}^2$; $p = [u, v] \mapsto [u, \sin(u) * v]$: (a) the VIEW of Domain2D of Hpc type; (b) the VIEWCOMPLEX of LAR(model) object of Lar type.

3. Backus, J., Williams, J.H., Wimmers, E.L., Lucas, P., Aiken, A.: FL LANGUAGE MANUAL. PARTS 1 AND 2. Tech. Rep. RJ 7100 (67163), IBM Almaden Research Center (1989) pages 40

4. Carlsson, C.: Syntax highlighting – OhMyREPL. URL https://kristofferc.github.io/OhMyREPL.jl/latest/. [retrieved may 22, 2024] pages 52

5. DiCarlo, A., Milicchio, F., Paoluzzi, A., Shapiro, V.: Chain-based representations for solid and physical modeling. IEEE Transactions on Automation Science and Engineering **6**(3), 454–467 (2009). URL https://ieeexplore.ieee.org/document/5071139 pages 47, 245

6. DiCarlo, A., Paoluzzi, A., Shapiro, V.: Linear algebraic representation for topological structures. Computer-Aided Design **46**, 269–274 (2014). DOI 10.1016/j.cad.2013.08.044. URL https://doi.org/10.1016/j.cad.2013.08.044 pages 47, 137, 184, 186, 218

7. Julia: Manual: The Julia REPL. URL https://docs.julialang.org/en/v1/stdlib/REPL/. [retrieved december 27, 2024] pages 51, 53

8. Julia: Manual: workflow-tips. URL https://docs.julialang.org/en/v1/manual/workflow-tips/. [retrieved December 27, 2024] pages 52

9. Paoluzzi, A.: Geometric Programming for Computer Aided Design. John Wiley Sons, Chichester, UK (2003). URL https://onlinelibrary.wiley.com/doi/book/10.1002/0470013885 pages 37, 38, 48, 73, 127, 130, 201, 206, 209, 244

10. Paoluzzi, A., Pascucci, V., Vicentino, M.: Geometric programming: a programming approach to geometric design. ACM Trans. Graph. **14**(3), 266–306 (1995). DOI 10.1145/212332.212349. URL http://doi.acm.org/10.1145/212332.212349 pages 37, 67, 245

11. Paoluzzi, A., Scorzelli, G.: Computational topology, boolean algebras, and solid modeling. Computer-Aided Design **181**, 103,839 (2025). DOI https://doi.org/10.1016/j.cad.2025.103839. URL https://www.sciencedirect.com/science/article/pii/S0010448525000016 pages 37, 47, 65, 143, 218, 221, 239

12. Paoluzzi, A., Scorzelli, G., Vicentino, M.: Securing the cultural heritage via geometric programming and modeling. Tech. rep., Dept of Computer Science and Engi-

neering, Roma Tre University (2009). URL https://www.academia.edu/47017676 pages 37

13. Paoluzzi, A., Shapiro, V., DiCarlo, A., Furiani, F., Martella, G., Scorzelli, G.: Topological computing of arrangements with (co)chains. ACM Trans. Spatial Algorithms Syst. **7**(1) (2020). DOI 10.1145/3401988. URL https://doi.org/10.1145/3401988 pages 47, 94, 137, 216, 218, 221, 224, 227, 228, 231, 233, 234, 235, 251, 258, 260, 278

14. Paoluzzi, A., Shapiro, V., DiCarlo, A., Scorzelli, G., Onofri, E.: Finite algebras for solid modeling using julias sparse arrays. Computer-Aided Design **155**, 103,436 (2023). DOI https://doi.org/10.1016/j.cad.2022.103436. URL https://www.sciencedirect.com/science/article/pii/S0010448522001695 pages 47, 65, 97, 137, 184, 218, 221, 254, 260, 270, 277

15. Pascucci, V., Ferrucci, V., Paoluzzi, A.: Dimension-independent convex-cell based lar: Representation scheme and implementation issues. In: C. Hoffmann, J. Rossignac (eds.) 3rd ACM/IEEE Symposium on Solid Modeling and Applications, pp. 17–19. ACM Press, New York, NY (1995). SMA 95 pages 48, 248

16. Scorzelli, G.: Pyplasm library (2023). URL https://libraries.io/pypi/pyplasm pages 37, 47

17. Scorzelli, G., Paoluzzi, A.: Plasm.jl: v0.1.0 (2023). URL https://github.com/PlasmLanguage/Plasm.jl pages 37

18. Whitaker, S.: Mastering the Julia REPL. URL https://blog.glcs.io/julia-repl#heading-starting-the-julia-repl. [retrieved may 22, 2024] pages 51

19. Williams, J.H., Wimmers, E.L.: Sacrificing Simplicity for Convenience: Where Do You Draw the Line? In: Proceedings of the 15th ACM SIGPLAN-SIGACT Symposium on Principles of Programming Languages, POPL '88, p. 169179. Association for Computing Machinery, New York, NY, USA (1988). DOI 10.1145/73560.73575. URL https://doi.org/10.1145/73560.73575 pages 37

Chapter 3
Topology primer

This chapter introduces basic topological methods for solid and geometric modeling computations in Building Information Modeling (BIM) and Computer-Aided Design (CAD). These methods include calculating space partitions, adjacency relations, and mesh cell orderings using algebraic topology and combinatorial techniques.

Cellular models, often called meshes in engineering, design, and graphics, use discrete cells to represent a geometric domain. In fields such as geospatial mapping, computer vision, robotics, graphics, finite element analysis, medical imaging, 3D printing, solid modeling, and geometric design, calculating incidences, adjacencies, and mesh cell orderings often depend on incompatible data structures and algorithms.

This chapter introduces the unifying topological concept of chain complex to compute 2D or 3D space partitions induced by collections of 1D, 2D, and 3D geometric objects, addressing these challenges. Such space decompositions, known as arrangements in combinatorial topology and algebra, form the foundation of our algebraic approach to geometric and solid modeling.

To efficiently manage large geometric datasets, we leverage fast, explicit expressions of topological relations via computational structures such as Julia's sparse matrices and vectors. These are employed in `Plasm` to compute the chain complex—a collection of linear spaces of cell chains and the linear transformations between them, spanning dimensions zero to three.

The last section illustrates `Plasm` functions for calculating domain integrals on piecewise-linear polyhedra, an essential solid modeling topic, using (co)chain complexes. In particular, we clarify the significance of cochains as tools for volumetric integration over coherently oriented triangulations of the boundary of the solid model. Special attention is given to integrating the monomial structure field to compute the mechanical properties of solids, including area, volume, centroid, products, and moments of inertia, in scalar and tensor forms.

© The Author(s), under exclusive license to Springer Nature Switzerland AG 2026
A. Paoluzzi and G. Scorzelli, *BIM Geometry with Julia Plasm—Functional Language for CAD Programming*, Digital Innovations in Architecture, Engineering and Construction, https://doi.org/10.1007/978-3-031-90244-4_3

3.1 Preliminaries

3.1.1 Motivation

The foundation of Plasm geometric language is built on key concepts from algebraic topology, which many readers may find unfamiliar and challenging to understand. Therefore, this chapter provides a simplified introduction to the topological ideas and methods needed to understand the language, especially its generalized approach to Constructive Solid Geometry (CSG). Plasm extends this to a Solid Boolean Algebra and its isomorphic Set Algebra.

The Plasm methods integrate generative, decomposition, enumerative, CSG, and boundary representation schemes. The book illustrates the language through numerous simple examples. It also explains how to use a complete solid algebra isomorphic to CSG, expanded with n-ary operations and explicitly considering external space, including holes within and between the elements of any assembly of solid models.

In particular, we have focused on the space arrangement generated by a set of geometric objects, represented by the boundaries of their cell decomposition. This allowed us to classify the whole Euclidean d-space into a collection of well-behaved d-cells and the set algebras of their subsets.

The atoms of this set algebra are the minimal (irreducible) elements of the space arrangement induced by the input. The quasi-disjoint union of specific subsets of such elementary components corresponds to the generator terms that produce the arrangement and is isomorphic to a set algebra.

The Domain Specific Language (DSL) Plasm supports all the examples and pictures in this book. In practice, any valid expression produces the Euclidean d-space arrangement and the related set of atoms generated by the expression terms. Applying a logical function (equivalent to the solid generating formula) to the binary representation of atoms, Plasm selects the subset of atoms solution of each solid expression. The evaluation process also classifies the inner and outer elements of linear spaces of chains.

In particular, computing a solid shape from a Boolean expression is like establishing a logical equivalence between a left side with solid geometric terms and point operations (union, intersection, difference, xor) and a right side that contains the disjoint union of the atom subset whose support space encloses the same Euclidean domain as the left side and is in canonical Boolean Sum Of Products. (SOP).

The Plasm Boolean method constructively derives this identity. While the left-hand side is represented in Plasm language as a combination of solid shapes, including hierarchical ones, the right-hand side is built step-by-step algorithmically, working with chain complexes to geometrically construct (1) all the atoms forming the space arrangement induced by the input and (2) the corresponding sequence of logical minterms. Each solid Boolean formula is ultimately converted into its canonical sum of minterms[1] and graphically exported or visualized.

3.1.2 Definitions

Let us begin with a note on notations: Greek letters denote the cells of a space partition, while Roman letters represent the chains of cells. These chains are encoded in coordinates as either signed integers or sparse arrays of signed integers.

By transforming back the binary representation of shape chains, the geometric representation of the objects is finally constructed so that it can be visualized or exported. The remainder of this chapter explains the different kinds of complexes the language uses.

A complex is a graded set $S = \{S_i\}_{i \in I}$ i.e. a family of sets, indexed here over $I = \{0, 1, 2, 3\}$. This section introduces two closely related types of complexes: the cell complex and the chain complex. The chains of cells are represented as either a vector or a matrix of signed or unsigned integers or as sparse arrays of signed or unsigned integers.

Definition 3.1 (d-Manifold) A manifold is a topological space that resembles a flat space locally, i.e., near all points. Each point of a d-dimensional manifold has a neighborhood that is homeomorphic[2] to $\mathbb{E}^d$, the Euclidean space of dimension d. Hence, this geometric object is often referred to as d-manifold.

Definition 3.2 (Cell) A p-cell σ is a p-manifold with boundary $(0 \leq p \leq d)$, which is linear, connected, possibly nonconvex, and not necessarily contractible. This definition refers to cellular complexes used in this paper and differs from others because a cell is neither necessarily simplicial, convex, or contractible.

In Plasm, cells are connected but may contain internal holes; conversely, cells of CW-complexes [12] are contractible to a point. In particular, we deal with Piecewise-Linear (PL) cells of dimensions 0, 1, 2, and 3, respectively. It should be noted that 2- and 3-cells may contain holes while remaining connected. In other words, Plasm cells are p-polyhedra, i.e. segments, polygons, and polyhedrons embedded in 2D or 3D space.

---

[1] In mathematics and computer science, a canonical form of a mathematical object is a standard way of presenting it as a mathematical expression. It often provides the most straightforward representation of an object and allows for its unique identification.

[2] A homeomorphic neighborhood refers to something that is "topologically equivalent," similar to a rubber patch that can be stretched without altering its topology.

Definition 3.3 (Cell complex) A cell p-complex or cellular p-complex is a finite set of cells with at most dimension p and all their r-dimensional boundary faces ($0 \leq r \leq p$). A face is an element of the PL boundary of a cell that satisfies the boundary compatibility condition. Two p-cells α, β are considered boundary compatible when their point set intersection contains the same r-faces ($0 \leq r \leq p$) for both α and β.

Definition 3.4 (Regular complex) A cell complex is said to be regular if the closure[3] of each cell is homeomorphic to a closed ball, and the boundary of each cell is a union of lower-dimensional cells.

In other words, every cell "fits together" nicely with its lower-dimensional faces, ensuring a well-behaved topological structure. Regularity guarantees that cell attachment is simple and follows the structure of a well-defined (boundary-compatible) chain complex.

Definition 3.5 (Skeleton) The s-skeleton of a p-complex Λ_p ($s \leq p$) is the set $\Lambda_s \subseteq \Lambda_p$ of all r-cells ($r \leq s$) of Λ_p. Every skeleton of a regular complex is a regular subcomplex. The difference $\Lambda_r - \Lambda_{r-1}$ of two skeletons is the set U_r of r-cells.

Definition 3.6 (Support space) The support space Λ of a cellular complex is the topological space that is "underlying" or "represented" by the complex, considering all the cells as glued pieces.

Definition 3.7 (Characteristic function) Given a subset A of a larger set S, the characteristic function $\chi_S(A)$, also called the indicator function of A, is defined as identically one on A and zero elsewhere. [23].

Geometric modeling

In computer graphics and solid modeling, the enumerative representation schemes [21] refer to a space partition embedding the object and associated with a cellular complex, where each model portion can be bijectively mapped to some (unique, even when composite) integer multi-index value. Consequently, the solid model remains identified by the indices of space cells assumed to be inner or on the boundary of the represented solid object.

While the cells of a decompositive representation are a partition of the object, those of an enumerative representation refer to a partition of the

---

[3] Closure of a set S in a topological space X is the smallest closed set containing S. It is denoted by $\overline{S}$ and consists of all the points in S along with its limit points (points that can be approached arbitrarily closely by points in S). Formally, $\overline{S} = S \cup S'$, where S' is the set of limit points of S. In other words, it is the union of S and its boundary.

space in which the object is embedded. In the following example, we first encounter the Cartesian (topological) product "*" of complexes.

It is easy to see that when given two low-dimensional spaces, represented through a space decomposition, their Cartesian product creates an enumerative scheme, where each cell corresponds one-to-one to two generating cells and their pair of indices.

Example 3.1.1 (Polyhedral Cartesian products) of 1-complexes. The reader should try the `VIEW` application expressions below:

```
grid1D = GRID(NN(10)([.1,-.1]));
grid2D = grid1D * grid1D;
grid3D = grid2D * grid1D;
VIEW(grid1D); VIEW(grid2D); VIEW(grid3D)
```

The `GRID` operator produces a 1-complex of line segments of specified sizes. `NN` repeats and concatenates number sequences. Negative numbers indicate empty segments. "*" is the Julia symbol for the `PROD` operator. Be careful when copying from `.pdf` files to the `REPL`: asterisks are replaced with a lowered character and must be manually replaced one-by-one with the correct character.

Remark 3.1 (Enumeration of generator's atoms) In solid modeling, we refer to the enumerative scheme when the space is partitioned and the list of its cells defines the object. In Chapter 7, we will see that geometric generators can produce the space partition (arrangement) into closed manifold cells known as atoms, and the results of a solid algebra formula are represented by a list of atoms [20, 18].

This book generalizes the CSG (Constructive Solid Geometry) scheme into a solid Boolean algebra. Here, geometric objects stem from Boolean expressions of any complexity. They are represented as characteristic functions of atom subsets from a set algebra that partitions the embedding space. Refer to Chapter 7. The Solid Boolean Algebra is a novel representation scheme [18] for solids situated halfway between decompositive and enumerative representations.

3.1.3 Basic geometry

Canonical representations in a Solid Boolean Algebra

Let's combine a collection of solid models to create new objects from the disjoint union of their unique parts (after the combination). Each element of this partition (of the generator objects and the embedding space), including the external space, is referred to as an "atom." Therefore, these "generators" are integrated, forming an arrangement of the embedding space such that each point of the space belongs to one and only one atom

A set of n atoms generates a set algebra with 2^n elements. Each element has a unique expression as a binary characteristic function, representing itself as a subset of the set of atoms and partitioning both the space and the generators. Every term in this algebra is canonically represented as a sum of minterms, and each atom serves as a minterm with only one bit set to 1.

The sum of minterms in a Boolean algebra is defined as follows:

1. Definition: The sum of minterms represents a Boolean function, expressed as a disjunction (OR operation) of all possible minterms that produce an output 1.

2. Minterm: A minterm is a product (AND operation) of all the variables in the function, with each variable appearing either in its proper form (x) or in its complemented form $(\bar{x})$.

3. Sum of Minterms: If a Boolean function F has n variables, the sum of minterms includes all terms where $F = 1$. It is often written in the following form:

$$F = m_1 + m_2 + \cdots + m_k$$

Where m_i represents a minterm corresponding to a specific binary combination of input variables that makes F equal to 1. [4]

Remark 3.2 (Characteristic function of solid objects) A binary string is the value of the characteristic function associated with a solid term in a solid algebra. The ones in the string represent the presence of the corresponding atom; the zeros represent the corresponding absence. The set of all such strings (minterms) is isomorphic to the Boolean algebra generated by a set of primitive geometric generators. The length of all strings is equal to the number of atoms in such algebra. The number of strings is 2^n if n is the number of atoms, which is also the cardinality of the basis in the C_3 chain space partitioning $\mathbb{E}^d$. Every solid object in this algebra has a canonical representation as a union of some atoms, i.e., as the sum of a subset of minterms. See Chapter 7.

3.2 Cellular models

A cellular model, or mesh, represents a domain using discrete cells. In many areas of geometric/numeric computational engineering, including geomapping, computer vision and graphics, finite element analysis, medical imaging, geometric design, and solid modeling, incidences, adjacencies, and order-

---

[4] It provides a canonical description of any Boolean function and is unique.

ing of mesh cells must be computed, generally using disparate and incompatible data structures and algorithms.

Coding 3.2.1 (Cartesian Product of 0/1 chain instances.)

```
grid1D = QUOTE(N(10)(.1));
grid2D = grid1D * grid1D;
grid3D = grid2D * grid1D;
grid11d = SKELETON(1)(grid2D);
VIEW(grid11d); VIEW(grid2D); VIEW(grid3D)
```

`N(10)(.1)` creates an array of 10 values `0.1`. The operator `QUOTE` transforms it into a 1D complex, i.e., a sequence of line segments. Figure 3.1 shows pictures of a 2D and 3D cuboidal complex.   □

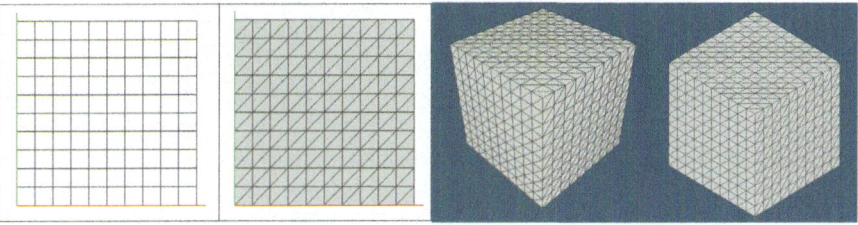

Fig. 3.1 Cellular complex: (a) 1-complex; (b) 2-complex; (c) 3-complex with perspective projection; (d) 3-complex with isometric parallel projection.

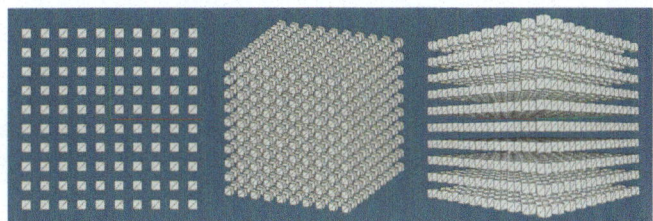

Fig. 3.2 disconnected cellular complex: (a) 2-complex; (b) 3-complex

Simplicial and cuboidal complexes, discussed in this section, are the most common types of cellular complexes. `Plasm` introduced the concept of multidimensional polyhedral complexes in [17], where cells are general polyhedra often decomposed into convex cells. Hierarchies of geometric objects were introduced simultaneously in [17] as hierarchical polyhedral complexes (HPC) to allow for the representation of object assemblies, also called structures, with components and substructures stored in local coordinate frames.

Chain complexes are different datasets that mathematicians use in combinatorial and/or algebraic topology that apply the tools of abstract algebra to study the number and type of holes in spaces.

Our geometric formulation, representation, and algorithms, based on chain complexes of boundary maps, may be applied to various geometric objects, from solid models to engineering meshes to biomedical images and 3D printing.

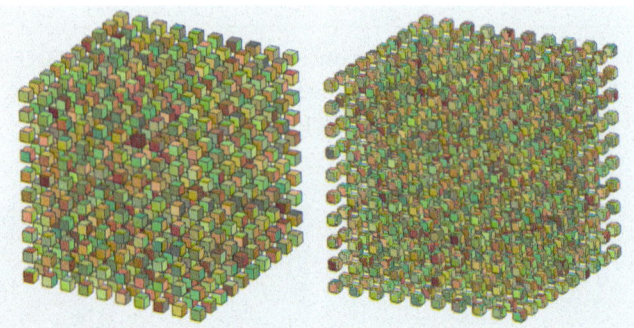

Fig. 3.3 Cellular complexes: (a) cellular 3-complex, with totally unconnected cells; (b) exploded 2-skeleton of the 3-complex.

Remark 3.3 (Chain complexes and numerical methods) Numerical methods that aim to integrate domain modeling, differential topology, and mathematical modeling with physical simulations are also based on chains and cochains, in particular in the mathematics of finite element methods that employ differential complexes to construct stable numerical schemes [3, 2].

Coding 3.2.2 (Grids as Cartesian Product of 1-products.)

The transformation of a geometric object from Hpc to Lar data structures is carried out by the LAR operator, which stores all the p-chains, where $0 \leq p \leq r$, of an r-chain within a single coordinate frame in d-space, with $r \leq d$.

```
largrid3D = LAR(grid3D);
largrid2D = LAR(SKELETON(2)(grid3D));
VIEWCOMPLEX(largrid3D);
VIEWCOMPLEX(largrid2D, explode=[1.3,1.3,1.3])
```

VIEWCOMPLEX is used to create graphics scenes interactively using Lar data structures that store the complete dataset of a cellular complex.

```
largrid3D.C
Dict{Symbol, Vector{Vector{Int64}}} with 5 entries:
  :CF => [[1, 2, 4, 9, 11, 13], [2, 3, 7, 10, 19, 125], [4, 5,…
  :CV => [[1, 2, 3, 4, 5, 6, 7, 8], [1, 2, 7, 8, 45, 46, 47, 4…
```

```
:FV => [[1, 2, 3, 4], [1, 2, 7, 8], [1, 2, 45, 46], [1, 3, 5…
:EV => [[1, 2], [1, 3], [1, 7], [1, 9], [1, 45], [2, 4], [2,…
:FE => [[1, 2, 6, 9], [1, 3, 7, 19], [1, 5, 8, 129], [2, 3, …
```

Conversely, the Hpc type stores only the (convex) cells of maximum dimension. The operator SKELETON(p)(obj) extracts the subset of p cells from obj :: Hpc.   □

Fig. 3.4 (a) 3-complex; (b) exploded 2-skeleton. The reader should guess the differences with the previous image.

3.2.1 Cubical complex

Let us now introduce the definition and use of the multidimensional grids of cuboidal, and the general Cartesian product $*$ of cellular complexes, also referred to as topological product, which is commutative and associative like the standard product of numbers. The $*$ operator, depending on the dimension of the inputs, can generate either full-dimensional (that is, solid) output complexes or lower-dimensional complexes of dimension d embedded in Euclidean n-space, with $d \leq n$. We say that an object is *solid* when $d = n$.

Regular cubical grid

Regular cubic grids are commonly used in finite element analysis, finite volume methods, finite difference methods, and generally for discretizing parameter spaces. Specifically, they create parametric curves or surfaces through MAP of a vector-valued function of d parameters over a grid discretization of the function's domain or the object space, as we will explore in a few pages.

Definition 3.8 (Regular cuboidal grid) A regular grid is a tessellation of n-dimensional Euclidean space formed by congruent parallelotopes (e.g., bricks). Its counterpart is an irregular grid. We refer to these objects as cuboidal complexes to emphasize their multidimensional nature in Plasm.

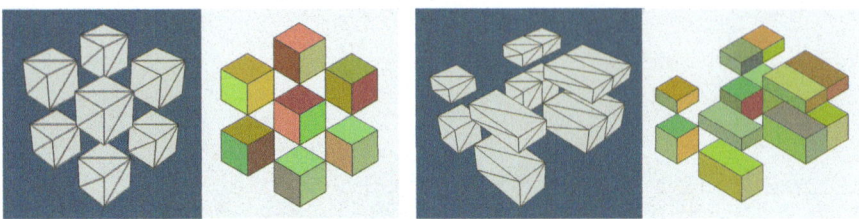

Fig. 3.5 Unconnected cubical grids: (a) regular cell complex; (b) irregular cell complex. See Script 3.2.3.

For example, think of a mesh of 3D cubes in three-dimensional space for the first case and the (non-manifold) framework of skeletal polygons of such cubic cells for the second case.

In particular, both n-dimensional solid grids of (hyper)-cuboidal cells and their d-dimensional skeletons ($0 \leq d \leq n$), embedded in $\mathbb{E}^n$, are generated by assembling the cells produced by a product of n either 0D or 1D cellular complexes that, in such lowest dimensions, coincide with simplicial complexes.

In Plasm, the cuboidal grids are not necessarily regular and can even be unconnected.

Coding 3.2.3 (3D Examples) Two simple examples of 3D cuboidal grids are shown in Figure 3.5. Each GRID instance generates a 1D cellular complex, and two Cartesian products, denoted in Plasm by the * infix symbol, produce the 3D cell complex.

```
VIEW(GRID([1,-1,1]) * GRID([1,-1,1]) * GRID([1,-1,1]))
VIEWCOMPLEX(LAR(GRID([1,-1,1])* GRID([1,-1,1])* GRID([1,-1,1])))

VIEW(GRID([1,1,-2,1]) * GRID([1,-1,2]) * GRID([1,-1,0.5]))
VIEWCOMPLEX(LAR(GRID([1,1,-2,1])*GRID([1,-1,2])*GRID([1,-1,.5]))
    )
```

It is easy to believe that negative parameters represent empty space, while positive parameters indicate full space. See Figure 3.5.   □

Regular domain of curved manifold

In Plasm, regular grids of cubical cells codify the domain of curved manifolds, including curves (1D), surfaces (2D), and curved solids (3D).

Let us start by discussing the internal structure of a minimal grid. Two adjacent 3-cells share only one common boundary, 2-cell, and not two, as might seem natural. Furthermore, in a well-formed d-complex, each $(d-1)$-cell (hence, 2-cells in 3D) cannot be shared by more than two d-cells. In particular, the topological properties tell us that this number is exactly two if the complex is closed, i.e., without boundary, like a 2-sphere, which is empty inside.

Example 3.2.1 (Small cellular complex) We recall that a convex 3D cell may be represented (a) as the convex hull of its vertices (as it is done in Julia Plasm by the `Hpc` type), or by its 2-skeleton, or better yet, by the 2-cells of its boundary (as we do in the `Lar` type), i.e., by its Brep:

```
grid = CUBOIDGRID([2,3,1]) # Lar
VIEWCOMPLEX(grid,explode=[1.2,1.2,2])
```

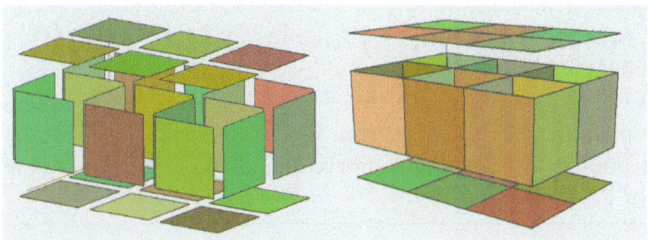

Fig. 3.6 Non-manifold internal brep, important with new materials models.

A small 3D `grid` is exploded in Figure 3.6. The complexes generated by `CUBOIDGRID` are strongly non-manifold, an essential issue in designing a new representation for solid modeling. The Plasm language and its data structures make the distinction between manifold and non-manifold objects and models obsolete. When using chain complexes, manifoldness and its reverse property are never considered.

Definition 3.9 (Bézier d-manifolds) We call Bézier d-manifold of degree n in $\mathbb{E}^m$ the `PL` (piecewise-linear) curved object obtained by mapping a vector function over the vertices of a `PL` cellular grid. The standard generative pattern in Plasm is with a `MAP`:

```
MAP(<Function>)(domain)
```

where the vector `<Function>` is the `CONS` (see Section 2.3) of a Julia `Vector`
`{Function}` whose m univariate (coordinate) functions $[0, 1] \subset \mathbb{R} \to \mathbb{R}$ are
usually[5] Bézier polynomials of degree n.

Definition 3.10 (Bézier polynomial of degrée n) The Bézier polyno-
mial of degree n in $\mathbb{E}^m$ is a function from $\mathbb{R} \to \mathbb{R}$, represented as a linear
combination of $n + 1$ real numbers (or real-valued functions, if transfinite)
using the $n + 1$ polynomials of the Bernstein basis, specifically:

$$B_k^n(x) = \binom{n}{k} x^k (1 - x)^{n-k}, \qquad 0 \le k \le n.$$

The curved surface of Figure 3.7b is obtained by mapping a vector `[C0,C1,`
`C2,C3]` of univariate functions $[0, 1] \subset \mathbb{R} \to \mathbb{R}$ which are Bézier polynomials
of various degrees.

   Examples of d-manifold curves and surfaces generated by Julia `Plasm`
scripts follow.

Example 3.2.2 (Bézier manifolds) The curve domain $[0, 1]$ of Figure 3.7a
is decomposed by `Plasm` in `32` segments (1-cells), i.e., is created as 1D cellular
complex embedded in $\mathbb{R}$.

```
VIEW(MAP(BEZIER(S1)([[0,0],[1,0],[1,1],[2,1],[3,1]])) (
    INTERVALS(1.0)(32)))
```

The functions `C0`, `C1`, `C2`, `C3` given below, generated by variable numbers
of control points, are 3D Bézier curve-generating functions, as asserted by
the selector `S1`. See Figure 3.7.

```
C0 = BEZIER(S1)([[0,0,0],[10,0,0]])
C1 = BEZIER(S1)([[0,2,0],[8,3,0], [9,2,0]])
C2 = BEZIER(S1)([[0,4,1],[7,5,-1], [8,5,1],[12,4,0]])
C3 = BEZIER(S1)([[0,6,0],[9,6,3], [10,6,-1]])
```

The control points here are $q \in \mathbb{E}^d$, with $d = 3$. A higher number of coordinate
functions would be produced by a higher dimension d of the manifold (via a
higher number of coordinates of control points). Later, we show examples of
dim $= 1, 2, 3$.

```
VIEW(MAP(BEZIER(S2)([C0,C1,C2,C3]))(INTERVALS(1.0)(10)*
    INTERVALS(1.0)(10)))
```

See the curve and surface generated in Figure 3.7.   □

---

[5] `Plasm` is able to create curved manifolds combining lower-dimensional curved manifolds
of various degrees. In other words, it is very flexible.

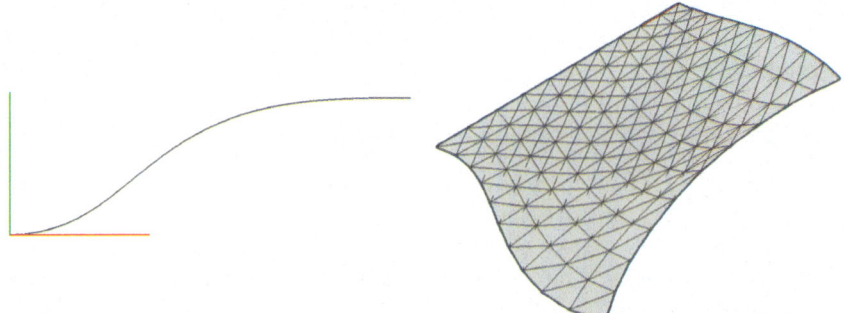

Fig. 3.7 Bézier manifolds. See Example 3.2.2: (a) Bézier curve of degree 4 in $\mathbb{E}^2$; (b) Bezier surface in $\mathbb{E}^3$ generated by four boundary curves. The MAP operator generates any finite-dimensional manifold in Plasm.

Example 3.2.3 Transfinite CUBICHERMITE patch generated by two boundary curves and two constant tangent fields. Note the different uses of S1 and S2.

```
c1 = CUBICHERMITE(S1)([[1,0,0],[0,1,0], [0,3,0],[-3,0,0]])
c2 = CUBICHERMITE(S1)([[0.5,0,0],[0,0.5,0], [0,1,0],[-1,0,0]])
surface = CUBICHERMITE(S2)([c1,c2, [1,1,1],[-1,-1,-1]])
dom1D = INTERVALS(1.0)(14)
VIEW(MAP(surface)(dom1D*dom1D))
```

For a definition of Transfinite Hermite and many other generation methods, all supported by Plasm.jl, interested users can refer to the chapter on Curves and Surfaces in Reference [16]. □

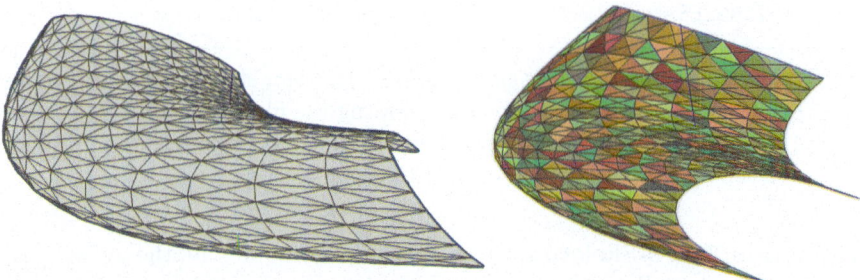

Fig. 3.8 Approximated Piecewise-Linear (PL) 2-manifold in $\mathbb{E}^3$. CUBICHERMITE airplane wing generated by C1, C2 curves and two opposite fields 1, −1 on opposite sides.

Definition 3.11 *p-Cell.* *p*-Manifold with boundary $(0 \leq p \leq d)$ which is piecewise-linear, connected, possibly nonconvex, and possibly noncontractible to a point, i.e., with holes.

This definition refers to cellular complexes used by Plasm and differs from others in the literature because a cell is neither simplicial, convex, or contractible. We call it polyhedral, and it may contain holes, so its boundary may be unconnected. In particular, every Hpc p-cell ($0 \leq p \leq d$) is connected, manifold, and convex. Lar objects are possibly non-convex and with holes.

Example 3.2.4 (Instancing scheme) The primitive instancing representation scheme [21] refers to specific code-generating parametric instances of named solid models. Plasm offers many primitive generators that users can extend. For example, a cube with an edge length of one is generated and represented as follows:

```
CUBE(1) #=
Hpc(MatrixNd(4), Hpc(MatrixNd(4), Geometry([[0.0, 0.0, 0.0],
    [1.0, 0.0, 0.0], [0.0, 1.0, 0.0], [1.0, 1.0, 0.0], [0.0,
    0.0, 1.0], [1.0, 0.0, 1.0], [0.0, 1.0, 1.0], [1.0, 1.0,
    1.0]], hulls=[[1, 2, 3, 4, 5, 6, 7, 8]]))) =#
```

Remark 3.4 (Primitive solid generators.) A current incomplete list of Plasm primitive generator functions follows. All functions generate a Julia object of Hpc type, which is required for the STRUCT assembly operator and many Plasm operators, including the VIEW function. To obtain the corresponding Lar value, which holds the entire cellular complex of the primitive, apply LAR(<name>).

```
CIRCLE, CIRCUMFERENCE, CONE, CROSSPOLYTOPE, CYLINDER,
    DODECAHEDRON, FINITECONE, ICOSAHEDRON, ICOSPHERE, MIRROR,
    NGON, OCTAHEDRON, PERMUTAHEDRON, POLYLINE, POLYMARKER,
    POLYPOINT, PRISM, RING, SPHERE, TETRAHEDRON, TORUS,
    TRIANGLEFAN, TRIANGLESTRIPE, TRUNCONE, TUBE.
```

To learn how to instantiate each function, the user should type their name in REPL after a ?. Julia will respond by showing the file name and line number where the function is defined. A quick look at the first lines of the source code will give the necessary details.

Definition 3.12 s-Skeleton. Given a p-complex Λ_p ($s \leq p$), the set $\Lambda_s \subseteq \Lambda_p$ consists of all r-cells ($r \leq s$) of Λ_p. Each skeleton of a regular complex is a regular subcomplex.

The difference $\Lambda_s - \Lambda_{s-1}$ of two skeletons is the set U_s of s-cells. An s-skeleton is regular when all r-cells ($r \leq s$) are faces of some s-cell.

Mapping 1-complexes to *d*-complexes

Example 3.2.5 (s-Skeleton) Skeleton extraction is used in `Plasm` to generate subcomplexes of objects with higher intrinsic dimension.

```
verts = [[0.,0.,0.],[3,0,0],[3,2,0],[0,2,0],
    [0,0,1.5],[3,0,1.5],[3,2,1.5],
    [0,2,1.5],[0,1,2.2],[3,1,2.2]]
```

Interesting uses of the `OFFSET` second-order operator are defined below and shown in Figure 3.2.1, where the `House` wireframe model is obtained as the 1-skeleton of the convex hull of a set of 3D points.

```
House = SKELETON(1)(CONVEXHULL(verts))
VIEWCOMPLEX(LAR(House))
VIEWCOMPLEX(LAR(OFFSET(kernel)(House)))
```

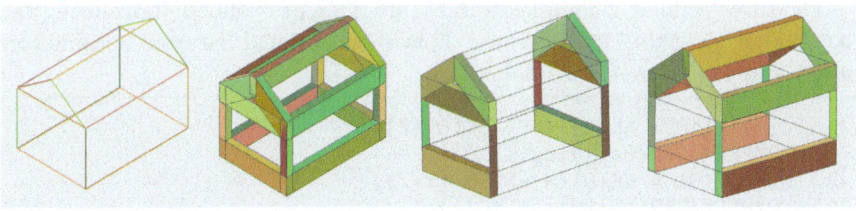

Fig. 3.9 Hpc 1-complex → Lar 3-complex, via **OFFSET** of 1-skeleton with kernels equal to [0.2, 0.2, 0.5], [0.2, 0., 0.5], [0., 0.2, 0.5], respectively.

In the following examples, we demonstrate how to automatically convert a wireframe (1D) model into a solid one, using the powerful `Plasm` operator `OFFSET`, which belongs to a family of similar operators [4] that are useful for transforming a schematic design into a more refined version of the idea. In particular, `OFFSET` operates by repeatedly extruding the object in a higher-dimensional space (up to $\mathbb{E}^6$) and then projecting it back appropriately.

Example 3.2.6 (Complex combinatorics) Fascinating subcomplexes can be easily extracted from a complex. Note the types: `LAR( :: Hpc) :: Lar`

```
X = GRID([2.4,4.5,-3,4.5,2.4])
Y = GRID([7,5]); Z = GRID([3,3]);
idea = X * Y * Z;
arrangement = ARRANGE3D(LAR(idea))
VIEWCOMPLEX(arrangement, explode = [1.2,1.2,2.0] )
VIEWCOMPLEX(arrangement, show=["CV"], explode = [1.2,1.2,2.0] )
```

Fig. 3.10 The exploded
arrangement of atoms
includes the two "outer"
atoms that bound the
"inner part" of the exterior
"unbounded" atom. The
total number of atoms
is 10, each depicted in
a single color. Refer to
Chapter 6 for details.

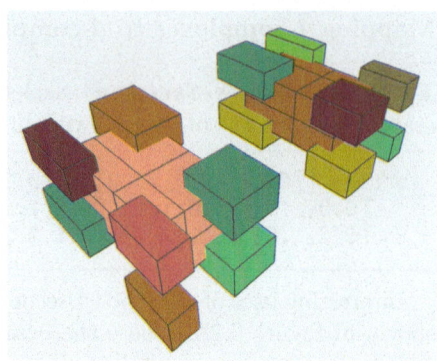

```
SK = SKELETON
building110 = X*Y*SK(0)(Z); VIEWCOMPLEX(LAR(building110))
building101 = X*SK(0)(Y)*Z; VIEWCOMPLEX(LAR(building101))
building011 = SK(0)(X)*Y*Z; VIEWCOMPLEX(LAR(building011))
```

The three cellular 2-complexes in Figure 3.2.6 are defined above using the
SK alias for the SKELETON operator. Lower-dimensional datasets obtained by
SK(0) and SK(1) are shown below.

```
building1_110 = SK(1)(X*Y)*SK(0)(Z);
VIEWCOMPLEX(LAR(building1_110));
building1_101 = SK(0)(X)*SK(1)(Y*Z);
VIEWCOMPLEX(LAR(building1_101));
building1_011 = R(2,3)(-π/2)(SK(1)(X*Z)*SK(0)(Y));
VIEWCOMPLEX(LAR(building1_011));
```

The 1-skeletons generated above produce via OFFSET operations the structural
frames of the building "idea" formulated in Example 3.2.6.   □

Example 3.2.7 (Complex combinatorics) In this example, we continue
the Plasm construction started in Example 3.2.6 by assigning appropriate
measures to various building subsystems, especially to the floors and different components in the x and y directions of the concrete building frame.
Notice the negative z component of the OFFSET kernel used to position the
longitudinal beams beneath the concrete frame, along with the curbs and
irons for attaching the above.

```
floors = OFFSET([.2,.2,.2])(building110);
framex = OFFSET([.2,.2,.2])(building1_011);
framey = OFFSET([.2,.2,-.4])(building1_101);
framexy = STRUCT(framex, framey);
framexyz = STRUCT(framex, framey, floors);
VIEWCOMPLEX(LAR(framexyz))
```

Fig. 3.11 Top: 2-complexes from the 3-complex of Figure 4.10a. Middle: wire-frame 1-complexes of z-layouts; y-sections; x-sections. Bottom: 1-skeleton of initial "idea"; final result with proper **SK** and **OFFSET** applications, shown in Examples 3.2.6 and 3.2.7.

3.2.2 Simplicial complex

In the current and next sections, we will continue alternating definitions of concepts with simple object examples and Julia **Plasm** coding of object properties.

Definition 3.13 (Join operation) The join of two compact sets of points $P, Q \subset \mathbb{E}^n$ is the set PQ of convex combinations of points in P and Q. The join operation is associative and commutative.

Definition 3.14 (Simplex) A d-simplex $\sigma_d \subset \mathbb{E}^n$ ($0 \leq d \leq n$) may be defined as the repeated join of $d+1$ affinely independent points, called vertices.

A d-simplex can be seen as a d-dimensional triangle: a 0-simplex is a point, a 1-simplex is a segment, a 2-simplex is a triangle, a 3-simplex is a tetrahedron, and so on. Two points are affinely independent when they are non-coincident, three points when they are non-collinear, four points when they are non-coplanar, etc.

Definition 3.15 (Simplicial Complex) A set Σ of simplices is called a triangulation. A simplicial complex, often simply denoted here as complex, is a triangulation Σ that verifies the following conditions:

1. if $\sigma \in \Sigma$, then any face of σ belongs to Σ;
2. if $\sigma, \tau \in \Sigma$, then either $\sigma \cap \tau = \emptyset$, or $\sigma \cap \tau$ is a face of both σ and τ.

A simplicial complex can be considered a well-formed multidimensional triangulation. Such triangulations are widely used in engineering analysis, e.g., topography or finite element methods.

Remark 3.5 (About REPL use and book typography) Let us return to our change of REPL appearance. All (quite) examples are written and evaluated in a REPL terminal. To save row space, the REPL prompt `julia>` is often not reported. Analogously, most vertical vector outputs are transformed into row output using the character "'". Unfortunately, Julia changes the output type from `::Vector` to `adjoint(::Vector)`. Do not care.

Coding 3.2.4 (Plasm d-dimensional simplex) A d-simplex is generated by the Plasm package using the unit SIMPLEX multi-dimensional function, as the convex hull of its vertices:

```julia
julia> using Plasm
julia> SIMPLEX(dim::Int)
SIMPLEX (generic function with one method)
```

Remark 3.6 (Plasm simplex vs standard simplex) In mathematics literature (see, e.g., [1]), the standard simplex Δ^n of dimension n is defined in the space $\mathbb{R}^{n+1}$ with vertices at the points $\mathbf{e}_i = (0 \ldots 1 \ldots 0)$, $i = 0 \ldots n$, where the 1 stands in the i-th place. Conversely, the Plasm simplex of dimension n is in $\mathbb{E}^n$ and contains the zero point to enforce the *solidity* of geometric objects, where the intrinsic dimension is equal to that of the embedding space

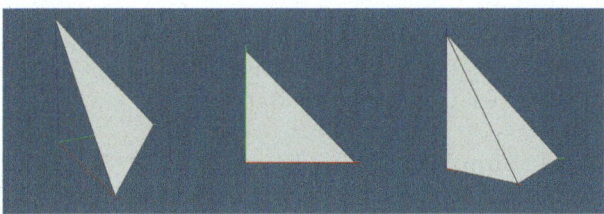

Fig. 3.12 (a) standard 2-simplex $\Delta^2 \subset \mathbb{E}^3$; (b) Plasm SIMPLEX(2) $\subset \mathbb{E}^2$; (c) Plasm SIMPLEX(3) $\subset \mathbb{E}^3$.

Definition 3.16 (Intrinsic dimension) The intrinsic dimension is the number of variables needed for a minimal representation of a datum, especially in a parametric form. For points, it is 0; for curves, it is 1; for surfaces, it is 2; and for solids in Euclidean space, it is 3, regardless of the n-dimensional space into which the object is embedded.

Example 3.2.8 (Comparison of standard and Plasm simplices) Let us note the Julia constructions of the coordinate matrices I and Δ, and the primitive MKPOL, to build an object of type Hpc, viewable with the function VIEW.

```
I = [1. 0. 0.; 0. 1. 0.; 0. 0. 1.]  # vertex coords in E³
Δ2 = MKPOL(I, [[1,2,3]])
VIEW(Δ2)
```

Note also the coordinate matrix built by blocks 1×3 and 3×3:

```
Δ = [0. 1. 0.; 0. 0. 1.]  # coords in E²
Σ2 = MKPOL(Δ, [[1,2,3]]) # SIMPLEX(2)
Σ3 = MKPOL([[0.;0.;0.] I], [[1,2,3,4]]) # SIMPLEX(3)
VIEW( Σ2 )
VIEW( Σ3 )
```

While the SIMPLEX primitive may build any simplex of finite dim $= d$, the Plasm visualization VIEW cannot (currently) go beyond $d = 3$.

```
julia> VIEW(SIMPLEX(2))
julia> VIEW(SIMPLEX(3))
```

Coding 3.2.5 (Dataset associated to simplices.) The simplex datasets follow for the first integer parameters. Note that SIMPLEX(d) contain $d + 1$ coordinate points of dimension d, and that hulls fields contain $d + 1$ indices of points, i.e. a single convex cell:

```
SIMPLEX(0)              #=
Hpc(MatrixNd(1), Geometry([Float64[]], hulls=[[1]])) =#
SIMPLEX(1)              #=
Hpc(MatrixNd(2), Geometry([[0.0], [1.0]], hulls=[[1, 2]])) =#
SIMPLEX(2)              #=
Hpc(MatrixNd(3), Geometry([[0.0, 0.0], [1.0, 0.0], [0.0, 1.0]],
    hulls=[[1, 2, 3]])) =#
SIMPLEX(3)              #=
Hpc(MatrixNd(4), Geometry([[0.0, 0.0, 0.0], [1.0, 0.0, 0.0],
    [0.0, 1.0, 0.0], [0.0, 0.0, 1.0]], hulls=[[1, 2, 3, 4]])) =#
SIMPLEX(4)              #=
Hpc(MatrixNd(5), Geometry([[0.0, 0.0, 0.0, 0.0], [1.0, 0.0, 0.0,
    0.0], [0.0, 1.0, 0.0, 0.0], [0.0, 0.0, 1.0, 0.0], [0.0,
    0.0, 0.0, 1.0]], hulls=[[1, 2, 3, 4, 5]])) =#
```

For visual simplicity, we remove the `julia>` prompt from `Plasm` scripts and use the multilinear comment `#= ... =#` to reduce the visual rumor. We remark that the expressions above return an `Hpc` value.

The order of a simplicial complex is the maximum order of its simplices. A complex Σ_d of order d is also called a d-complex. A d-complex is said to be regular or pure if each simplex is a face of a d-simplex[6]. A regular d-complex is a homogeneously d-dimensional complex.

Definition 3.17 (Combinatorial boundary of simplex) The combinatorial boundary $\Sigma_{d-1} = \partial\sigma_d$ of a simplex σ_d is a simplicial complex consisting of all proper s-faces ($s < d$) of σ_d.

Two simplices σ and τ in a complex Σ are called s-adjacent if they have a common s-face. Hereafter, when we refer to adjacencies into a d-complex, we intend to refer to the maximum order adjacencies, i.e., to $(d-1)$-adjacencies. $\mathcal{K}_s$ ($s \leq d$) denotes the set of s-faces of Σ_d.

Definition 3.18 Skeleton and Faces. The set $\langle \mathbf{v}_0, \mathbf{v}_1, \ldots, \mathbf{v}_d \rangle$ of vertices of σ_d is called the 0-skeleton of σ_d. The s-simplex generated from any subset of $s + 1$ vertices ($0 \leq s \leq d$) of σ_d is called an s-face of σ_d.

Remark 3.7 Let us notice, from the definition, that a simplex may be considered both as a purely combinatorial object and as a geometric object, i.e. as the compact (closed and bounded) point-set defined by the convex hull of a discrete set of points.

Coding 3.2.6 (Extract all faces from a simplex)

```
LAR(SIMPLEX(3)).C #=
Dict{Symbol, Vector{Vector{Int64}}} with 4 entries:
  :CV => [[1, 2, 3, 4]]
  :FV => [[1, 2, 3], [1, 2, 4], [1, 3, 4], [2, 3, 4]]
  :EV => [[1, 2], [1, 3], [1, 4], [2, 3], [2, 4], [3, 4]]
  :FE => [[1, 2, 4], [1, 3, 5], [2, 3, 6], [4, 5, 6]] =#
```

The `.C` attribute of `Lar` value generated by the expression `LAR(SIMPLEX(3))` contains the 3-cell indexed by `:CV` (solid), the four 2-cells by `:FV` (faces), and the six 1-cells by `:EV` (edges). `LAR` cannot currently go beyond $d = 3$.   □

Remark 3.8 (higher-dimensional simplices) When the user needs to generate the boundary complex of a simplex of higher dimension, she may evaluate `SIMPLEX(5,boundary=true)` to get a pair `V, Cells` where `Cells` is the array with all dimensional skeletons from 0 to 5. The reader is invited to try. It is easy to show the cardinality of boundary cells of the simplex of dimension n is 2^n by summing the binomial numbers on the $n + 1$ row of the Pascal (better: Tartaglia's) triangle.

---

[6] It is assumed that any simplex σ has also itself and the empty set $\emptyset$ as faces.

Coding 3.2.7 (LARSIMPLEX and HPCSIMPLEX) Currently, the
SIMPLEX operator serves as a synonym for LARSIMPLEX, and HPCSIMPLEX is
also available, with both being multidimensional. For two examples, see the
following script:

```
HPCSIMPLEX(6)
Hpc(MatrixNd(7), Geometry([[0.0, 0.0, 0.0, 0.0, 0.0, 0.0], [1.0,
    0.0, 0.0, 0.0, 0.0, 0.0], [0.0, 1.0, 0.0, 0.0, 0.0, 0.0],
    [0.0, 0.0, 1.0, 0.0, 0.0, 0.0], [0.0, 0.0, 0.0, 1.0, 0.0,
    0.0], [0.0, 0.0, 0.0, 0.0, 1.0, 0.0], [0.0, 0.0, 0.0, 0.0,
    0.0, 1.0]], hulls=[[1, 2, 3, 4, 5, 6, 7]]))
```

```
LARSIMPLEX(4,boundary=true)
([0.0 1.0 … 0.0 0.0; 0.0 0.0 … 0.0 0.0; 0.0 0.0 … 1.0 0.0; 0.0
    0.0 … 0.0 1.0], [[[1], [2], [3], [4], [5]], [[1, 2], [1, 3],
    [1, 4], [1, 5], [2, 3], [2, 4], [2, 5], [3, 4], [3, 5], [4,
    5]], [[1, 2, 3], [1, 2, 4], [1, 2, 5], [1, 3, 4], [1, 3,
    5], [1, 4, 5], [2, 3, 4], [2, 3, 5], [2, 4, 5], [3, 4, 5]],
    [[1, 2, 3, 4], [1, 2, 3, 5], [1, 2, 4, 5], [1, 3, 4, 5], [2,
    3, 4, 5]], [[1, 2, 3, 4, 5]]])
```

For a higher-dimensional LARSIMPLEX dataset, a pair V, Vector{Vector{
Chains}} is returned instead of a Lar dataset object.   □

Simplicial Complexes

Coding 3.2.8 (binomial numbers and simplex faces) The boundary
cells of a simplex are a good example of a simplicial complex. This script
computes the number of cells of all intrinsic dimensions. There are

$$\sum_{k=0}^{n} \binom{n}{k} = 2^n$$

k-faces in a n-simplex. Testing this statement is simple and very elegant with
Julia. You may verify, e.g., that simplex(5) faces are a set of 2^5 elements,
including the empty and "improper" cells, coinciding with the simplex itself.
An efficient Julia's generator expression is used in the last two rows:

```
[binomial(5,k) for k=0:5]' #=
1×6 adjoint(::Vector{Int64}) with type Int64:
  1   5   10   10   5   1      =#
sum(binomial(5,k) for k=0:5) == 2^5 == 32
sum(binomial(50,k) for k=0:50) == 2^50 == 1125899906842624
```

Coding 3.2.9 (Convex hull of given points) A simple coding example creates the simplicial complex, offering the triangulation of the boundary of a convex hull. Note the input type here, which features random points organized by rows:

```
julia> p = rand(6,3)                    #=
6×3 Matrix{Float64}:
 0.456121  0.340689  0.523394
 0.670731  0.920846  0.810581
 0.511325  0.83709   0.765548
 0.27295   0.344676  0.246891
 0.155611  0.262588  0.372059
 0.997037  0.689132  0.594624          =#
```

The Matrix p holds six random 3D points by rows. The q Julia variable below receives a Vector{Vector{Float64}} with 6 elements:

```
julia> q = [p[k,:] for k=1:size(p,1)]              #=
6-element Vector{Vector{Float64}}:
 [0.4561213293752391, 0.34068894716553066, 0.5233939444809501]
 [0.6707310911896536, 0.9208461259235498, 0.8105811405026019]
 [0.5113250218604997, 0.8370897242704682, 0.7655476328700838]
 [0.2729499977250678, 0.3446764528053594, 0.24689130455109154]
 [0.15561059083290985, 0.26258788197490757, 0.37205895443742]
 [0.9970371401112009, 0.6891321979374976, 0.5946242627157257]
   =#
```

The convex hull of these points is calculated using the Plasm function CONVEXHULL and stored in the Hpc data structure:

```
julia> CONVEXHULL(q)              #=
Hpc(MatrixNd(4), Geometry([[0.4561213293752391,
     0.34068894716553066, 0.5233939444809501],
     [0.6707310911896536, 0.9208461259235498,
     0.8105811405026019], [0.5113250218604997,
     0.8370897242704682, 0.7655476328700838],
     [0.2729499977250678, 0.3446764528053594,
     0.24689130455109154], [0.15561059083290985,
     0.26258788197490757, 0.37205895443742], [0.9970371401112009,
      0.6891321979374976, 0.5946242627157257]], hulls=[[1, 2, 3,
     4, 5, 6]]))            =#
```

Finally, the last dataset is transformed into a Lar data structure containing all the boundary polygons :FV and edges :EV.

```
julia> LAR(CONVEXHULL(q))              #=
Lar(3, 3, 6, [0.27294999772507 0.67073109118965 …
     0.15561059083291 0.5113250218605; 0.34467645280536
     0.92084612592355 … 0.26258788197491 0.83708972427047;
     0.24689130455109 0.8105811405026 … 0.37205895443742
     0.76554763287008], Dict{Symbol, AbstractArray}(:CV => [[1,
```

```
2, 3, 4, 5, 6]], :FV => [[1, 2, 3], [2, 3, 4], [1, 4, 5],
[1, 3, 4], [1, 5, 6], [1, 2, 6], [4, 5, 6], [2, 4, 6]], :EV
=> [[2, 3], [1, 3], [1, 2], [3, 4], [2, 4], [1, 4], [4, 5],
[1, 5], [5, 6], [1, 6], [2, 6], [4, 6]]))                    =#
```

We remark that the `Hpc` describes only the convex cells of its dataset. Conversely, the `Lar` data structure explicitly gives all cells of any dimension. □

Exercise 3.1 (Boundary of points convex hull) Readers should visualize the boundary of a vector q of n random points in $[0, 1]^3$, notice the number of boundary vertices, and argue how it varies, with $n = 10, 100, 1000, 10000, ...$

```
julia> VIEW(CONVEXHULL(q))
julia> VIEWCOMPLEX(LAR(CONVEXHULL(q)))
```

Simplex orientation

Simplices can be considered as either oriented or non-oriented. The ordering of 0-skeleton of a simplex implies an *orientation* of it.

Definition 3.19 (Ordering of simplex vertices) The n-simplex can be oriented according to the even or odd permutation class of its 0-skeleton. Therefore, when $\sigma = \langle \mathbf{v}_0, \mathbf{v}_1, \ldots, \mathbf{v}_n \rangle$ is represented by indices $\langle 0, 1, , 2, \ldots, n \rangle$ of its vertices, a single swap of two indices exchanges its permutation class and hence its orientation.

The two opposite orientations of a simplex are denoted as $+\sigma$ and $-\sigma$.

Definition 3.20 (Coherent orientation) Two simplices are coherently oriented when their common faces have opposite orientations. A complex is orientable when all its simplices can be coherently oriented.

It is assumed that:

1. the two orientations of a simplex represent its relative interior and exterior;
2. the two orientations of an orientable simplicial complex analogously represent the relative interior and exterior of the complex, respectively;
3. the boundary of a complex maintains the same orientation of the complex.

The volume associated with an orientation of a simplex (or complex) is positive, while the one related to the opposite orientation has the same absolute value and opposite sign. It is assumed that the bounded object has a positive volume. It is also assumed that either a minus sign or a multiplying factor -1 denotes complementation, i.e., an opposite orientation of the simplex, which can be explicitly obtained by swapping two vertices in its ordered 0-skeleton. For example:

$$+\sigma_3 = \langle \mathbf{v}_0, \mathbf{v}_1, \mathbf{v}_2, \mathbf{v}_3 \rangle, \; -\sigma_3 = \langle \mathbf{v}_1, \mathbf{v}_0, \mathbf{v}_2, \mathbf{v}_3 \rangle.$$

Fig. 3.13 Coherent orientation of the 2-faces of 3-simplex $\langle \mathbf{v}_0, \mathbf{v}_1, \mathbf{v}_2, \mathbf{v}_3 \rangle$ whith indices $\langle 0, 1, 2, 3 \rangle$: $\langle 1, 2, 3 \rangle, -\langle 0, 2, 3 \rangle,$ $\langle 0, 1, 3 \rangle, -\langle 0, 1, 2 \rangle =$ $\langle 1, 2, 3 \rangle, \langle 2, 0, 3 \rangle, \langle 0, 1, 3 \rangle, \langle 1, 0, 2 \rangle.$

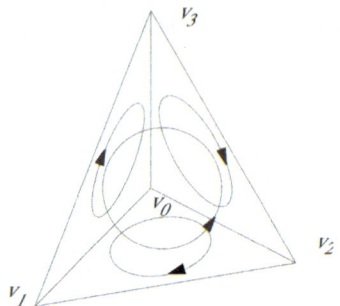

Definition 3.21 (Volume of a 3-simplex) The volume of a 3-simplex σ is a 3-cochain (see Section 3.5.2) from the 2-cycle of σ 2-faces (boundary triangles) $\rightarrow \mathbb{R}$. A cochain is a $\phi(\sigma)$ map from a chain space to real numbers $\phi : C_3 \rightarrow \mathbb{R}$. Analogously, the σ surface area is a map $\psi : C_2 \rightarrow \mathbb{R}$.

Facet extraction

Since no ambiguity may arise, the $(d-1)$-faces of a d-dimensional simplex or complex are also called facets. It is possible to see [25] what follows:

Theorem 3.1 (Whitney, 1957) *The oriented facets $\sigma_{d-1}^{(i)}$ $(0 \le i \le d)$ of the oriented d-simplex $\sigma_d = +\langle \mathbf{v}_0, \mathbf{v}_1, \ldots, \mathbf{v}_d \rangle$ are obtained [25] by removing the i-th vertex $\mathbf{v}_i$ from the 0-skeleton of σ_d:*

$$\sigma_{d-1}^{(i)} = (-1)^i (\sigma_d - \langle \mathbf{v}_i \rangle), \qquad 0 \le i \le d. \tag{3.1}$$

The 0-skeleton of σ_{d-1}^i is therefore obtained by removing the i-th vertex from the 0-skeleton of σ_d and either by swapping a pair of vertices or, better, by inverting the simplex sign when i is odd (see Figure 3.13).

Coding 3.2.10 (Facets of a d-simplex.)
The Julia function SIMPLEXFACETS is provided in Plasm to computationally generate all the oriented facets of a simplices array of d-simplexes. A precondition is that all Vector{Int64} elements in the input sequence have the same length, i.e., dimension. We may start with an array of simplices of the same length and iterate over their facets to obtain a simplicial complex.

```
CV = SIMPLEX(3)[2]    #=
```

```
[[1, 2, 3, 4]]       =#
FV = SIMPLEXFACETS(CV)   #=
[[1, 2, 3], [1, 2, 4], [1, 3, 4], [2, 3,4]] =#
EV = SIMPLEXFACETS(FV)   #=
[[1, 2], [1, 3], [1, 4], [2, 3], [2, 4], [3, 4]] =#
VV = SIMPLEXFACETS(EV)   #=
[[1], [2], [3], [4]]     =#
SIMPLEX(3,boundary=true)[2]   #=
4-element Vector{Vector{Vector{Int64}}}:
  [[1], [2], [3], [4]]
  [[1, 2], [1, 3], [1, 4], [2, 3], [2, 4], [3, 4]]
  [[1, 2, 3], [1, 2, 4], [1, 3, 4], [2, 3, 4]]
  [[1, 2, 3, 4]]     =#
```

Coding 3.2.11 (Generation of boundary complex)   We can directly generate the boundary complex of a standard simplex `obj` in any dimension, and obtain it all by evaluating the expression `@show obj;`:

```
obj = SIMPLEX(4,boundary=true);
```

The reader should evaluate the expression in REPL: a `Pair` dataset is produced by `Plasm`, with components `obj[2]` reporting a Julia `Array` of boundary `Cells` of all dimensions:

```
obj[2]         #=
5-element Vector{Vector{Vector{Int64}}}:
[[1],[2],[3],[4],[5]]
[[1,2],[1,3],[1,4],[1,5],[2,3],[2,4],[2,5],[3,4],[3,5],[4,5]]
[[1,2,3],[1,2,4],[1,2,5],[1,3,4],[1,3,5],[1,4,5],
   [2,3,4],[2,3,5],[2,4,5],[3,4,5]]
[[1,2,3,4],[1,2,3,5],[1,2,4,5],[1,3,4,5],[2,3,4,5]]
[[1,2,3,4,5]]   =#
```

and `obj[1]` reporting the vertices coordinates by columns:

```
obj[1]       #=
4x5 Matrix{Float64}:
 0.0  1.0  0.0  0.0  0.0
 0.0  0.0  1.0  0.0  0.0
 0.0  0.0  0.0  1.0  0.0
 0.0  0.0  0.0  0.0  1.0           =#
```

Coding 3.2.12 (Simplex generation in Hpc)   Conversely, the `Plasm` data structure `Hpc` may store a single convex cell `HPCSIMPLEX(d)` with $d+1$ vertices for any finite d:

```
HPCSIMPLEX(4)           #=
```

```
Hpc(MatrixNd(5), Geometry([[0.0, 0.0, 0.0, 0.0], [1.0, 0.0, 0.0,
    0.0], [0.0, 1.0, 0.0, 0.0], [0.0, 0.0, 1.0, 0.0], [0.0,
    0.0, 0.0, 1.0]], hulls=[[1, 2, 3, 4, 5]]))          =#
```

Definition 3.22 (Prism by simplicial extrusion) The prism over a simplex $\sigma_d = \langle \mathbf{v}_0, \ldots, \mathbf{v}_d \rangle \subset \mathbb{E}^d$, is defined as the set

$$P_{d+1} := \sigma_d \times [a, b], \qquad [a, b] \subset \mathbb{R}, \quad P_{d+1} \subset \mathbb{E}^{d+1}$$

and will be called simplicial $(d + 1)$-prism.

An oriented complex which triangulates P_{d+1} can be defined combinatorially, by using a closed form formula for its $\mathcal{K}_{d+1}$ skeleton:

$$\mathcal{K}_{d+1} = \sigma_{d+1,i} = (-1)^{id} \langle \mathbf{v}_i^a, \mathbf{v}_{i+1}^a, \ldots, \mathbf{v}_d^a, \mathbf{v}_0^b, \mathbf{v}_1^b, \ldots, \mathbf{v}_i^b \rangle, \quad 0 \le i \le d \quad (3.2)$$

where $\mathbf{v}_i^a = (\mathbf{v}_i, a) \in \mathbb{E}^{d+1}$ and $\mathbf{v}_i^b = (\mathbf{v}_i, b) \in \mathbb{E}^{d+1}$ are Euclidean points.

Closed formulas to triangulate the $(d + 1)$-prism over a d-complex in a time linear with the size of the output while also computing the d-adjacencies between the resulting $(d + 1)$-simplices, can be found in [5, 11, 10].

Property 3.1 (Cube tetrahedrization) We prove that a unit cube can be decomposed into six well-assembled unit tetrahedra (3-simplices), looking at Figure 3.14, where a standard 2-simplex in $\mathbb{E}^2$ is extruded orthogonally in $\mathbb{E}^3$ to generate three standard 3-simplices, thus producing a half-cube partition.

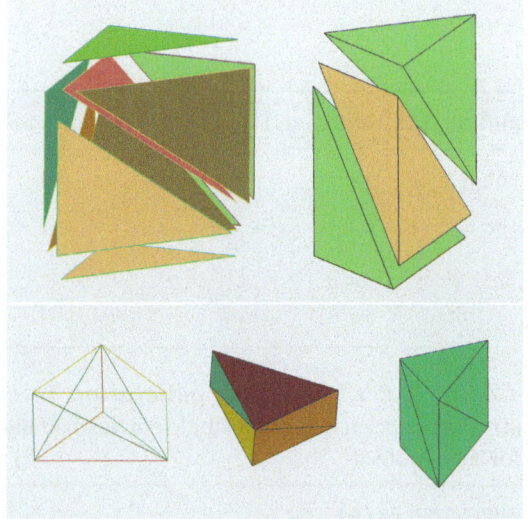

Fig. 3.14 Simplicial complex extrusion of the 2D triangle in $\mathbb{E}^2$, producing the half-cube in $\mathbb{E}^3$ as a complex with three tetrahedra.

Top: (a) 2-skeleton of exploded half-cube; (b) exploded 3-cells (tetrahedra).

Bottom: (a) 1-skeleton (set of 1-cells); (b) 2-skeleton (set of 2-cells); (b) 3-skeleton (set of 3-cells).

Coding 3.2.13 (Unit 2-simplex extrusion)

In Plasm, we have a generalized version EXTRUDESIMPLICES of the extrusion Formula 3.2, used as an algorithm for multidimensional extrusion of simplicial d-complexes. Of course, (currently) it makes only sense to go 1D→2D or 2D→3D.

We can show the previous Property 3.1 by extruding, via Equation 3.2, the unit 2-simplex. The combinatorial formula is implemented in the Plasm function EXTRUDESIMPLICES:

```
halfcube = MKPOL(EXTRUDESIMPLICES(([0. 1 0; 0 0 1], [[1,2,3]]),
    [1]))
VIEW( halfcube )
```

The operator EXTRUDESIMPLICES can be applied to a 0-, 1-, 2-, ... simplicial model, to get a 1-, 2-, 3-, ... dimensional model. The optional parameter pattern::Array{Array{Int}} of Chains type specifies how to decompose the added dimension. □

Example 3.2.9 (Highly non-manifold multiextrusion)

A classic PLaSM object model is a Pair (vertices, cells) to be extruded, whereas pattern is an array of numbers for lateral measures of the **extruded model**, which enjoys one added dimension. The pattern's elements are assumed as either solid or empty measures, according to their (+/-) sign.

```
V = [[0.,0] [1,0] [2,0] [0,1] [1,1] [2,1] [0,2] [1,2] [2,2]];
FV = [[1,2,4],[2,3,5],[3,5,6],[4,5,7],[5,7,8],[6,8,9]];
pattern = repeat([1,.2,-2],outer=4);
model = (V, FV);
VIEW(MKPOL(V,FV))
```

Fig. 3.15 Simplicial complex extrusion of the 2D triangle in $\mathbb{E}^2$. (a) the input 2-complex model=(V,FV) in $\mathbb{E}^2$; (b) the multi-extrusion solid 3-complex model=(W,FW) output in $\mathbb{E}^3$. Note that both input and output are locally non-manifold in several places!

```
W,FW = EXTRUDESIMPLICES(model, pattern);
VIEW(MKPOL(W,FW))
VIEWCOMPLEX(LAR(MKPOL(W,FW)))
```

Fig. 3.16 A highly no-
manifold **Plasm** model:
(a) projective view of
extruded 3-complex; (b)
parallel view of 3-complex
in $\mathbb{E}^3$.

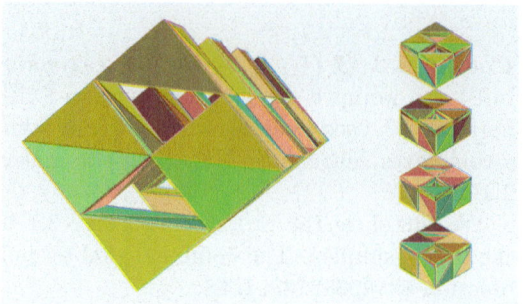

Definition 3.23 (Delaunay triangulation) A Delaunay triangulation of
a set of random points of $\mathbb{E}^2$ is a simplicial complex subdividing its convex
hull into triangles whose circumcircles do not contain any other points.

In $\mathbb{E}^3$, we have Delaunay tetrahedralization with the same property using
circumspheres. See Figure 3.17.

Fig. 3.17 Simplicial complex tetrahedralizing the
set of 20 random 3D points.

Top: (a) exterior view;
(b) wire-frame view of 1-
skeleton.
Bottom: (a) exploded view
of 2-skeleton; (b) exploded
view of 3-complex.

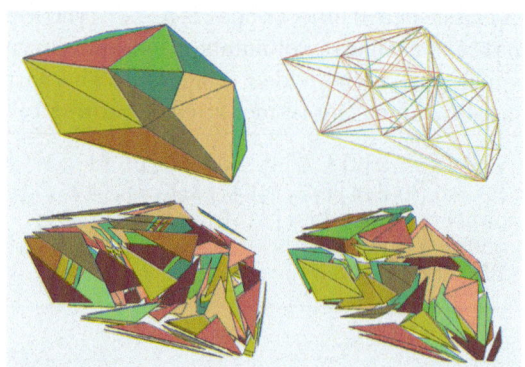

*Remark 3.9 (***About non-manifoldness***)*
The class of models in the **Plasm** representation scheme, defined as the category of chain complexes, does not impose any requirements regarding the
non-manifoldness or its reverse property for mathematical models within the
scheme.

Therefore, our **Plasm** programs do not need to consider non-manifoldness
in data structures (cellular complexes) or algorithms that manipulate them.
Conversely, this property previously required highly complex, specific data
structures and algorithms in the field of solid modeling.

Simplicial grids via extrusion

In layout design, a grid system provides a design framework of intersecting vertical and horizontal lines. Designers use this framework to place and align text, images, and other graphic elements.

Similarly, a simplicial grid is a geometric grid structure composed of simplicial cells of the same dimension aligned on a cuboidal layout grid of 1D, 2D, 3D, or higher elements. This construction is mainly used for domain decomposition, where coordinate functions are mapped to create curved structures, such as proper curves, curved surfaces, and curved solids.

The curved complexes are easily created by applying the `MAP` operators to a `Plasm` grid, so ensuring that all vertex points are moved consistently without changing the grid's topology.

```
VIEW(MKPOL(SIMPLEXGRID([10])))
VIEW(MKPOL(SIMPLEXGRID([10,10])))
VIEW(MKPOL(SIMPLEXGRID([10,10,2])))
```

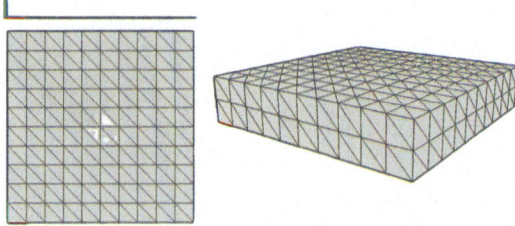

Fig. 3.18 Plasm-generated 1D, 2D, and 3D simplicial grids. The drawings are redundant, since the cells are convex.

```
VIEW(MKPOL(SIMPLEXGRID([10])), properties=DICT("background_color"=WHITE))
VIEW(MKPOL(SIMPLEXGRID([10,10])))
VIEW(MKPOL(SIMPLEXGRID([10,10,2])))
```

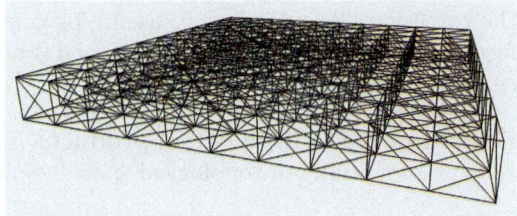

Fig. 3.19 Spatial structural lattice: a strongly rigid structure

```
grid::Hpc = SK(1)(MKPOL(SIMPLEXGRID([10,10,1])))
```

Above is a structure generated by Plasm, consisting of a $10 \times 10 \times 1$ grid, where each cell is the one-skeleton (the network of edges and vertices) of a sextuple of 3-simplices. In other words, each cell is built from two groups of three simplices (the simplest 3-dimensional building blocks) that form a rigid network.

Because the structure is based on a triangulated framework (with triangles being inherently stable and resistant to deformation), and because these rigid cells are assembled in a regular grid, the load is efficiently distributed throughout the entire assembly. This repeated, interlocking pattern of triangular elements across the grid significantly contributes to the structure's overall rigidity.

Fig. 3.20 Exploded simplicial grids of dimensions 1 (lines), 2 (surfaces), and 3 (solids). This simplicial 2-complex is generated by SIMPLEXFACETS from the 3-complex SIMPLEXGRID([10,10,1]).

3.2.3 Product of complexes

Here, we introduce a multidimensional modeling operation that generalizes the extrusion [4] and is highly useful for generating 3D solids from 2D surfaces, 2-complexes from 1-complexes, and producing their skeletons using one or more 0-complexes.

Definition 3.24 (Topological product) A product space is the Cartesian product of a family of topological spaces equipped with a natural topology.

The natural topology on a subset of a topological space, in our case $\mathbb{E}^d$, is the relative topology (or subspace topology). The product of two cell complexes can be assembled into a cell complex, as shown below.

Definition 3.25 (Product of cell complexes) If X and Y are m- and n-complexes in $\mathbb{E}^m$ and $\mathbb{E}^n$, respectively, $X \times Y \subset \mathbb{E}^{m+n}$ is a cell complex in which each cell is a product of a cell in X and a cell in Y, endowed with the relative topology in $\mathbb{E}^{m+n}$.

The `Plasm POWER` function, also denoted as infix $*$, takes a pair of `Plasm` models of `Hpc` type as input and returns the value `Hpc` of their Cartesian product. The operation is associative. Hence, no parentheses are needed, as seen in Coding 3.2.3.

3.3 From cells to chains and cochains

A minimal set of definitions is provided here. Some are repeated to highlight their importance, encouraging the reader to consider them carefully and remember their meaning throughout this book. Reading [?] would be helpful for a complete understanding of the `Lar` data structures.

Definition 3.26 Chains. Given a cellular d-complex U_p, the elements of the chain powersets $C_p = \mathcal{P}(U_p)$, $0 \leq p \leq d$, are subsets of cells of the same dimension. They are called chains, i.e., subsets of cells. Their sets C_p are p-chain spaces.

Linearity of chain spaces

Chain spaces are a dataset used in combinatorial and algebraic topology. In these areas, we use the tools of abstract algebra to examine the number and types of holes in spaces.

The graded collections C_p of chains are closed w.r.t. modular addition $(+ \mod 2)$ of chains and multiplication by a scalar in $\{0, 1\}$ or $\{0, 1, -1\}$. Because they fulfill the typical properties of associativity, commutativity, distributivity, and identity operations, chain spaces form linear (i.e., vector) spaces.

Since C_0, C_1, C_2, C_3 are linear spaces, all their elements may be (uniquely) represented by linear combination of a (given) chain basis with scalar numbers. The simpler (or minimal, i.e., irreducible) basis for each C_p is the ordered set of singletons $\{\gamma_k \in U_p\}$, with $k = 1, 2, \ldots, m$, where m_p is the cardinality of cell collection: $m_p := \#U_p$. This number gives the dimension of the C_p space, where $u_k = \{\gamma_k\}$ is represented in coordinates by a tuple of $m - 1$ zeros and one unit in k position. Of course, each vector in C_p is represented by a binary m_p-vector.

This vector is an image of the characteristic function $\chi(C_p)$ evaluated upon a chain. Each chain vector $c \in C_p$ has coordinate representation as such an m_p-vector with values in the field of scalars.

3.4 Chain complex

This section deals with algebraic topological tools to compute boundaries and adjacencies of chains of cells and to generate the discrete solid primitives needed by a computational modeler. Basic geometric and algebraic topology provides the mathematical concepts needed to compute and explore the cells of the space partition induced by a set of geometric objects and the related incidence/neighborhood relations.

Gradation

Graded vector spaces are commonly used in algebra and geometry to arrange complex systems into manageable layers or levels. A graded vector space can be understood as a collection of vector spaces, each assigned to a specific "level" or "degree," similar to shelves in a library where each shelf holds a different type of book. In simple terms:

- The "grading" organizes the vector space into sections, often indexed by numbers (e.g., degree 0, degree 1, degree 2, etc.).
- Each level (or degree) contains its own vector space with elements that "belong" to that degree.
- This structure enables you to work with these vector spaces individually or collectively, depending on the context.

In particular, a graded vector space is a vector space that possesses the additional structure of a gradation, which is a decomposition of the space into a direct sum of vector subspaces generally indexed by the integers. Formally:

Definition 3.27 (Graded vector space) A graded vector space is a vector space V expressed as a direct sum of subspaces V_p indexed by integers in $[0, d] := \{p \in \mathbb{N} \ : \ 0 \le p \le d\}$:

$$V = \oplus_{p=0}^{d} V_p. \tag{3.3}$$

Definition 3.28 (Graded linear maps) A linear map $f : V \to W$ between graded vector spaces is a graded map of degree k if $f(V_p) \subset W_{p+k}$ for each p.

Let X be a cellular complex. We are interested in working with subsets of cells with the same dimension and their 1-graded maps, which entirely specify the topology of X.

Chains

In algebraic topology, the chain concept studies the structure of topological spaces by breaking them down into simpler pieces, such as points, edges, faces, and higher-dimensional analogs. Chains formally represent these pieces, enabling the application of algebraic methods. In essence, the chain concept provides algebraic tools for geometric and topological analysis, enabling the study of spaces in terms of their structure and the relationships among their parts.

Definition 3.29 (p-chain) A p-chain refers to any subset of p-cells within a cellular complex X

Definition 3.30 (Unit p-chain) We denote the set of unit (or elementary) p-chains, where $0 \leq p \leq d$, as $U_p = \Lambda_p - \Lambda_{p-1}$. Here, Λ_p refers to the p-skeleton of X, and $\Lambda_{-1} = \emptyset$.

Definition 3.31 (Linear p-space) $C_p = \mathcal{P}(U_p)$ is the space of p-chains, where $\mathcal{P}$ is the power set.

The set $C = \bigoplus C_p$, the direct sum of chain spaces, can be given the structure of a graded vector space (see Definition 3.34 and [2]) by defining the sum of chains with the same dimension and the product times scalars in a field, with the usual properties.

Definition 3.32 (Chain bases) As a linear space, each C_p contains a set of irreducible generators. The natural basis $U_p \subset C_p$ is the set of independent (or elementary) chains $u_p \in C_p$, given by singleton elements.

Consequently, every chain $c \in C_p$ can be expressed as a linear combination of this basis with field elements and is uniquely generated.

Once the basis is fixed, i.e., U_p is ordered, the unique unsigned coordinate representation of each $\{\lambda_k\} =: u_k \in C_p$ is a binary array with one non-zero element in position k, and all other elements 0. The ordered sequence of scalars may be drawn either from $\{0, 1\}$ (unsigned representation) or from $\{0, 1, -1\}$ (oriented representation). With abuse of language, we often call p-cells the elements of U_p.

Definition 3.33 (Chain complex) A chain complex is a vector space V provided with a graded linear map $\partial : V \to V$ of degree -1 called boundary operator, which satisfies $\partial^2 = 0$.

In simpler words, a chain complex is a sequence of vector spaces C_p and linear maps $\partial_p : C_p \to C_{p-1}$, such that the composition of two consecutive boundary maps is zero, i.e., $\partial_{p-1} \circ \partial_p = 0$, ensuring the existence of cycles and boundaries.

In this book, the notation $C_\bullet$ is used for the chain complex over the binary field $\{0, 1\}$, and $C_\bullet^\circlearrowleft$ for the oriented chain complex over the ternary field $\{0, 1, -1\}$, which is used to define the oriented boundaries. Figure 3.21 illustrates the concept.

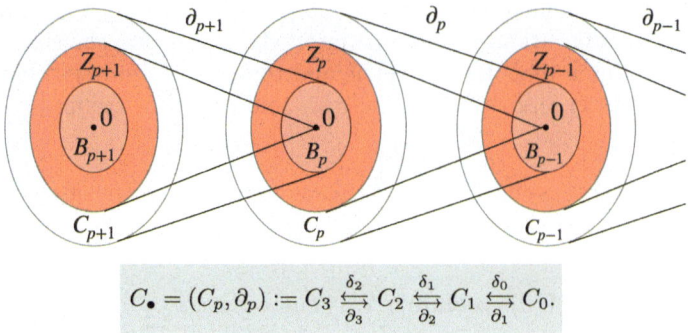

$$C_\bullet = (C_p, \partial_p) := C_3 \underset{\partial_3}{\overset{\delta_2}{\rightleftarrows}} C_2 \underset{\partial_2}{\overset{\delta_1}{\rightleftarrows}} C_1 \underset{\partial_1}{\overset{\delta_0}{\rightleftarrows}} C_0.$$

Fig. 3.21 The graphical representation of a chain complex, with boundary operators ∂_p, their domains (the gray sets) and codomains (the pink sets), and the sets of cycles Z_p (the red sets) that the boundary operators map to zero.

Homology studies the cycles and boundaries in a chain complex C. The homology groups $H_n(C)$ are the quotient of the cycles (kernel of ∂_n) by the boundaries (image of ∂_{n+1}). Homology captures topological features of a space, such as holes and voids, by examining the non-boundary cycles that are not boundaries of higher-dimensional chains.

The diagram of a complex $C_\bullet = (C_p, \partial_p)$, $0 \le p \le d$, is shown in Eq. 3.5, where a short ordered sequence of chain spaces is intertwined with linear operators of boundary $\partial_p : C_p \to C_{p-1}$ and coboundary $\delta_p : C_{p-1} \to C_p$.

We don't distinguish between chain (C_p) and cochain (C^p) spaces since we are only interested in topological properties [19]. In this case, we have

$$\partial_p = \delta_{p-1}^\top.$$

As a topological operator, δ_p maps a p-chain c_k to the subset of $(p+1)$-chains that share c_k as a common element. For example, in graphical terms, δ_1 maps an edge γ to the subset of faces with γ as a common edge.

3.4.1 Linear chain spaces

In mathematics, a graded vector space has the extra structure of a gradation, which decomposes the space into a direct sum of vector subspaces, generally indexed by integers.

Definition 3.34 (Graded vector space) A graded vector space is a vector space V expressed as a direct sum of subspaces V_p indexed by integers in $[0, d] := \{p \in \mathbb{N} : 0 \le p \le d\}$:

$$V = \oplus_{p=0}^d V_p. \tag{3.4}$$

Definition 3.35 (Graded linear maps) A linear map $f : V \to W$ between graded vector spaces is a graded map of degree k if $f(V_p) \subset W_{p+k}$ for each p.

Let be given a cellular complex X. We are interested in working with subsets of cells with the same dimension and their 1-graded maps, which entirely specify the X topology.

Definition 3.36 (p-chain) With some abuse of language, a p-chain can be seen as any subset of p-cells in a cellular complex X.

Definition 3.37 (Unit p-chain) In this sense we write $U_p = \Lambda_p - \Lambda_{p-1}$ for the set of unit (or elementary) p-chains ($0 \le p \le d$), where Λ_p is the p-skeleton of X and $\Lambda_{-1} = \emptyset$.

Definition 3.38 (Linear p-space) $C_p = \mathcal{P}(U_p)$ is the space of p-chains, where $\mathcal{P}$ is the power set.

The set $C = \oplus C_p$, the direct sum of chain spaces, can be given the structure of a graded vector space (see 3.34 and [2]) by defining sums of chains with the same dimension, and products times scalars in a field, with the usual properties.

Definition 3.39 (Chain bases) As a linear space, each C_p contains a set of irreducible generators. The natural basis $U_p \subset C_p$ is the set of independent (or elementary) chains $u_p \in C_p$, given by singleton elements.

Consequently, every chain $c \in C_p$ can be written as a linear combination of this basis with field elements and is uniquely generated.

Definition 3.40 (Chain coordinates) Once the basis is fixed, i.e., U_p is ordered, the unique unsigned coordinate representation of each $\{\lambda_k\} =: u_k \in U_p$ is a binary array with one non-zero element in position k, and all other elements 0.

Since C_0, C_1, C_2, C_3 are linear spaces, all their elements may be (uniquely) represented by linear combination of the chain basis with scalar numbers. The simpler (or minimal) basis for each C_p is the ordered set of singletons $\{\gamma_k\} \in U_p$, and $k = 1, 2, \ldots, m$, where m is the cardinality of cell collection $\#U_p$. This number m gives the dimension of the C_p space, where $u_k = \{\gamma_k\}$ is represented in coordinates by a tuple of $m - 1$ zeros and one non-zero in k position.

Such ordered sequences of scalars may be drawn either from $\{0, 1\}$ (unsigned representation) or from $\{0, 1, -1\}$ (oriented representation). Of course, each p-chain in C_p is represented by a unique m-vector of coordinates.

Remark 3.10 (We identify cells and unit chains) With abuse of language, we often call p-cells the elements of U_p. Depending on the context, we make the same use of "cell basis" and "chain basis," identifying the single cells with the unit chains (their singletons).

Definition 3.41 (Chain complex) A chain complex is a vector space V provided with a graded linear map $\partial : V \to V$ of degree -1 called boundary operator, which satisfies $\partial^2 = 0$.

In other words, a chain complex is a sequence of vector spaces C_p and linear maps $\partial_p : C_p \to C_{p-1}$, such that $\partial_{p-1} \circ \partial_p = 0$. The notation $C_\bullet$ is used in this paper for the chain complex over the binary field $\{0, 1\}$, and $C_\bullet^\circlearrowleft$ for the oriented chain complex over the ternary field $\{0, 1, -1\}$, used to get oriented boundaries.

Definition 3.42 (Cochain complex) A cochain complex is a graded vector space V provided with a graded linear map $\delta : V \to V$ of degree $+1$ called coboundary operator, which satisfies $\delta^2 = 0$.

That is to say, a cochain complex is a sequence of vector spaces C^p and linear maps $\delta^p : C^p \to C^{p+1}$, such that $\delta^{p+1} \circ \delta^p = 0$.

Remark 3.11 The reader should note, as a rule of dumb to remember the meaning of boundary and coboundary operators, that boundaries map cells in decreasing dimensions and coboundaries in increasing dimensions. So, e.g., it is $\partial_2 : C_2 \to C_1$ and $\delta_1 : C_1 \to C_2$. Note also that every operator index is the one of its domain.

3.4.2 Linear chain operators

We can also say that a chain complex is a sequence of linear spaces C_p of chains and linear boundary/coboundary maps ∂_p, δ_p between them:

Given a collection S of geometric objects[7], we shall compute the topology of their space arrangement $\mathcal{A}(S)$ as a chain complex (see Section 6)

$$C_\bullet = (C_p, \partial_p) := C_3 \underset{\partial_3}{\overset{\delta_2}{\leftrightarrows}} C_2 \underset{\partial_2}{\overset{\delta_1}{\leftrightarrows}} C_1 \underset{\partial_1}{\overset{\delta_0}{\leftrightarrows}} C_0 \ ; \tag{3.5}$$

$$\delta_p = \partial_{p+1}^\top \qquad 3 \le p \le 0 \tag{3.6}$$

---

[7] Examples include but are not limited to: line segments, quads, triangles, polygons, meshes, pixels, voxels, volume images, B-reps, etc. In mathematical terms, a geometric object is a topological space embedded in some $\mathbb{E}^d$ [8].

In the remainder of this book, we identify chains and cochains, as in [20]. In this way, we express that we are only interested in the combinatorial topology of cell complexes, not in their differential aspects.

A standard basis, also called a natural basis, is an special orthonormal vector basis in which each basis vector has a single non-zero coordinate entry with value 1 [22]. We are only interested in topological properties, so we have identified the natural bases of the corresponding chain and cochain spaces elementwise.

Consequently, the boundary and coboundary matrices are related by transposition. Given the coordinate vector of a p-cell, both the incident $p-1$ and $p+1$ cells are provided respectively.

Operator matrices

The matrices of boundary and coboundary operators (their transpose) are very sparse, with sparsity growing linearly with the number n of rows (sparse columns in Julia). With common data structures [7] for sparse matrices, the storage cost $O(n)$ is linear with the number of cells, with $O(1)$ small cost per cell that depends on the storage scheme.

Remark 3.12 The standard notation of matrices of operators is given by the operator symbol enclosed between squared brackets. For example, $[\partial_3]$ and $[\delta_2]$ stand for the matrices of 3-boundary and 2-coboundary, respectively.

According to what was discussed in Section 3.4.2, we can identify the U_0 basis of 0-chains with the ordered sequence V of vertices (0-cells) of a cellular complex, and the U_1, U_2 bases of 1- and 2-chains with ordered sequences E of edges (1-cells) and F of faces (2-cells), respectively. Analogously, we use the symbol C for the ordered 3-cells.

The natural bases of cells remain identified by the ordinal numbers of their cells (1, 2, 3, ...), and the chain spaces dimension (for example, dim V) by #V, #E, #F, and #C.

Therefore, we select a pair of characters from V, E, F, C, such as EF, to represent the matrix of the mapping F → E from 2-chains ∈ F → 1-chains ∈ E. In mathematical language $\partial_2 : C_2 \to C_1$ (see Eq. 3.5). Therefore, for clarity and simplicity, we often use the following notations.

$$\delta_2 : C_2 \to C_3 \equiv \text{CF: F} \to \text{C}$$
$$\delta_1 : C_1 \to C_2 \equiv \text{FE: E} \to \text{F}$$
$$\delta_0 : C_0 \to C_1 \equiv \text{EV: V} \to \text{E} \ .$$

$$\partial_3 : C_3 \to C_2 \equiv \text{FC: C ;} \to \text{F}$$
$$\partial_2 : C_2 \to C_1 \equiv \text{EF: F} \to \text{E}$$
$$\partial_1 : C_1 \to C_0 \equiv \text{VE: E} \to \text{V} \ .$$

Remark 3.13 (Matrix and vector formats) The actual content of vectors V, E, F, C is the coordinate representation of a chain (subset) of topological entities, and hence the characteristic function image (binary vector) of such a subset of cells.

Remark 3.14 (Linear operator vs matrix) Linear operators between chain spaces act between two linear spaces. Applying the operator's matrix produces a vector in the range from a vector in the domain, both represented in coordinates. It is important to note that any matrix column corresponds to an element of the domain's basis, expressed in the coordinates of the target space, specifically by the coefficients of a linear combination of the codomain basis.

Numerical examples

It may be helpful for our readers to review the meanings of left- and right-multiplication of a vector by a matrix.

Left-multiplying a matrix A by a row vector x, i.e., xA, results in a linear combination of A's rows using the elements of x as scalars, producing a row vector. Similarly, right-multiplying a matrix A by a column vector x, or Ax, results in a linear combination of A's columns and produces a column vector.

Coding 3.4.1 (Cube cell complex) Let's begin by creating a single 3-cube cellular complex to use in the following examples.

```
obj = LAR(CUBE(3))    #=
Lar(3, 3, 8, [3.0 0.0 … 3.0 0.0; 3.0 3.0 … 3.0 3.0; 0.0 0.0 …
    3.0 3.0], Dict{Symbol, AbstractArray}(:CV => [[1, 2, 3, 4,
    5, 6, 7, 8]], :FV => [[1, 2, 3, 4], [3, 4, 5, 6], [1, 3, 5,
    7], [2, 4, 6, 8], [1, 2, 7, 8], [5, 6, 7, 8]], :EV => [[3,
    4], [2, 4], [1, 2], [1, 3], [5, 6], [4, 6], [3, 5], [5, 7],
    [1, 7], [6, 8], [2, 8], [7, 8]])) =#

V  = obj.V
EV = obj.C[:EV]
FV = obj.C[:FV]
```

Coding 3.4.2 (Construction of an operator matrix) With some abuse of language, we build the Plasm operator EF :: ChainOP $\equiv \partial_2$: F $\rightarrow$ E.

When copying text from a source PDF and merging it into REPL to test the book examples, readers should pay close attention to a few characters (like * and '), as some might be changed on different platforms compared to the original. If changes occur, correct them manually.

```
KEV = lar2cop(EV); KFV = lar2cop(FV);  KVF = KFV';

∂_2 = KEV * KVF .÷ 2
12×6 SparseArrays.SparseMatrixCSC{Int64, Int64} with 24 stored
    entries:
 1  1  .  .  .  .
 1  .  1  .  .  .
 .  1  1  .  .  .
 1  .  .  1  .  .
 .  1  .  1  .  .
 1  .  .  .  1  .
 .  .  1  .  1  .
 .  .  .  1  1  .
 .  .  .  .  1  1
 .  .  1  .  .  1
 .  .  .  1  .  1
 .  1  .  .  .  1
```

We see that the generated sparse matrix $[\partial_2] =$ FE is $12 \times 6 = $ #F $\times$ #E. □

It may be worth discussing this matrix. Remember that, by right multi-plication, any mapping matrix sends the space of columns (2-chains: faces here) to the space of rows (1-chains: edges here). With any cubical complex, we have two 1s for each row and four 1s for each column. They characterize the two faces as incident sets on the edges, and the four edges as incident sets on the faces.

Coding 3.4.3 (*How to use the linear operators.*)
A boundary operator—a linear map—sends each vector in the domain space (the span of the matrix columns) to its image in the target space (the span of the matrix rows).

```
face = zeros(length(FV)); face[4] = 1

(∂_2 * face)'  #=
1×12 adjoint(::Vector{Float64}) with eltype Float64:
 0.0  0.0  0.0  1.0  1.0  0.0  0.0  1.0  0.0  0.0  1.0  0.0  =#
```

In particular, the matrix of a linear operator—its coordinate representa-tion relative to the chosen bases—maps the coordinate vector of a domain element to its image in the target space. Of course, both are binary vectors. For example, face[4] has boundary edges 4, 5, 8, and 11 (look for 1s in the last row above). □

Coding 3.4.4 (*Cell with a hole*) Figure 3.22 illustrates an example of 2D cellular complex $X = X_2$, consisting of 8 unit 0-chains (0-cells) u_0^h, 8 unit 1-chains (1-cells) u_1^k, and 2 unit 2-chains (2-cells) u_2^j.
A user-readable representation of the geometric complex $(X_2, V :: \text{Points})$ is below. V :: Points is the array of vertices that provides the embedding map

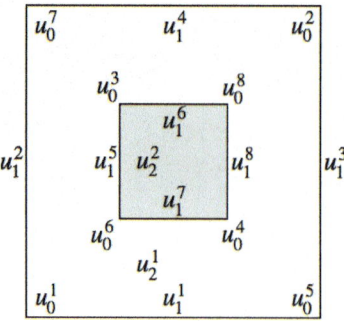

Fig. 3.22 Cellular 2-complex with two unit 2-cells (u_2^p), eight unit 1-cells (u_1^q), and eight unit 0-cells (u_0^r).

$C_0 \to \mathbb{E}^2$, implemented as array $\mathbb{N} \to \mathbb{R}^2$. EV and FV respectively provide the canonical (sorted) $X_2 :: \mathtt{Lar}$ of 1-cells and 2-cells as lists of lists of 0-cell indices. These can be interpreted as user-readable CSR (Compressed Sparse Row) characteristic matrices M_1 and M_2 of the generators of 1-cells and 2-cells, respectively, according to [9].

```
V = [[0.,0.],[3,3],[1,2],[2,1],[3,0],[1,1],[0,3],[2,2]]
FV = [[1,2,3,4,5,6,7,8],[3,4,6,8]]
EV = [[1,5],[1,7],[2,5],[2,7],[3,6],[3,8],[4,6],[4,8]]
```

The unsigned matrix of the boundary operator $\partial_2 : C_2 \to C_1$, computed by filtering elements of value 2 in the matrix $M_1 M_2^t$, is:

$$[\partial_2] = \mathrm{filter}(M_1 M_2^t, 2) = \mathrm{filter} \left(\begin{pmatrix} 1\,0\,0\,0\,1\,0\,0\,0 \\ 1\,0\,0\,0\,0\,0\,1\,0 \\ 0\,1\,0\,0\,1\,0\,0\,0 \\ 0\,1\,0\,0\,0\,0\,1\,0 \\ 0\,0\,1\,0\,0\,1\,0\,0 \\ 0\,0\,1\,0\,0\,0\,0\,1 \\ 0\,0\,0\,1\,0\,1\,0\,0 \\ 0\,0\,0\,1\,0\,0\,0\,1 \end{pmatrix} \begin{pmatrix} 1\,0 \\ 1\,0 \\ 1\,1 \\ 1\,1 \\ 1\,0 \\ 1\,1 \\ 1\,0 \\ 1\,1 \end{pmatrix}, 2 \right) = \begin{pmatrix} 1\,0 \\ 1\,0 \\ 1\,0 \\ 1\,0 \\ 1\,1 \\ 1\,1 \\ 1\,1 \\ 1\,1 \end{pmatrix}$$

The first column represents the non-convex 2-cell with the hole, and the second represents the convex cell within the hole. The reader may quickly check that the four ones in positions from fifth to eighth in the second column of $[\partial_2]$ correspond to the last four unit 1-chains in EV array. By multiplication (mod 2) of $[\partial_2]$ times the coordinate representation $[c]$ of the 2-complex in Figure 3.22, i.e., times the total 2-chain $c = u_2^1 + u_2^2 = \begin{pmatrix} 1 & 1 \end{pmatrix}^t$, we get the coordinate representation

$$[\partial_2][c] = \begin{pmatrix} 1\,1\,1\,1\,0\,0\,0\,0 \end{pmatrix}^t$$

of the 1-boundary of c, i.e., the cycle $u_1^1 + u_1^2 + u_1^3 + u_1^4$ made by the first four 1-cells in EV. $\qquad\qquad\qquad\qquad\qquad\qquad\qquad\qquad\qquad\qquad\qquad\qquad\square$

The $[\partial_2] \equiv$ KEF computation in Plasm may be expressed as:

```
KFV = lar2cop(FV); KEV = lar2cop(EV)

(KEV * KFV') .÷ Int8(2)
```

3.5 Cochain complex

Definition 3.43 (Cochain complex) A cochain complex is a graded vector space V furnished with a graded linear map $\delta : V \to V$ of degree $+1$ called coboundary operator, which satisfies $\delta^2 = 0$.

That is to say; a cochain complex is a sequence of vector spaces C^p and linear maps $\delta^p : C^p \to C^{p+1}$, such that $\delta^{p+1} \circ \delta^p = 0$. See Equation 3.5.

Remark 3.15 (Boundary vs coboundary operators) As a general rule, readers should remember the definitions of boundary and coboundary operators: boundaries map cells to lower dimensions, while coboundaries map in increasing dimensions. For example, it is $\partial_2 : C_2 \to C_1$ and $\delta_1 : C_1 \to C_2$. Note that each operator's index corresponds to its domain index.

The space C^p of cochains on X is the dual space $(C_p)^* = \{\phi : C_p \to \mathbb{R}\}$, consisting of real-valued linear functions that integrate over chains, that is, subsets of cells of the same dimension that decompose X.

Definition 3.44 (Domain integral) A domain integral is the value obtained when a given cochain is evaluated on a specific chain.

This value depends bilinearly on the pair (cochain, chain), meaning that it is a linear function of the cochain when the chain is fixed and a linear function of the chain when the cochain is fixed.

Thus, the cochain is not the integral's value but the operation that calculates the integral, i.e., the linear map that assigns the integral's value to each chain. The canonical example of cochain use is domain integration on a curve, surface, or volume, expressed as the sum of discrete integral values (real numbers) over the 1-, 2-, or 3-dimensional cells that decompose the domain.

Remark 3.16 (Plasm integration)
Since the early development of classic PLaSM as a language for solid modeling, we have developed methods for integrating monomials into geometric shapes. Specifically, boundary triangulation and PL curve integration enable surface and volume integration. The integral over monomials can be easily extended to polynomials by leveraging the linearity of both the integral and the integrand functions.

The remainder of this section explains how they are used as Plasm primitives to integrate over surfaces and boundary representations of solid models.

Cochain and integration

In algebraic topology, the connection between integration and cochains emerges in the context of duality between chains and cochains. This duality links geometric ideas, such as integration over manifolds, with algebraic structures, like cohomology[8] which can be regarded as a higher-dimensional version of integration. For simplicity, we refer to simplices instead of cells.

Chains and Cohomology

Chains are formal sums of simplices or cells (e.g., vertices, edges, faces, etc.) used to represent the "building blocks" of a space.

Cochains are dual to chains. They assign values (often real numbers, integers, or other coefficients) to chains, typically thought of as functions on chains. Cochains provide a way to measure or "weigh" these building blocks in a way that is useful in cohomology theory.

Integration as a Dual of Boundary Operator

The boundary operator maps a chain (a combination of simplices) to its boundary (a more straightforward set of simplices). This gives a way of representing topological features like loops and surfaces.

The integration of a cochain over a chain can be interpreted as measuring how the cochain "interacts" with the chain, much like how integration measures the total of a function over a geometric domain.

Specifically, the cochain evaluates the chain by summing over the simplices of the chain. The integration of a cochain is the sum of the contributions of the cochain to each of the simplices in the chain. In this sense, the cochain gives a "measure" of the space that can be integrated.

Cochain Complex and Integration

The cochain complex is closely related to the chain complex, where cochains are used to define cohomology (which is the dual of homology). In cohomology, we integrate cochains over chains to detect topological features such as holes and other geometric structures in a space.

For example, in the case of de Rham cohomology (where cochains are differential forms), integration corresponds to evaluating the form (cochain) over the manifold (chain), giving a topological invariant. Cochains can be considered differential forms[9] integrated over the space's cells in this setting.

---

[8] Cohomology is a mathematical tool that assigns algebraic invariants to a topological space, capturing its global properties. It is the dual concept to homology, where cochains replace chains and measure how cochains fail to be exact.

[9] Differential forms provide a unified approach to defining integrands over curves, surfaces, solids, and higher-dimensional manifolds.

In summary, Chains represent the underlying topological space as sums of cells. Cochains provide a way to measure or evaluate chains, and the integration of cochains over chains offers a dual, algebraic method for studying the geometric features of a space. The relationship between integration and cochains is central to cohomology theory, where cochains represent topological features, and the integration over chains computes topological invariants.

3.5.1 Cochain integration with Plasm

Evaluating the integral properties of solid structures (e.g., volume, center of gravity, moments of inertia, etc.) is often necessary to plan or control the static or dynamic behavior of mechanical or civil models in CAD and BIM projects. Therefore, this section addresses the precise evaluation of the inertial properties of homogeneous polyhedral objects.

Since [6], a finite integration method is employed in Plasm to calculate monomial integrals of various orders over solids and polyhedral surfaces in 3D and 2D Euclidean space. This method precisely evaluates domain integrals of trivariate polynomials. We use this finite solution in Plasm for surface and volume integration of polynomials over a domain's boundary triangulation. In other words, this section focuses on cochain evaluation over our chain-based boundary models.

In particular, the surface and volume integrals are evaluated in Plasm by transforming them into line integrals along the boundary of each 2-simplex in a triangulation of the domain boundary [6]. A different approach to finite integration, which uses a decomposition into volume elements induced by a boundary triangulation, is presented in [15], where a closed formula for volume integration over polyhedral volumes is introduced by decomposing the solid into a collection of tetrahedra; however, this method is not suitable for surface integration.

Volume integration

Triple (volumetric) integrals determine the volume, moments of inertia, and other properties of rigid homogeneous solids within 3D Euclidean space. Automating the calculation of these integral properties for geometrically complex solids is essential in CAD/CAM, robotics, and related fields. The papers [13, 14] summarize the established methods for computing the integral properties of solids.

As previously noted, closed formulas for integrals over polyhedral volumes are presented in [6]. These formulas are easily derived by converting volume integrals into surface integrals using the Gauss Divergence Theorem, which states that the outward flux of a vector through a closed surface equals the volume integral of its divergence over the region. The divergence theorem reflects the physical principle that, in the absence of matter creation or destruction, the density within a given space can only change through matter flowing in or out across its boundary [24].

It is possible to show [6] that such integrals are computable in polynomial time, and in particular, that inertia moments are computable in $O(E)$ time, with E representing the number of edges on the boundary of a solid model of the domain of integration.

Surface integration

Surface integral values (numbers) are computed as a summation of monomial integrals (cochains) over a triangulation (chain) of the surface. Any triangle is mapped into the unit triangle in implementing the Plasm integration module; we based many unit tests on the standard coherently oriented 3-simplex below:

```
V = [0.0 1.0 0.0 0.0; 0.0 0.0 1.0 0.0; 0.0 0.0 0.0 1.0];
FV = [[1, 2, 4], [1, 3, 2], [4, 3, 1], [2, 3, 4]];
tetra = V, FV;
VOLUME(tetra) # => 0.16666666666666666
```

because it is minimal, easy, and produces results comparable to those of many external sources. For example, the relation

```
0.16666666666666666 == 1/6 # => true
```

is correct since the unit standard CUBE(1), clearly of unit volume, can be partitioned into six such tetrahedra (see Remark 3.1).

But later, we realized that other tests, even over the standard cube, gave incorrect results. The reason? You must remember to compute or check a coherently oriented boundary triangulation of the solid correctly.

The best solution was to use the output of a BREP operator, which reports the coherently oriented triangles of the BOUNDARY of a solid expression, meaning the 3D domain defined by an input expression. In the case of a cube (single atom), its six faces are coherently oriented, resulting in such a triangulation.

BREP interface for Plasm integrals

Here, we introduce the API and FL-based implementation for a minimal Plasm interface that converts Hpc and Lar data objects into the model data structure recognized by the native PlaSM language and integration methods. Readers will value the notable compactness of this classic FL-based language.

```
BREP(obj::Hpc) = CONS([S1,CAT∘S2])(get_oriented_triangles(obj))
BREP(obj::Lar) = CONS([S1,CAT∘S2])(get_oriented_triangles(MKPOL(
    obj.V, obj.C[:CV])))
```

Coding 3.5.1 (CUBE(2)) The simplest test examples are given here, as usual in writing unit tests for software development.

```
obj = CUBE(2)        #=
Hpc(MatrixNd([[1.0, 0.0, 0.0, 0.0], [0.0, 2.0, 0.0, 0.0], [0.0,
    0.0, 2.0, 0.0], [0.0, 0.0, 0.0, 2.0]]), Hpc(MatrixNd(4),
    Geometry([[0.0, 0.0, 0.0], [1.0, 0.0, 0.0], [0.0, 1.0, 0.0],
    [1.0, 1.0, 0.0], [0.0, 0.0, 1.0], [1.0, 0.0, 1.0], [0.0,
    1.0, 1.0], [1.0, 1.0, 1.0]], hulls=[[1, 2, 3, 4, 5, 6, 7,
    8]])))   =#
```

Now compute some integrals on Hpc dataset:

```
VOLUME(BREP(obj::Hpc)) # => 8
SURFACE(BREP(obj::Hpc)) # => 24
CENTROID(BREP(obj::Hpc))' #=
1×3 adjoint(::Vector{Float64}) with eltype Float64:
    1.0  1.0  1.0      =#
```

Note that the operator BREP only functions with Hpc objects. A Lar object must be converted to Hpc first. Refer to the script below. □

Coding 3.5.2 (Cube with coherently-oriented BREP())
The cubegrid object consists of one, two, and three unit cubes in the first, second, and third dimensions, respectively. BREP extracts the coherently oriented boundary triangulation:

```
lar = CUBOIDGRID([1,2,1])
cubegrid = MKPOL(lar.V, lar.C[:CV])::Hpc
VOLUME(BREP(cubegrid)) # => 2.0
SURFACE(BREP(cubegrid)) # => 10.0
```

We may scale the V coordinates by 2.0 in all three directions to increase the volume by a factor of 2^3 and the surface area to 10×2^2.

```
V,TV = BREP(cubegrid)
scaledgrid = (V * 2, TV)
VOLUME(BREP(MKPOL(scaledgrid))) # => 16.0
SURFACE(BREP(MKPOL(scaledgrid))) # => 40.0
CENTROID(BREP(MKPOL(scaledgrid)))    #=
3-element Vector{Float64}:
 1.0
 2.0
 1.0      =#
```

For the original, non-scaled cubegrid, we have:

```
CENTROID(BREP(cubegrid))          #=
3-element Vector{Float64}:
 0.5
 1.0
 0.5        =#
```

The language operators used in script 3.5.2 utilize a pair (V,TV) generated by BREP, where V is a coordinate matrix and TV represents the coherently-oriented boundary triangulation. It is important to note that the common (non-boundary) face of the two unit cubes in cubegrid is not considered.

To test the numerical robustness of integration formulas with more approximated coordinates, for example, within a hierarchical STRUCT operator, we (currently) need to convert the geometric value to Lar format and extract its vertices and boundary triangulation.

Coding 3.5.3 (Integration tests with transformations)

We present additional test examples by transforming the domain model with Plasm affine tensors or by directly changing its coordinates via V transformations. Remember that R()() can be applied only to Hpc values.

```
julia> obj = R(2,3)(π/3)(CUBE(1))    #=
Hpc(MatrixNd([[1.0, 0.0, 0.0, 0.0], [0.0, 1.0, 0.0, 0.0], [0.0,
    0.0, 0.5000000000000001, -0.8660254037844386], [0.0, 0.0,
    0.8660254037844386, 0.5000000000000001]]), Hpc(MatrixNd(4),
    Hpc(MatrixNd(4), Geometry([[0.0, 0.0, 0.0], [1.0, 0.0, 0.0],
    [0.0, 1.0, 0.0], [1.0, 1.0, 0.0], [0.0, 0.0, 1.0], [1.0,
    0.0, 1.0], [0.0, 1.0, 1.0], [1.0, 1.0, 1.0]], hulls=[[1, 2,
    3, 4, 5, 6, 7, 8]])))) =#
```

The pair "rotated coords, triangles" is therefore used, where it is possible to see that now some approximation errors appear:

```
julia> VOLUME(BREP(obj)) # => 1.0000000192425853
```

We like to remark that the applied rotation was around the x axis so that the x coordinates of our model points cannot mutate, and so cannot mutate the first centroid coordinate:

```
julia> centre = CENTROID(BREP(obj))          #=
3-element Vector{Float64}:
  0.5
 -0.18301270190371720
  0.6830127145649264        =#
```

Let's note we can check the correctness of centroid position by directly computing the Hpc value made by a single vertex:

```
MKPOL([centre],[[1]])    #=
Hpc(MatrixNd(4), Geometry([[0.5, -0.1830127019037172,
    0.6830127145649264]], hulls=[[1]]))      =#
```

Coding 3.5.4 (Rotated test object) The last series of unit tests are focused on the rotated unit cube of $\pi/4$ around the z axis.

```
obj = CUBE(1)
VOLUME(BREP(R(1,2)(π/4)(obj)))    # => 0.9999999657714582
SURFACE(BREP(R(1,2)(π/4)(obj)))   # => 5.999999863085833
VOLUME(BREP(obj))                 # => 1.0
SURFACE(BREP(obj))                # => 6.0
```

3.5.2 Mechanical properties with structure products

In BIM and CAD applications, we are interested in integrals:

$$I_S = \iint_S f(\mathbf{p})\, dS; \quad I_P = \iiint_V f(\mathbf{p})\, dV,$$

where S and P are linear and regular 2- or 3-polyhedra in 3D, dS and dV are the differential surface and the differential volume.

The integrating function $f(\mathbf{p})$ is a trivariate polynomial of indetermined $x, y, z \in \mathbb{R}$:

$$f(\mathbf{p}) = \sum_{\alpha=0}^{m} \sum_{\beta=0}^{n} \sum_{\gamma=0}^{p} a_{\alpha\beta\gamma} x^\alpha y^\beta z^\gamma, \qquad \alpha, \beta, \gamma \in \mathbb{N} \tag{3.7}$$

whose integration is implemented in Plasm through integrating the trivariate monomial $x^\alpha y^\beta z^\gamma$, known as the structure product. Due to domain and integration function linearity, the language calculates any polynomial integral over any PL surface S and any PL volume V.

The computation through the Plasm integration subsystem relies on a fundamental lower-level group of functions M, TTT, II, and III, [6] from the Plasm module integr.jl, which is not documented here. These functions are used to implement any integral structure property as described below.

Coding 3.5.5 (Unit tetrahedron and cube)
The two straightforward polyhedral datasets created by BREP and used in the unit tests below are called tetra and cube.

Let's remember that the Julia input data type required by `Plasm` for integration methods is a 2–element `Vector{Array}` containing vertices by columns and consistently oriented boundary triangles.

```
tetra = BREP(MKPOL(SIMPLEX(3)))        #=
2-element Vector{Array}:
 [0.0 0.0 0.0 1.0; 0.0 1.0 0.0 0.0; 1.0 0.0 0.0 0.0]
 [[1, 2, 3], [1, 3, 4], [1, 4, 2], [2, 4, 3]]        =#
cube  = BREP(CUBE(1))      #=
2-element Vector{Array}:
 [0.0 0.0 … 1.0 1.0; 1.0 1.0 … 0.0 1.0; 1.0 0.0 … 1.0 0.0]
 [[1, 2, 3], [5, 1, 3], [4, 5, 3], [2, 4, 3], [2, 8, 4], [8, 6,
    4], [6, 7, 4], [7, 5, 4], [1, 5, 7], [6, 1, 7], [1, 6, 8],
    [2, 1, 8]]      =#
```

Surface and volume

The surface area of a solid is the sum of all its outside faces or curved surfaces. It is important for calculating properties such as heat transfer, required materials, or fluid interactions. For polyhedral solids (e.g., prisms), the surface area is the total of all face areas, while for curved solids (e.g., spheres), you might use polyhedral models to estimate.

A solid's volume measures its three-dimensional space. Knowing the volume is essential for determining an object's capacity, weight, or mass, especially in engineering and manufacturing.

Coding 3.5.6 (Surface)

```
function SURFACE(P)::Float64
    return II(P, 0, 0, 0)
end
```

```
SURFACE(tetra) = 2.3660254037844384
SURFACE(cube) = 6.0
```

While the unit cube's surface is precise, how can we verify the accuracy of the value returned by `SURFACE(tetra)`? When applied to any solid, the function returns the total of signed areas of all triangles from the solid shells. See Figure 3.23.

Coding 3.5.7 (Volume)

```
function VOLUME(P)::Float64
    return III(P, 0, 0, 0)
end
```

Fig. 3.23 In a unit SIM-
PLEX(3) model, we have
three boundary triangles
of equal area, contributing
a total of $3 \times 0.5 = 1.5$
to the surface area. Therefore, the oblique triangle
adds 2.3660254037844384 -
$1.5 = 0.8660254037844384$
= sqrt(3)/2, representing
the area of the standard
oblique 2-simplex in $\mathbb{E}^3$.

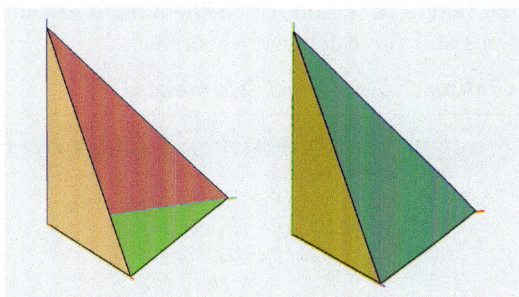

```
VOLUME(tetra) = 0.16666666666666666
VOLUME(cube) = 1.0
```

We know that (1) the unit cube can be generated by the union of two
triangular prisms of height one and basis the unit 2-simplex; (2) each such
prism is decomposable into three 3-simplices, so we get that the unit cube
is decomposable into 6 unit 3-simplices. Hence, we have: VOLUME(tetra) =
VOLUME(cube)$/6 = 0.16666666666666666$.

First and Second Moments and Products of Inertia

The first moment and second moment of an object differ in their descriptions
of how mass, area, or other properties are distributed in relation to a reference
point, axis, or plane. We give below their meaning in 2D, easier to visualize.

First Moment   The first moment measures the distribution of mass or
  area relative to an axis. In 2D it is calculated as $M_x = \iint_S y\, dA$ (about the
  x-axis) or $M_y = \iint_S x\, dA$ (about the y-axis). The first moment is physically
  used to locate an object's centroid.

Second Moment   The second moment in 2D measures the distribution of
  mass or area relative to an axis but emphasizes points farther from the
  axis (squares their distance). It is calculated as $I_x = \iint_S y^2\, dA$ (about
  the x-axis) or $I_y = \iint_S x^2\, dA$ (about the y-axis). The second moment is
  also known as the moment of inertia and reflects the object's resistance to
  bending or rotation.

In summary, the first moments relate to the centroid's position, while the
second describes the resistance to rotational or bending forces. Of course, in
3D, the integral involves the differential volume dV, and the vector integrals
have three components.

In solid mechanics, the moment of inertia about an arbitrary axis through
the origin is the volume integral of the squared (orthogonal) distances from

the points of the object to the axis. It is calculated using space transformations and the Euclidean theorem.

Coding 3.5.8 (First Moments)

```
function FIRSTMOMENT(P)::Array{Float64,1}
    out = zeros(3)
    out[1] = III(P, 1, 0, 0)
    out[2] = III(P, 0, 1, 0)
    out[3] = III(P, 0, 0, 1)
    return out
end
```

Let us check the implementation against the `cube` and `tetra` objects:

```
FIRSTMOMENT(tetra)     #=
3-element Vector{Float64}:
 0.04166666666666663
 0.041666666666666664
 0.041666666666666685     =#
FIRSTMOMENT(cube)      #=
3-element Vector{Float64}:
 0.5
 0.5
 0.5   =#
```

Coding 3.5.9 (Second Moments)

```
function SECONDMOMENT(P)::Array{Float64,1}
    out = zeros(3)
    out[1] = III(P, 2, 0, 0)
    out[2] = III(P, 0, 2, 0)
    out[3] = III(P, 0, 0, 2)
    return out
end
```

```
SECONDMOMENT(tetra)       #=
3-element Vector{Float64}:
 0.01666666666666668
 0.016666666666666663
 0.016666666666666663     =#
SECONDMOMENT(cube)      #=
3-element Vector{Float64}:
 0.3333333333333333
 0.33333333333333326
 0.33333333333333326     =#
```

The products of inertia describe how the mass of a body is distributed relative to two perpendicular axes in a chosen coordinate system. They are crucial in mechanics and engineering, particularly when analyzing rotational motion, rigid bodies, and moments of inertia.

Coding 3.5.10 (Products of Inertia)

```
function INERTIAPRODUCT(P)::Array{Float64,1}
    out = zeros(3)
    out[1] = III(P, 0, 1, 1)
    out[2] = III(P, 1, 0, 1)
    out[3] = III(P, 1, 1, 0)
    return out
end
```

```
INERTIAPRODUCT(tetra)    #=
3-element Vector{Float64}:
 0.008333333333333361
 0.008333333333333331
 0.008333333333333325    =#
INERTIAPRODUCT(cube)    #=
3-element Vector{Float64}:
 0.25
 0.25
 0.25    =#
```

Centroids and Moments of Inertia

The centroid of a geometric object is the center of mass (assuming uniform density) and acts as the "balance point" of the shape. For a 2D shape, the centroid coordinates $(\bar{x}, \bar{y})$ are found as the weighted average of the vertices or area distribution. In 3D, it involves volume-based calculations. The centroid plays a key role in structural analysis, computer graphics, and robotics.

In mechanics, the centroid and moment of inertia help analyze balance, stability, and rotational dynamics. They are used in algorithms for shape analysis, modeling, and simulations in computational geometry.

Coding 3.5.11 (Products of Inertia)

The products of inertia measure how mass (or area) is distributed relative to two perpendicular axes, quantifying cross-coupling effects in rotational motion. For a 2D object, the product of inertia I_{xy} is given by $I_{xy} = \iint_S xy\, dA$, where x and y are the coordinates of the area element dA.

Nonzero products of inertia indicate asymmetry and are essential in analyzing principal axes, rotational stability, and dynamic behavior of objects.

```
function CENTROID(P)::Array{Float64,1}
    return FIRSTMOMENT(P) ./ VOLUME(P)
end
```

```
CENTROID(tetra)    #=
3-element Vector{Float64}:
 0.25
 0.25
 0.25    =#
CENTROID(cube)     #=
3-element Vector{Float64}:
 0.5
 0.5
 0.5     =#
```

By definition, for each vector component we have `CENTROID(tetra)` = `FIRSTMOMENT(tetra)` / `VOLUME(tetra)`, which equals $0.041\overline{6}/0.1\overline{6} = 0.25$.

The moment of inertia measures how mass is spread out relative to an axis of rotation, indicating an object's resistance to rotational motion. For a 2D object, the moments of inertia I_x and I_y are calculated about the x- and y-axes by integrating $y^2 dA$ or $x^2 dA$ over the area S of the object.

Coding 3.5.12 (Moments of Inertia)

```
function INERTIAMOMENT(P)::Array{Float64,1}
    out = zeros(3)
    result = SECONDMOMENT(P)
    out[1] = result[2] + result[3]
    out[2] = result[3] + result[1]
    out[3] = result[1] + result[2]
    return out
end
```

```
INERTIAMOMENT(tetra)    #=
3-element Vector{Float64}:
 0.033333333333333326
 0.03333333333333334
 0.03333333333333334     =#
INERTIAMOMENT(cube)     #=
3-element Vector{Float64}:
 0.6666666666666665
 0.6666666666666665
 0.6666666666666665      =#
```

Coding 3.5.13 (Body inertia moments w.r.t. any axis)

The vector moments of inertia can be easily computed with respect to any axis not passing through the origin. Let be given a `Plasm` model named body,

If $\mathbf{p} = (p_x, p_y, p_z)$ and $\mathbf{d} = (d_x, d_y, d_z)$ are a 3D point and the direction of an axis for $\mathbf{p}$ and not passing for the origin, respectively, then the (scalar) inertial moment of the body w.r.t. the $(\mathbf{p}, \mathbf{d})$ axis may be computed as follows:

1. translation of space such that **p** is moved to the origin;
2. general rotation such that the axis is made coincident with the z coordinate direction;
3. computation od `body` moments of inertia about the z axis. □

Exercise 3.2 After reading the following book chapter, the reader should write and test a `Plasm` function named `axial_inertia`, which has parameters `body :: Hpc`, `p :: Vector`, and `d :: Vector`. □

While the moment of inertia about a single axis is a scalar, the complete description of rotational inertia for a 3D object in free rotation is a more complex quantity called the inertia tensor, which is a rank-2 tensor.

References

1. Standard simplex. ncyclopedia of Mathematics (2024) pages 78
2. Arnold, D.N.: Finite Element Exterior Calculus, CBMS-NSF Regional Conference Series in Applied Mathematics, vol. 93. Society for Industrial and Applied Mathematics (SIAM), Philadelphia, PA (2018) pages 68, 93, 95
3. Arnold, D.N., Falk, R.S., Winther, R.: Finite element exterior calculus: from Hodge theory to numerical stability. Bull. Amer. Math. Soc. (N.S.) **47**, 281–354 (2010) pages 68
4. Bernardini, F., Ferrucci, V., Paoluzzi, A., Pascucci, V.: Product operator on cell complexes. In: SMA '93: Proceedings on the second ACM symposium on Solid modeling and applications, pp. 43–52. ACM Press, New York, NY, USA (1993). DOI http://doi.acm.org/10.1145/164360.164378 pages 75, 90, 248
5. Cattani, C., Paoluzzi, A.: Solid modeling in any dimension. Tech. Rep. 02-89, Dip. di Informatica e Sistemistica, Università 'La Sapienza' (1989) pages 86
6. Cattani, C., Paoluzzi, A.: Boundary integration over linear polyhedra. Computer-Aided Design **22**(2), 130–135 (1990). DOI https://doi.org/10.1016/0010-4485(90)90007-Y. URL https://www.sciencedirect.com/science/article/pii/001044859090007Y pages 103, 104, 107
7. Cimrman, R.: Sparse matrices in scipy. In: G. Varoquaux, E. Gouillart, O. Vahtras, P. deBuyl (eds.) Scipy lecture notes, release: 2022.1 edn., p. Section 2.5. Zenodo (2015). DOI 10.5281/zenodo.594102. URL https://scipy-lectures.org/advanced/scipy_sparse/index.html pages 97
8. Delfinado, C., Edelsbrunner, H.: An incremental algorithm for betti numbers of sinplicial complexes on the 3-sphere. Computer Aided Geometric Design **12**, 771–784 (1995) pages 96, 222
9. DiCarlo, A., Paoluzzi, A., Shapiro, V.: Linear algebraic representation for topological structures. Comput. Aided Des. **46**, 269–274 (2014). DOI 10.1016/j.cad.2013.08.044. URL http://dx.doi.org/10.1016/j.cad.2013.08.044 pages 100
10. Ferrucci, V.: Generalised extrusion of polyhedra. In: Proceedings on the Second ACM Symposium on Solid Modeling and Applications, SMA '93, p. 3542. Association for Computing Machinery, New York, NY, USA (1993). DOI 10.1145/164360.164376. URL https://doi.org/10.1145/164360.164376 pages 86, 248
11. Ferrucci, V., Paoluzzi, A.: Extrusion and boundary evaluation for multidimensional polyhedra. Comput. Aided Des. **23**(1), 4050 (1991). DOI 10.1016/0010-4485(91)90080-G. URL https://doi.org/10.1016/0010-4485(91)90080-G pages 86
12. Hatcher, A.: Algebraic topology. Cambridge University Press (2002) pages 63

13. Lee, Y.T., Requicha, A.A.G.: Algorithms for computing the volume and other integral properties of solids. i. known methods and open issues. Commun. ACM **25**(9), 635–641 (1982). DOI 10.1145/358628.358643. URL https://doi.org/10.1145/358628.358643 pages 103

14. Lee, Y.T., Requicha, A.A.G.: Algorithms for computing the volume and other integral properties of solids. ii. a family of algorithms based on representation conversion and cellular approximation. Commun. ACM **25**(9), 642–650 (1982). DOI 10.1145/358628.358648. URL https://doi.org/10.1145/358628.358648 pages 103

15. Lien, S.l., Kajiya, J.T.: A symbolic method for calculating the integral properties of arbitrary nonconvex polyhedra. IEEE Computer Graphics and Applications **4**(10), 35–42 (1984). DOI 10.1109/MCG.1984.6429334 pages 103

16. Paoluzzi, A.: Geometric Programming for Computer Aided Design. John Wiley Sons, Chichester, UK (2003). URL https://onlinelibrary.wiley.com/doi/book/10.1002/0470013885 pages 37, 38, 48, 73, 127, 130, 201, 206, 209, 244

17. Paoluzzi, A., Pascucci, V., Vicentino, M.: Geometric programming: a programming approach to geometric design. ACM Trans. Graph. **14**(3), 266–306 (1995). DOI 10.1145/212332.212349. URL http://doi.acm.org/10.1145/212332.212349 pages 37, 67, 245

18. Paoluzzi, A., Scorzelli, G.: Computational topology, boolean algebras, and solid modeling. Computer-Aided Design **181**, 103,839 (2025). DOI https://doi.org/10.1016/j.cad.2025.103839. URL https://www.sciencedirect.com/science/article/pii/S0010448525000016 pages 37, 47, 65, 143, 218, 221, 239

19. Paoluzzi, A., Shapiro, V., DiCarlo, A., Furiani, F., Martella, G., Scorzelli, G.: Topological computing of arrangements with (co)chains. ACM Trans. Spatial Algorithms Syst. **7**(1) (2020). DOI 10.1145/3401988. URL https://doi.org/10.1145/3401988 pages 47, 94, 137, 216, 218, 221, 224, 227, 228, 231, 233, 234, 235, 251, 258, 260, 278

20. Paoluzzi, A., Shapiro, V., DiCarlo, A., Scorzelli, G., Onofri, E.: Finite algebras for solid modeling using julias sparse arrays. Computer-Aided Design **155**, 103,436 (2023). DOI https://doi.org/10.1016/j.cad.2022.103436. URL https://www.sciencedirect.com/science/article/pii/S0010448522001695 pages 47, 65, 97, 137, 184, 218, 221, 254, 260, 270, 277

21. Requicha, A.: Representations for rigid solids: Theory, methods and systems. ACM Computing Surveys **12**(4), 437–464 (1980). URL https://doi.org/10.1109/TC.1980.1675470 pages 64, 74, 138, 218, 244

22. Roth, A., Weisstein, E.W.: Standard basis. In: MathWorld, p. Algebra > Linear Algebra > Linear Systems of Equations. A Wolfram Web Resource (2005). URL https://mathworld.wolfram.com/StandardBasis.html pages 97

23. Rowland, T.: Characteristic function. In: From MathWorld, p. Foundations of Mathematics > Set Theory > Sets. A Wolfram Web Resource (2005). URL ttps://mathworld.wolfram.com/CharacteristicFunction.html pages 64

24. Weisstein, E.W.: Quadrature. MathWorld–A Wolfram Web Resource URL https://mathworld.wolfram.com/DivergenceTheorem.html. [retrieved October 10, 2024] pages 104

25. Whitney, H.: Geometric Integration Theory. Princeton University Press, Princeton (1957). DOI doi:10.1515/9781400877577. URL https://doi.org/10.1515/9781400877577 pages 84

Chapter 4
Geometric models

In mathematics and science, a space is a set of objects with some additional structure. In this chapter, we explore the geometrical spaces used to represent actual objects from natural, artificial, or virtual environments, or, better yet, the geometric models used to study, visualize, and/or simulate their physical behavior in computer memory.

Geometric models determine the physical appearance of architecture, engineering, and construction (AEC) products at any scale, from structural and envelope parts to entire buildings and built environments. They are used for design, collaboration, tenders, and contracts.

We ported the functional language `Plasm` to Julia to enhance the design, modeling, and visualization of geometric objects. We consider Julia `Plasm` to be the best option because of its simplicity and compactness in creating symbolic geometric models for Building Information Modeling (BIM) and Computer-Aided Design (CAD).

The chapter highlights the expressive capabilities of Plasm geometric types and parametric functions. We also cover simple methods for constructing parametric assemblies, where other objects can serve as actual parameters. Each model is usually created in its reference frame and then moved through elementary rotations, scalings, translations, and other techniques to assemble component models accurately.

Therefore, this chapter demonstrates the simple `Plasm` syntax for geometric transformations and assemblies. The language also offers a general mechanism (Julia dictionaries) for exporting models characterized by colors, textures, materials, and more, including boolops among lower-level components. Additionally, `Plasm` can be embedded in the Jupyter platform to document design choices progressively within digital notebooks.

A. Paoluzzi and G. Scorzelli, *BIM Geometry with Julia Plasm—Functional Language for CAD Programming*, Digital Innovations in Architecture, Engineering and Construction, https://doi.org/10.1007/978-3-031-90244-4_4

4.1 Geometric Transformations

This section mainly covers linear, affine, and convex spaces as collections of vectors and/or points with specific properties. In particular, geometric transformations are functions that move, resize, or rotate objects.

4.1.1 Vector space

The concept of a vector space is undoubtedly the most useful tool invented by mathematicians for scientists, engineers, and architects to represent and study, both formally and graphically, the primary and complex arrangements and behaviors of our natural and artificial environments.

Definition 4.1 (Vector space) or linear space $\mathcal{V}$ over a field $\mathcal{F}$ is defined as a set, closed under two composition rules: addition and product by a scalar. Closed means that the composition output belongs to the set of inputs.

The elements $\mathbf{v} \in \mathcal{V}$ are called vectors and are often represented by an oriented arrow with a given direction, orientation, and length. The elements $\alpha \in \mathcal{F}$ are called scalars. The sum of two non-zero vectors $\mathbf{u}, \mathbf{v} \in \mathcal{V}$ is a third vector $\mathbf{w} = \mathbf{u} + \mathbf{v} \in \mathcal{V}$ with direction and length different from both $\mathbf{u}$ and $\mathbf{v}$. We may write $\mathbf{v} + \mathbf{v} = 2\mathbf{v}$, and see here both the operations of a vector space: (a) addition of vectors and (b) multiplication times a scalar.

The product of a vector by a scalar $\alpha \mathbf{v} = \mathbf{v} \alpha \in \mathcal{V}$ is a vector collinear with $\mathbf{v}$ and with different length if $\alpha \neq 1$. This explains the name "scalar" since a number "scales" (changes the length of) the vector it multiplies.

The length of $\mathbf{u} = \alpha \mathbf{v}$ shrinks or grows w.r.t. $\mathbf{v}$ according to a positive $0 < \alpha < 1$. If $\alpha < 0$, then $\mathbf{u}$ has orientation opposite to $\mathbf{v}$.

Linear independence

A linear combination of vectors is a new vector defined as a sum of scalar multiples of other vectors. Let $\mathbf{v}_1, \mathbf{v}_2, ..., \mathbf{v}_n \in \mathcal{V}$ and $\alpha_1, \alpha_2, ..., \alpha_n \in \mathcal{F}$, with $\mathcal{V}$ a vector space on the field $\mathcal{F}$ of scalar numbers. The vector

$$\mathbf{w} = \alpha_1 \mathbf{v}_1 + \alpha_2 \mathbf{v}_2 + \cdots + \alpha_n \mathbf{v}_n = \sum_{i=1}^{n} \alpha_i \mathbf{v}_i \in \mathcal{V}$$

is called a linear combination of vectors $\mathbf{v}_1, \mathbf{v}_2, ..., \mathbf{v}_n$.

- Two or more vectors are said linearly independent if none of them can be written as a linear combination of the others; in other terms, if none belongs to the span (generated subspace) of the others;

- if at least one of them can be written as a linear combination of the others, then they are said linearly dependent.

Given a set of vectors, you can determine if they are linearly independent by writing the vectors as the columns of a matrix $\mathbf{A}$ and solving $\mathbf{A}\mathbf{x} = \mathbf{0}$.

If there are non-zero solutions, then the vectors are linearly dependent. If the only solution is $\mathbf{x} = \mathbf{0}$, then they are linearly independent.

The typical vector spaces in this book will be (a) the numeric field $\mathbb{R}$ for scalars and $\mathbb{R}^n$ for vector coordinates, with (b) dim $= d$, for $0 \leq d \leq 3$.

Examples

Coding 4.1.1 (Matrices $m \times n$ are vector spaces)
Generate a random 100×100 matrix `A = rand(100,100)`. The default of `rand` function are values of type `Float64` in `[0.0,1.0]`, so we ask the compiler to compute the matrix multiplication by the scalar π, denoted by the Greek symbol. The generated value is shown commented and simplified for printing in a small space and with a small visual rumor:

```
A = rand(100,100) * π          #=
100×100 Matrix{Float64}:
  1.86139  1.35513  1.31263   …   2.11579  1.34978
  2.62044  2.96187  2.73373       1.2714   1.0682
  ⋮
  2.10896  1.56324  1.62982       1.73493  0.769944  =#
```

Coding 4.1.2 (Random vector generation.) Then, we generate a 100-element vector of random integers within the interval `[0,100]`, by using the Julia `0:100` iterator:

```
b = rand(0:100, 100)      #=
100-element Vector{Int64}:
  61
  85
  ⋮
  51              =#
```

Coding 4.1.3 (Multiplication matrix-vector.) This operation is implemented natively and very efficiently for big matrices in Julia, as well as the product and the sum of dense matrices (the detail of printing depends on available space on the output device):

```
A * b             #=
100-element Vector{Float64}:
  2786.383421064984
```

```
2554.1378635659034
  ⋮
2735.1205994270754   =#
```

Coding 4.1.4 (Addition matrix-vector.) Conversely, the sum of a matrix with a vector is not a linear operation, so we need to broadcast ("." operator) the vector `b` on all columns of `A`:

```
A .+ b                                                              #=
100x100 Matrix{Float64}:
  15.8614   15.3551   15.3126   …   16.1158   15.3498
  19.6204   19.9619   19.7337       18.2714   18.0682
  ⋮                                          ⋱
  41.109    40.5632   40.6298       40.7349   39.7699 =#
```

Summary: While the `Matrix` by `Vector` multiplication is a native operation on the linear space of `Numbers`, adding a `Matrix` and a `Vector` is not, and the broadcast operator "." must be used. The same applies to adding a matrix and a single scalar value. The reader should try.

Subspace

Let $\mathcal{V}$ be a vector space on the field $\mathcal{F}$. We say that $\mathcal{U} \subset \mathcal{V}$ is a subspace of $\mathcal{V}$ when $\mathcal{U}$ is a vector space about the same operations. In particular, $\mathcal{U} \subset \mathcal{V}$ is a subspace of $\mathcal{V}$ if and only if:

1. $\mathcal{U}$ includes the $\mathbf{0}$ vector;
2. for each $\alpha \in \mathcal{F}$ and $\mathbf{u}_1, \mathbf{u}_2 \in \mathcal{U}$, $\alpha\, \mathbf{u}_1 + \mathbf{u}_2 \in \mathcal{U}$

The codimension of a subspace $\mathcal{U} \subset \mathcal{V}$ is defined as $\dim \mathcal{V} - \dim \mathcal{U}$.

It may be helpful to note that the intersection of subspaces is a subspace. In particular, if $\mathcal{U}_1, \mathcal{U}_2$ are subspaces of $\mathcal{V}$, then $\mathcal{U}_1 \cap \mathcal{U}_2$ is a subspace of $\mathcal{V}$.

Generators, Bases, and Coordinates

Span and generators

The smallest subset of vectors that can be generated by linear combinations of a subset $S \subset \mathcal{V}$ of linearly independent vectors is called *span S*. Therefore, the subset S is a set of generators.

The *span S* is closed concerning addition and multiplication, and hence is a subspace including the zero vector, which is contained in every subspace.

Bases and coordinates

The basis and dimension of the linear space V are a minimum set of generators for V, and its number $d = \dim V$ of elements, respectively.

Every basis of a linear space V has the same number d of elements.

When a basis for V has been fixed, that is, an ordered minimal subset of generators $B \subset V$ has been chosen, every vector $\mathbf{v} \in V$ can be expressed uniquely as a linear combination of elements of B with scalars. The ordered tuple of such scalars is called the coordinate tuple, or the coordinates, of vector $\mathbf{v} \in V$, and denoted as $[\mathbf{v}]$.

Remark 4.1 Representing vectors by coordinates requires a minimum set of space generators, and their ordering has already been chosen. The space has been parameterized since every vector is uniquely identified by the linear combination of the basis elements with a unique tuple of scalars.

Remark 4.2 The basis is often denoted by the ordered sequence $(\mathbf{e}_1, \ldots, \mathbf{e}_d)$ of vector elements or by the matrix $[\mathbf{e}_1 \cdots \mathbf{e}_d]$ of their coordinates by columns, where $\mathbf{e}_i = [0, 0, 1, \ldots, 0]^t$ is a (column) tuple of zeroes, with only one element 1 in position i. The standard basis of a coordinate vector space is the set of vectors whose components are all zero, except one of value 1.

Examples of vector spaces

The usual geometric example of vector space has the oriented arrows[1] as elements, summed with the parallelogram rule, and scaling related to elongation or shortening. Other examples are the linear spaces of matrices $\mathcal{M}_m^n(\mathbb{R})$ with real elements, m rows and n columns, and $\mathcal{M}_m(\mathbb{R})$ and $\mathcal{M}^n(\mathbb{R})$ of column and row vectors, respectively.

Linear space is also the space $\mathbb{P}_n(\mathbb{R})$ of real polynomial functions $p : \mathbb{R} \to \mathbb{R}$ such that $x \mapsto p^n(x)$ of degree $\leq n$. In particular, $p^n(x) = a_0 + a_1\,x + a_2\,x^2 + \cdots + a_{n-1}\,x^{n-1} + a_n\,x^n$ is exactly a linear combination of a tuple of $n+1$ scalars a_k, $0 \leq k \leq n$, with the power basis x^k of polynomials, $0 \leq k \leq n$.

The Bernstein bases of polynomial space $\mathbb{P}_n(\mathbb{R})$

Polynomials' Bernstein-Bézier basis is fundamental in geometric modeling and computer graphics. The $n+1$ Bernstein polynomials of degree n, defined as

$$B_k^n(x) = \binom{n}{k}\,x^k\,(1-x)^{n-k}, \qquad 0 \leq k \leq n$$

---

[1] Formally: the equivalence classes of equipollent oriented arrows. In Euclidean geometry, equipollence is a binary relation between directed line segments.

form a basis for the vector space $\mathbb{P}_n(\mathbb{R})$ of polynomials of degree at most n with real coefficients. Therefore, to implement the basis and to draw a graph of each basis function, we have to consider the degree n we are interested in, the ordinal number k of each function $(0 \leq k \leq n)$, and finally, the independent variable x such that $x \mapsto p^n(x)$.

Coding 4.1.5 (Pure functional style in Julia) The Julia function B (for basis) given here returns the whole range of Bézier-Bernstein polynomial bases of any degree and is very useful in curved geometric modeling.

```
B = n -> k -> u -> binomial(n,k) * u^k * (1-u)^(n-k)
B(3)   # => 4-element Vector{Function}
B(3)(2)   # => generic function with 1 method
B(3)(2)(0.5)    # => 0.375
```

This functional programming style is available in Julia and is adopted when useful to obtain curried function applications. [2]   □

Let us make some checks: B(1)(0)(0.5) == B(1)(1)(0.5) == 0.5 #=> true, for the degree-1 basis made by two polynomials B(1)(0) and B(1)(1). The higher-level function generates the whole basis of degree n Bernstein(n) that we test in the quadratic case, where it returns a 3−element array of Vector{Function}:

Coding 4.1.6 (Generating Bernstein polynomial bases) The function Bernstein(n) is defined by mapping the (partial) vector function B(n) over the integer array [0, 1, ..., n] indexing the component functions.

```
Bernstein(n) = map(B(n), collect(0:n))
# => #7 Bernstein (generic function with one method)
Bernstein(2)
# => 3-element Vector{Function}
Bernstein(2)[2]
# => #9 (generic function with one method)
Bernstein(2)[2](0.1)
# => 0.18000000000000002
```

Let's create a discrete sequence of functions for the Vector function Bernstein(2) to test the implementation. Above, we also see the second function of Bernstein's basis of degree n=2, and finally, its value 0.18 computed for u = 0.1.

Coding 4.1.7 (Sampling of third quadratic polynomial) The function Bernstein(2)[3] denotes the third function of the vector Bernstein(2) of

---

[2] Currying transforms a function of multiple arguments into a series of function calls. Partial application, connected to currying, fixes the value of some of a function's arguments without fully evaluating the function. The term honors the mathematician and logician Haskell Curry (1900-1982).

type :: `Vector{Function}`, and we compute the three function values for `u` = `0.5`, that you can check in Figure 4.1.

Exercise 4.1 (Sampling of third basis polynomial of degree 4) . The reader may build the sequence of pairs `(u, Bernstein(2)[3](u))`. A sampling of x values is created by `collect(0:0.1:1)`.

The Bernstein basis enjoys interesting properties. They are ≥ 0 for every independent variable value $x \in [0, 1]$. For every u, the elements of each basis sum to 1. Such bases are said partition of unity. Other properties also hold.

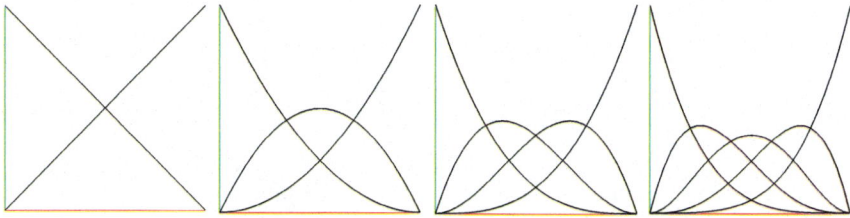

Fig. 4.1 The four images give the graphs, each in $[0, 1] \times [0, 1] \subset \mathbb{E}^2$, of Bernstein's bases of linear, quadratic, cubic, and quartic polynomials in linear spaces $\mathbb{P}_n(\mathbb{R})$, with degree $1 \leq n \leq 4$, and order $\dim \mathbb{P}_n(\mathbb{R}) = n + 1 = 2, 3, 4$, and 5, respectively.

Coding 4.1.8 (Example of `FL` & `Plasm` Combinators)
We must highlight the impressive expressive power of `FL` programming inherited from `Plasm.jl`, as demonstrated below.

Analyzing goals and types of subexpressions is very useful, as it allows one to step through the development process and understand the language expressions. First, we define the 1D model `dom1D` of the function domain $[0, 1]$ subdivided with 36 intervals, which is a geometric value of `Hpc` type:

```
dom1D = INTERVALS(1.0)(36)        #=
Hpc(MatrixNd([[1.0, 0.0], [0.0, 1.0]]), BuildMkPol[BuildMkPol(
    points=[[0.0], [0.027777777777777776], ...  =#
```

Then, consider the $\mathbb{E} \to \mathbb{E}^2$ vector function `CONS(...)∘S1` of one variable applied to the vertices of `dom1D :: Hpc` by the `MAP` operator.

```
F(k) = CONS([ID, Bernstein(4)[k]]) ∘ S1        #=
#122 (generic function with one method)    =#
```

The selector function `S1` extracts the first (and unique) coordinate value from the internal data structure `point` array. The `CONS` operator, acting on the argument of `Vector{Function}` type, transforms a function array into a vector function of coordinate functions `x(u)`, `y(u)`, etc.

Hence, we have $\mathsf{F} = \mathsf{CONS(::Vector\{Function\})} \circ \mathsf{S}d : \mathbb{E}^d \to \mathbb{E}^n$, where d is the number of coordinates of domain vertices (one in this case), and n is the dimension of the embedding space, i.e., the number of coordinate functions inside the array argument, i.e. $\mathsf{CONS([,])}$, two in this case:

Finally, we note that expressions like $\mathsf{MAP(F)(dom1D)}$ apply the F vector function to an object of Hpc type, bending it accordingly. Remember that we generated an array of Hpc curves for $\mathsf{k = 0:n}$ — with n=4.

For this purpose, we finally need to apply the STRUCT operator to transform the array of five curves into a single Hpc value. We finally visualize:

```
quarticbasis = STRUCT([ MAP(F(k))(dom1D) for k=1:5 ])
VIEWCOMPLEX(LAR(quarticbasis))
```

Change of basis

Given a basis in a linear space, it is often necessary to find a new way to describe the space using a different, possibly simpler, or more useful basis. In other words, it might be required to determine new coordinates for the vectors of the space relative to this new basis. This process is known as change of basis.

For concreteness, let us have vector data in a 3D linear space, where we need to compute their representation differently. Let us denote the new basis as $(\mathbf{u}_i)$, and the standard one as $(\mathbf{e}_i)$, with $i \in \{1, 2, 3\}$. We may write the change of coordinates as a matrix map $\mathbf{T} : \mathcal{V} \to \mathcal{V}$, such that $[u_{ij}] \mapsto [e_{ij}]$, so transforming the old coordinates of the new basis into the standard basis.

Therefore we set $\mathbf{T}[u_{ij}] = [e_{ij}]$ and, since it is $[e_{ij}] = \mathbf{I}_3$, i.e., the 3×3 identity matrix, the solution of this equation is $\mathbf{T} = [u_{ij}]^{-1}$. Of course, this matrix exists provided that $\det [u_{ij}] \neq 0$. In other words, the n vectors must be linearly independent. To obtain a Cartesian basis, the vectors must be orthonormal.[3]

4.1.2 Affine space

In geometric modeling and computer graphics, it is helpful to distinguish between a space of vectors and a space of points (supported by vectors). A space of points that provides for an operation of displacement is called an affine space and is represented here by the $\mathcal{A}$ symbol.

---

[3] A set of vectors is considered orthogonal if every pair of vectors in the set is orthogonal (the dot product is 0). The set is orthonormal if it is orthogonal, and each vector is a unit vector (norm equals 1).

Definition 4.2 (Affine action on a point space) We call affine action the function:

$$\mathcal{A} \times \mathcal{V} \to \mathcal{A},$$

so that the displacement from a point $\mathbf{a} \in \mathcal{A}$ to a point $\mathbf{b} \in \mathcal{A}$ is given in $\mathcal{V}$.

Definition 4.3 (Difference of points) An affine space $\mathcal{A}$ is so endowed with an operation of difference of points $\mathcal{A} \times \mathcal{A} \to \mathcal{V}$ where

$$\mathbf{a} + \mathbf{v} = \mathbf{b}, \quad \text{and hence} \quad \mathbf{b} - \mathbf{a} = \mathbf{v} \in \mathcal{V}$$

Remarks

We note that a **vector space** provides for an internal operation of (a) sum of two vectors and the external (b) product of a vector by a scalar, both returning a vector. Conversely, in an **affine space**, we have an external operation of (i) difference of points, returning a vector, and a (ii) sum of a point and a vector, called displacement, returning a point. The zero vector $\mathbf{0}$ is contained in all subspaces. Conversely, in an affine space $\mathcal{A}$ all points are equivalent, with no distinguished elements.

The space of points has been parameterized when a Cartesian system (origin and basis) has been chosen. To use coordinates to make it easier to work with points, we have to choose :

1. a point, called origin, to associate with the zero vector, and
2. an orthonormal basis of vectors.

The above steps consent to associate each point with the tuple of coordinates of its displacement vector from the origin.

In a vector space, all the subspaces have at least one common element, the zero vector. In contrast, two affine subspaces may not have common elements. In such a case, they are said parallel.

The dimension n of an affine space $\mathcal{A}$ corresponds to that of its supporting vector space $\mathcal{V}$. A common term for affine subspaces of dimension d is d-hyperplane or d-hyperspace when $d > 3$. Lines and planes are affine subspaces of dimension 1 and 2, respectively.

Affine independence and local parameterization

In an affine space $\mathcal{A}$ of sufficiently high dimension n, we say that two points are affinely independent when noncoincident, three points when non aligned, four points when noncoplanar, and so on.

In general, $d+1$ points ($d \leq n$) are affinely independent when the d vectors defined by the differences $\mathbf{p}_k - \mathbf{p}_0$ of the points from one of them are linearly independent.

The affine independence of a subset of $d+1$ points is often used to establish local coordinate systems on lines, planes, and higher dimensional subsets of points. Choose two noncoincident points $\mathbf{a}, \mathbf{b}$ on a 3D line. Any other point $\mathbf{p}$ of the line remains parameterized by the α scalar in the expression

$$\mathbf{p} = \mathbf{a} + \alpha(\mathbf{b} - \mathbf{a}).$$

Note that $\mathbf{b} - \mathbf{a}$ is a vector, $\alpha(\mathbf{b} - \mathbf{a})$ is a vector, $\mathbf{a} + \alpha(\mathbf{b} - \mathbf{a})$ is a point plus a vector, which is a point. The typing of our expression looks correct!

Analogously, let us choose three noncolinear points $\mathbf{a}, \mathbf{b}, \mathbf{c}$ in 3D space. They are undoubtedly noncoincident and fix a plane, i.e., a unique affine subspace of dimension 2 embedded in space. This plane is parameterized by the pairs (α, β) of scalars in the expression.

$$\mathbf{q} = \alpha(\mathbf{b} - \mathbf{a}) + \beta(\mathbf{c} - \mathbf{a}),$$

As the reader may immediately check. Similar local coordinates will hold in every affine d-subspace in n-dimensional space.

The typical affine space of points used in this book is the Euclidean space $\mathbb{E}^d$, $1 \leq d \leq 3$, usually equipped with a Cartesian system, with coordinates in the `Float64` number system.

4.1.3 Convex space

Let's consider the Euclidean space, $\mathbb{E}^d$, $1 \leq d \leq 3$, which is the fundamental space of geometry meant to represent physical space.

Affine subspaces become convex sets when a numerical constraint is imposed on the possible parameter values of an affine combination of points. Two points $\mathbf{a}, \mathbf{b} \in \mathcal{A}$ become the extreme elements of a line segment $\mathbf{p} = \alpha \mathbf{a} + (1-\alpha)\mathbf{b}$ as set of points, by adding the further constraint $\alpha + \beta = 1$ to the parameter values $\alpha, \beta \geq 0$, so posing $\beta = 1 - \alpha$. Analogously, setting $\alpha + \beta + \gamma = 1$ and $\alpha, \beta, \gamma \geq 0$ constraints the set of points combination of three nonaligned points $(\mathbf{a}, \mathbf{b}, \mathbf{c})$.

Positive, Affine and Convex Combination of Points

Let $\mathbf{p}_1, \mathbf{p}_2, \ldots, \mathbf{p}_n$ be affinely independent points in $\mathbb{E}^d$, and $\alpha_1, \alpha_2, \ldots, \alpha_n$ be scalar in $\mathbb{R}$. Their combination $\alpha_1\mathbf{p}_1 + \alpha_2\mathbf{p}_2 + \cdots + \alpha_n\mathbf{p}_n$ is said to be positive, affine, and convex, respectively, when

1. $\alpha_1, \alpha_2, \ldots, \alpha_n \geq 0$ (positive combination)
2. $\alpha_1 + \alpha_2 + \cdots + \alpha_n = 1$ (affine combination)
3. $\alpha_1 + \alpha_2 + \cdots + \alpha_n = 1$ and $\alpha_1, \alpha_2, \ldots, \alpha_n \geq 0$ (convex combination)

Verifying that the affine combination of points is a point may be interesting. Let us eliminate the α_1 scalar using the unitary sum of scalars constraint:

$$\mathbf{p} = (1 - \alpha_2 - \cdots - \alpha_n)\mathbf{p}_1 + \alpha_2\mathbf{p}_2 + \cdots + \alpha_n\mathbf{p}_n \qquad (4.1)$$
$$= \mathbf{p}_1 + \alpha_2(\mathbf{p}_2 - \mathbf{p}_1) + \cdots + \alpha_n(\mathbf{p}_n - \mathbf{p}_1) \qquad (4.2)$$

which, of course, is a point in $\mathbb{E}^d$.

A convex combination is positive and affine.

The set of all convex combinations of points $C \subset \mathbb{E}^n$ is the convex hull of C. The convex hull is the smallest compact set containing the points in C. It is the intersection of all compact sets of $\mathbb{E}^d$ that contain hull C.

Convex coordinates of a point $\mathbf{c} \in$ hull C are the scalars whose convex combination with C elements produces $\mathbf{c}$. If $C \subset \mathbb{E}^n$ has $n + 1$ affinely independent elements, that is, it is a simplex (see Definition 3.13). The convex coordinates of each $\mathbf{c} \in$ hull C are unique.

Characterization of affine pyramids

Regarding the local parameterization of affine subspaces, we did not place any restrictions on the signs of the scalar parameters (usually in the real numbers field, $\mathbb{R}$). If all the scalars are positive, the points generated stay within the interior of a planar or solid angle contained inside the lower-dimensional affine subspaces that are formed. It becomes clear that the entire subspace will be divided by an arrangement of subspaces centered at the fixed point of the set of linearly independent vectors. For example, think about how 3D space is divided into $8 = 2^3$ octants by 4 noncoplanar points, especially those where three points are at a unit distance from each other, with the three difference vectors being pairwise orthogonal.

4.1.3.1 Positive, Affine, and Convex coordinates

Positive Coordinates

Positive coordinates represent a point in a coordinate system where all coefficients or weights are non-negative. These are commonly encountered in settings such as barycentric coordinates, where the weights express the contribution of basis points (e.g., vertices of a triangle) to a point within a geometric structure. Positive coordinates ensure the point lies within or on the convex hull of the basis points. For instance, in a triangle, a point $\mathbf{p}$ can be expressed as $\mathbf{p} = \alpha_1\mathbf{p}_1 + \alpha_2\mathbf{p}_2 + \alpha_3\mathbf{p}_3$, where $\alpha_1, \alpha_2, \alpha_3 \geq 0$, and $\mathbf{p}_1$, $\mathbf{p}_2$, and $\mathbf{p}_3$ are the triangle's vertices. Positive coordinates are crucial in computational geometry, computer graphics, and interpolation, ensuring physically plausible interpolations.

Affine Coordinates

Affine coordinates describe a point in terms of a weighted sum of basis points, where the weights sum to one but are not necessarily non-negative. This property allows affine coordinates to represent points within and outside the convex hull of the basis points, making them valuable for affine transformations and general geometric modeling. For a point $\mathbf{p}$ in terms of vertices $\mathbf{p}_1$, $\mathbf{p}_2$, and $\mathbf{p}_3$, the representation $\mathbf{p} = \alpha_1\mathbf{p}_1 + \alpha_2\mathbf{p}_2 + \alpha_3\mathbf{p}_3$ requires $\alpha_1 + \alpha_2 + \alpha_3 = 1$. Affine coordinates are essential in applications like affine transformations; refer to the following pages, where scaling, rotation, and translation maintain parallelism and barycentric properties.

Convex Coordinates

Convex coordinates are a subset of affine coordinates, where the weights are non-negative and sum to one. This dual constraint ensures that the point lies strictly within the convex hull of the given basis points. Convex coordinates are essential in convex analysis and optimization, where problems often involve points confined to convex regions. For example, solutions lie on the convex hull defined by the feasible vertices in simplex methods for linear programming. Convex coordinates are also key in defining convex combinations, which are crucial for blending, interpolation, and shape approximation in geometric modeling and data analysis.

4.1.4 Space Transformations

The user interface for affine coordinate transformation of geometric objects is provided through standard Julia functions and matrices. Internally, `Plasm` performs such mapping of local coordinates using its multidimensional matrix type, called `MatrixNd`, and the field `T::MatrixNd` within the recursive datatype `Hpc`. See Section 4.3.3.

Definition 4.4 (Geometric transformation) A geometric transformation is a bijective function, i.e., a one-to-one (injective) and onto (surjective) mapping $\mathbb{E}^d \to \mathbb{E}^d$.

By definition, geometric transformations of plane or space are invertible and represented by invertible square matrices. We will see that rotation, scaling, and shearing are linear transformations; translation is affine.

Homogeneous Coordinates

In computer graphics, the homogeneous coordinates are often used instead of Cartesian coordinates. In the homogeneous plane or space, lines are mapped to lines, but parallel lines are not maintained as parallel. The primary reason for this shift is the ability to treat affine transformations (such as translation) as linear and to combine them seamlessly with linear transformations (like rotation, scaling, etc.).

In homogeneous coordinates, the Euclidean plane $\mathbb{E}^2 \setminus \{0\}$ is in bijective correspondence with the bundle of lines through the origin in $\mathbb{E}^3 \setminus \mathbb{E}^2$ (a model for the projective plane), so that each point $(x, y) \in \mathbb{E}^2$ corresponds to a line $\lambda(W, X, Y)$ such that

$$(x, y) \equiv \left(\frac{X}{W}, \frac{Y}{W} \right) = (x, y) \leftrightarrow (W, X, Y) = (1, x, y).$$

The same construction applies to each $\mathbb{E}^d$, for $d \geq 2$. After division, homogeneous coordinates are said to be normalized.

Remark 4.3 (Lines and Plane at Infinity) The statement above holds for projective (or, more generally, homogeneous) transformations. In projective geometry, every line is mapped to a line, since these transformations are linear in homogeneous coordinates. However, parallelism—a notion tied to the Euclidean concept of the line at infinity—is not preserved under general projective transformations. In contrast, affine transformations (a special subclass of projective transformations that preserve the hyperplane at infinity) do preserve parallelism: parallel lines remain parallel.

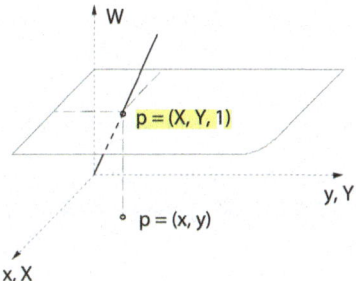

Fig. 4.2 The homogeneous plane is a model of a projective plane, where all finite points have a homogeneous coordinate equal to one, and the points at infinity have it equal to zero. All the points at infinity form the line at infinity, and all the lines at infinity form the plane at infinity. Commonly seen in graphics books, see, e.g [5].

In Plasm, by design choice to make the multidimensional approach to geometric design more accessible, the added homogeneous coordinate is the first, not the last, as we may see in many computer graphics books.

In addition, for the sake of clarity, we can use the HOMO operator to transform a $d \times d$ matrix in a $(d + 1) \times (d + 1)$ matrix, i.e., a 3×3 on the 2D plane and 4×4 on 3D space. The returned matrix type is `MatrixNd`, which supports dimension-independent programming.

Remark 4.4 (Homogeneous coordinates) allow all transformations to be combined linearly, using the products of their matrices in homogeneous normalized coordinates. In the remainder of this section, we describe the geometric effect of each transformation and the structure of the corresponding matrices.

*Remark 4.5 (*Plasm *maps are* **tensors***)* The reader should note that our maps or transformations are invertible functions of space into itself (automorphisms), represented (even translation, as we will see) by square matrices, i.e., are rank two tensors. Since they are also `Plasm` functions that can be applied to geometric objects, they multiply the object coordinates by matrices.

4.1.4.1 2D rotation

In a planar rotation, all points of the 2D plane move along an arc of a circle, with the same angle at the center, while the center is the only fixed point. In a space rotation, a straight line of fixed points (the axis) passes through the origin. All the other 3D points describe a circle arc with the same angle along the plane (orthogonal to the rotation axis) they belong to.

Let us consider how unit vectors $\mathbf{e}_1 = \begin{pmatrix} 1 \\ 0 \end{pmatrix}$ and $\mathbf{e}_2 = \begin{pmatrix} 0 \\ 1 \end{pmatrix}$, columns of the matrix $(\mathbf{e}_1 \ \mathbf{e}_2)$ are transformed by the (yet unknown) $\mathbf{R}(\alpha)$ rotation matrix into the columns of the matrix at right-hand side:

$$\begin{pmatrix} \cos \alpha & -\sin \alpha \\ \sin \alpha & \cos \alpha \end{pmatrix} = \mathbf{R}(\alpha) (\mathbf{e}_1 \ \mathbf{e}_2). \quad \text{Hence we have:} \quad \mathbf{R}(\alpha) = \begin{pmatrix} \cos \alpha & -\sin \alpha \\ \sin \alpha & \cos \alpha \end{pmatrix}$$

There is only one class of planar rotations, parameterized by α, the rotation angle about the origin. Conversely, we will see three classes of elementary space rotations, parameterized by $alpha_x$, $alpha_y$, and α_z, the rotation angles about each coordinate axis.

Coding 4.1.9 (Plasm notation for rotation) The plane rotation function in `Plasm` is: `R(1,2)(`α`)` because its effect is to change x, y. The plane rotation function in `Plasm` is `R(1,2)(`α`)` because it changes the first and second coordinates of the 2D model to which it is applied. Internally, these are applied to a planar geometric object of `Hpc` type using the `STRUCT` operator (see Section 4.3.3) that contains `Hpc` values and transformation tensors: the first and second coordinates of the 2D model it is applied to.

```
SQUARE(d) = CUBOID([d,d])        #=
SQUARE (generic function with one method)     =#
```

```
obj = R(1,2)(π/4)(SQUARE(1))        #=
Hpc(MatrixNd([[1.0, 0.0, 0.0], [0.0, 0.7071067811865476,
    -0.7071067811865475], [0.0, 0.7071067811865475,
    0.7071067811865476]]), Hpc(MatrixNd(3), Hpc(MatrixNd(3),
    Geometry([[0.0, 0.0], [1.0, 0.0], [1.0, 1.0], [0.0, 1.0]],
    hulls=[[1, 2, 3, 4]])))) =#
VIEW(obj)
```

Remark 4.6 CUBOID(shape) :: Hpc is the generator of multidimensional hyper-parallelopipeds, depending on length and content of shape vector.

   CUBOID([1,1]) is the unit square; CUBOID([1,2,3]) is the parallelopiped of sides 1, 2, and 3; CUBOID([1,1,1,1]) is the 4D unit hypercube.

4.1.4.2 Elementary rotations (about the Cartesian axes)

The multidimensional Plasm language has elementary rotations about each $d-1$-dimensional subspace, changing two coordinates according to the pattern of a planar rotation about the origin using *sin* and *cos*. It allows the rotation of any r-model ($r \le d$) in any dimension $d \ge 2$

Remark 4.7 (Number of elementary rotations in d-space) It is clear that in dimension d, there are binomial($d, 2$) elementary rotations, representing the number of ways to choose 2 coordinates from d. Hence we have *one* for $d = 2$, *three* for $d = 3$, *six* for $d = 3$, and so on.

Definition 4.5 (Elementary rotations in 3D)
   Assume that the rotation axes are $\mathbf{e}_1, \mathbf{e}_2, \mathbf{e}_3$, with rotation angles α, β, γ respectively. The corresponding elementary matrices, derivable as before by change of coordinates, are:

$$\mathbf{R}_x(\alpha) = \begin{pmatrix} 1 & 0 & 0 \\ 0 & \cos\alpha & -\sin\alpha \\ 0 & \sin\alpha & \cos\alpha \end{pmatrix}, \ \mathbf{R}_y(\beta) = \begin{pmatrix} \cos\beta & 0 & \sin\beta \\ 0 & 1 & 0 \\ -\sin\beta & 0 & \cos\beta \end{pmatrix}, \ \mathbf{R}_z(\gamma) = \begin{pmatrix} \cos\gamma & -\sin\gamma & 0 \\ \sin\gamma & \cos\gamma & 0 \\ 0 & 0 & 1 \end{pmatrix}$$

$$(4.3)$$

   Readers should note that the elementary rotation is defined in any dimension d such that only 2 coordinates are changed by the rotation.

   In Plasm, the 3D elementary rotations are represented, respectively, by the tensors: R(2,3)(α), R(1,3)(β), and R(1,2)(γ).

Coding 4.1.10 (Elementary rotation) We use a Platonic solid, a permutahedron, to show the application of the tensor R(1,2)(pi/2) to it:

```
obj1 = R(2,3)(pi/2)( PERMUTAHEDRON(3) ); # rotation about x
obj2 = R(1,3)(pi/2)( PERMUTAHEDRON(3) ); # rotation about y
obj3 = R(1,2)(pi/2)( PERMUTAHEDRON(3) ); # rotation about z
VIEW( obj1 ); VIEW( obj2 ); VIEW( obj3 );
```

Fig. 4.3 Three parallel
projections of a unit cube:
V = VIEWCOMPLEX ∘ LAR;
(a) V(CUBE(1));
(b) V(R(1,2)(π/4)(CUBE(1)));
(c) V(R(1,2)(π/2)(CUBE(1))).

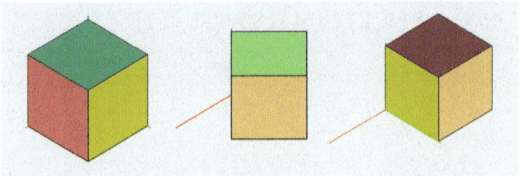

Remark 4.8 (Reduction of visual noice) Just note that in all `Plasm` geometric operators, the constraint of using functions as unary has been relaxed to make it possible to write, e.g., `obj = R(1,2)(pi/2)(obj)` instead of `obj = R([1,2])(pi/2)(obj)`. In the remainder, we often use this style.

Coding 4.1.11 (Permutahedron) The reader might be curious to see how such an important and beautiful polyhedron [9] whose vertex coordinates are the permutations of the first d natural numbers. It is generated in `Plasm` as follows:

```
function PERMUTAHEDRON(d)
    vertices = ToFloat64(PERMUTATIONS(collect(1:d+1)))
    center = MEANPOINT(vertices)
    cells = [collect(1:length(vertices))]
    object = MKPOL(vertices, cells, [[1]])
    object = T(INTSTO(d))(-center)(object)
    for i in 1:d
        object = R(i,d+1)(pi/4)(object)
    end
    object = PROJECT(1)(object)
    return object
end
```

The `Plasm` function `INTSTO`(d) (integers to d) is used to generate the sequence `[1,2,...,d]`, extremes included. The other functions are easy to understand. Let's note the multidimensionality. It is possible to evaluate the **PERMUTAHEDRON** function for *small* d = {1,2,3,4,..} (from [5]).   □

4.1.4.3 General rotation in 3D

A rotation of 3D space has a fixed line of points (the rotation axis) passing through the origin. We may compute the corresponding matrix as a function of a direction vector along the axis and a real rotation angle. For this purpose, we can compose three linear transformations by multiplying their matrices. Therefore, we have:

Definition 4.6 (General 3D rotation with axis d and angle α) The ordering of transformations is from right to left:

$$\mathbf{R}(\mathbf{d}, \alpha) = \mathbf{Q}^{-1}(\mathbf{d}) \, \mathbf{R}_z(\alpha) \, \mathbf{Q}(\mathbf{d})$$

First, a space rotation that brings the vector $\mathbf{d}$ on a coordinate axis say $\mathbf{e}_3$; second, a space rotation $\mathbf{R}_z(\alpha)$ about the z-axis; third, the inverse of the first transformation, so to bring the rotation axis in its original direction.

$\mathbf{Q}(\mathbf{d})$ must transform the unit vector $\mathbf{d}$ to the $\mathbf{e}_3$ unit vector. Therefore, we can compute the coordinate transformation that maps three orthonormal vectors $(\mathbf{u}_1, \mathbf{u}_2, \mathbf{u}_3)$ to the standard basis $(\mathbf{e}_1, \mathbf{e}_2, \mathbf{e}_3)$. We can select the triple:

$$
\begin{aligned}
\mathbf{u}_3 &= \mathbf{d}/\|\mathbf{d}\|, \\
\mathbf{u}_2 &= (\mathbf{u}_3 \times \mathbf{e}_3)/\|\mathbf{u}_3 \times \mathbf{e}_3\|, \\
\mathbf{u}_1 &= \mathbf{u}_2 \times \mathbf{u}_3,
\end{aligned}
\tag{4.4}
$$

To write the transformation of coordinates:

$$(\mathbf{e}_1, \mathbf{e}_2, \mathbf{e}_3) = \mathbf{Q}(\mathbf{d}) \, (\mathbf{u}_1, \mathbf{u}_2, \mathbf{u}_3)$$

Therefore, since the right-hand side is the identity, we have:

$$\mathbf{Q}(\mathbf{d}) = (\mathbf{u}_1, \mathbf{u}_2, \mathbf{u}_3)^{-1}$$

But $\mathbf{Q}(\mathbf{d})$ maps orthonormal vectors to orthonormal vectors; hence, it is a normal transformation, so its inverse is equal to the transpose. So, we can write:

$$\mathbf{R}(\mathbf{d}, \alpha) = \mathbf{Q}^t(\mathbf{d}) \, \mathbf{R}_z(\alpha) \, \mathbf{Q}(\mathbf{d}). \tag{4.5}$$

Coding 4.1.12 (3D General Rotation matrix) Let's use a test-driven programming style, with parameter values easy to test. Rotate 45 degrees about the diagonal axis of the unit cube with a vertex on the origin, i.e., the model generated by the CUBE(1) expression in Plasm. We may follow this procedure using a functional approach:

```
using Plasm, LinearAlgebra
d = [1,1,1];
u₃ = normalize(d);
u₂ = normalize(u₃ × [0,0,1]);
u₁ = u₂ × u₃;
```

and write the following matrix for the transformation of coordinates that maps the $\mathbf{e}_3$ axis to the direction of the d vector.

```
Q(d) = [u₁ u₂ u₃]'
```

The single quote stands for the Julia transpose of a matrix.   □

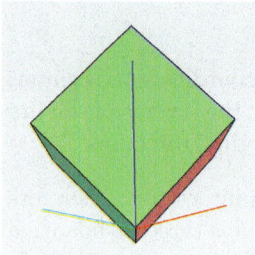

```
rotated = GR([1,1,1],π/3)(CUBE(1))
VIEWCOMPLEX(LAR(rotated))
```

Fig. 4.4 General rotation (GR) of angle $\pi/3$ about the axis [1,1,1]. Note that the rotation was about the main cube diagonal, supported by the axis x = y = z. The parallel projection of the model view is from the back.

Coding 4.1.13 (General 3D rotation tensor) In what follows, MAT transforms a Julia Matrix into a Plasm tensor applicable to Hpc values. The HOMO function applies to a square matrix, by adding a new unitary first row and column for homogeneous coordinates (see Section 4.1.4).

```
GR(d,α) = MAT(HOMO(Q(d)')) ∘ R(1,2)(α) ∘ MAT(HOMO(Q(d)))
```

The GR (general rotation) is a Plasm tensor depending on the axis d and the angle α. Our geometric model is therefore rotated and viewed as above. □

4.1.4.4 Scaling

In a scaling transformation, all points are moved along the line through the origin. The scaling is said elementary when only one of the coordinates changes. There are two scaling parameters s_x, s_y in 2D geometry and three scaling parameters s_x, s_y, s_z in 3D, to be used in scalar products by the point coordinates. The transformation can be a dilatation of space when scaling parameters are more significant than one or a contraction of space when scaling parameters are less than one.

Definition 4.7 (Scaling transformations in 3D)

$$S(s_x, s_y, s_z) = \begin{pmatrix} s_x & 0 & 0 \\ 0 & s_y & 0 \\ 0 & 0 & s_z \end{pmatrix}; \; S_x = \begin{pmatrix} s_x & 0 & 0 \\ 0 & 1 & 0 \\ 0 & 0 & 1 \end{pmatrix}, \; S_y = \begin{pmatrix} 1 & 0 & 0 \\ 0 & s_y & 0 \\ 0 & 0 & 1 \end{pmatrix}, \; S_z = \begin{pmatrix} 1 & 0 & 0 \\ 0 & 1 & 0 \\ 0 & 0 & s_z \end{pmatrix}$$

The scaling matrices are diagonal. The origin remains fixed. In fact:

$$S\begin{pmatrix} 0 & 0 & 0 \end{pmatrix}^t = \begin{pmatrix} 0 & 0 & 0 \end{pmatrix}^t.$$

Hence, a scaling transformation is linear. It is easy to see that scale transformations are multiplicative, commutative, and associative because the matrix is diagonal:

$$\mathbf{S}(s_x, s_y, s_z) = \mathbf{S}_x(s_x)\,\mathbf{S}_y(s_y)\,\mathbf{S}_z(s_z).$$

Coding 4.1.14 (How to scale a Plasm model?) Similarly to the previous coding example, we will utilize `CUBE(1)` as our model object and appropriately scale it.

```
beam = S(1,2,3)(.2,6,.4)(CUBE(1))
wall = S(3)(8)(beam)
floor = S(1,2,3)(4,6,.2)(CUBE(1))
VIEWCOMPLEX(UNION( wall, T(1)(4), beam, T(1,3)(-3.9,.4), floor )
    )
```

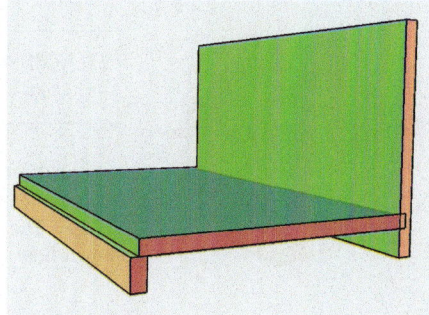

Fig. 4.5 In this case, scale maps are composed with translations in the context of a **STRUCT** tensor (see Ch. 4.3.3) with "compositional" semantics. In such a case, the global coordinate system is the local system of first **STRUCT**'s object (**wall**). The last instance is substituted by **BOOL(UNION())** operator (see Section 7.4.3.)

Coding 4.1.15 (How to scale a Plasm model?) Note that the effect of transformations impacts only the homogeneous matrices ahead of **Hpc** values.

```
scaledcube1 = S(1,2,3)(.1,.1,10)(CUBE(1))      #=
Hpc(MatrixNd([[1.0, 0.0, 0.0, 0.0], [0.0, 0.1, 0.0, 0.0], [0.0,
    0.0, 0.1, 0.0], [0.0, 0.0, 0.0, 10.0]]), Hpc(MatrixNd(4),
    Hpc(MatrixNd(4), Geometry([[0.0, 0.0, 0.0], [1.0, 0.0, 0.0],
    [0.0, 1.0, 0.0], [1.0, 1.0, 0.0], [0.0, 0.0, 1.0], [1.0,
    0.0, 1.0], [0.0, 1.0, 1.0], [1.0, 1.0, 1.0]], hulls=[[1, 2,
    3, 4, 5, 6, 7, 8]])))) =#
scaledcube2 = S(2)(100)(SQUARE(1))   #=
Hpc(MatrixNd([[1.0, 0.0, 0.0], [0.0, 1.0, 0.0], [0.0, 0.0,
    100.0]]), Hpc(MatrixNd(3), Hpc(MatrixNd(3), Geometry([[0.0,
    0.0], [1.0, 0.0], [1.0, 1.0], [0.0, 1.0]], hulls=[[1, 2, 3,
    4]])))) =#
```

Of course, `S(1,2,3)(.1,.1,10)` and `S(2)(100)` are tensor objects.   □

Coding 4.1.16 (Construction of octahedron model) As an exciting coding example, we show a simple construction of an octahedron model, starting from the 3D **SIMPLEX** model. In particular, we double the simplicial parts of the generated object twice, and finally we double again and impose the **UNION** of all the 8 octahedron parts.

```
tetra = HPCSIMPLEX(3);
twotetra = STRUCT( tetra, S(1)(-1), tetra );
fourtetra = STRUCT( twotetra, S(2)(-1), twotetra );
octahedron = BOOL(UNION( fourtetra, S(3)(-1), fourtetra ))
```

The cellular complex associated with the solid model octahedron::Lar is worth examining. In particular, the generating Boolean operation has resulted in a decompositional representation into height standard tetrahedra (see Figure 4.6).

```
VIEWCOMPLEX(octahedron, show=["CV"],
            explode=[2,2,2])
```

Fig. 4.6 Plasm viewing the octahedron object's atoms explosed. Remember that **VIEWCOMPLEX** applies to Lar values. The height atoms of the **BOOL** operation are shown here exploded (see 7.4.3 for details).

4.1.4.5 Shearing

In a 2D elementary shearing tranformation, all points of each line (each plane in 3D) orthogonal to a coordinate axis move by summing one (fixed) vector. The coordinate line (plane in 3D) remains fixed and the translation vector changes linearly with the distance of its line (plane) from the origin. Each of the two elementary planar shearings depends on a single scalar parameter (the translation of the line at a unit distance from the coordinate line).

Definition 4.8 (Elementary shearing in 2D)

$$\mathbf{H}_x(h_x) = \begin{pmatrix} 1 & 0 \\ h_x & 1 \end{pmatrix}, \qquad \mathbf{H}_y(h_y) = \begin{pmatrix} 1 & h_y \\ 0 & 1 \end{pmatrix};$$

Each of the three elementary space shears in 3D depends on two scalar parameters—the coordinates of the planar translation vector of the plane at a unit distance from the coordinate plane.

Definition 4.9 (Elementary shearing in 3D)

$$\mathbf{H}_x(h_y, h_z) = \begin{pmatrix} 1 & 0 & 0 \\ h_y & 1 & 0 \\ h_z & 0 & 1 \end{pmatrix}, \mathbf{H}_y(h_x, h_z) = \begin{pmatrix} 1 & h_x & 0 \\ 0 & 1 & 0 \\ 0 & h_z & 1 \end{pmatrix}, \mathbf{H}_z(h_x, h_y) = \begin{pmatrix} 1 & 0 & h_x \\ 0 & 1 & h_y \\ 0 & 0 & 1 \end{pmatrix}.$$

An elementary shearing differs from the identity matrix only for the elements of a single column, both in 2D and 3D, and in homogeneous 4D coordinates.

Plasm **implementation of shearing tensor**

In Plasm, the shearing tensor is named H and has the following syntax, corresponding to the semantics above: first, indicate the column index; then give the $d - 1$ ordered transformation parameters, i.e., one in 2D and two in 3D, some of which are possibly zeros. Therefore, we have H(col)(pars).

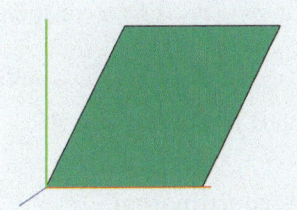

```
SQUARE(d) = CUBOID([d,d])
shearedsquare = H(2)(.5)(SQUARE(1))
VIEWCOMPLEX(LAR(shearedsquare))
```

Fig. 4.7 Unit square sheared on the (second) coordinate y. The y of model points does not change.

Typically, shearing is used in 2D by typesetting systems of computerized typography to get italic versions of character fonts defined with Bézier curves or splines. Note how we define a parametric square (with a vertex on the origin). See Figure 4.7. Look also at the coordinate frame seen from back in Figure 4.8.

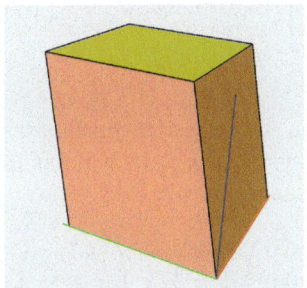

```
shearedcube = H(3)(.2,.3)(CUBE(1))

VIEWCOMPLEX(LAR(shearedcube))
```

Fig. 4.8 Unit cube sheared on the (third) coordinate z. The z values of the points do not change.

Remark 4.9 (Dimensional independence) It is important to note that the H mapping, like R, GR, S, T, MAT, and HOMO, is dimension independent, meaning it can be applied to the space of any embedding dimension d for geometric models. Homogeneous normalized matrices are used for implementation purposes.

4.1.4.6 Translation

Definition 4.10 (Translation transformation) a translation is an invertible transformation of Euclidean space $\mathbb{E}^d$ generated by summing a fixed vector to all points.

A translation of a plane $\mathbb{E}^2$ (or space $\mathbb{E}^3$) is not a linear transformation since it moves the origin, but it is an affine transformation since all $\mathbb{E}^2$ (or $\mathbb{E}^3$) mapped points change by sum with a fixed vector (the affine action).

Definition 4.11 (Affine translation) A translation in 2D (or 3D) depends on two (or three) scalar parameters, that is, on the coordinates of a translation vector, displacing the points of the affine space. We may therefore translate, using coordinates, by summing to the generic vector $\mathbf{v} = (v_i) \in \mathbb{E}^d$ a constant vector:

$$\mathbf{v'} = \mathbf{v} + \mathbf{t} \tag{4.6}$$

where $\mathbf{t} = (t_i)$ is called the translation vector, applied to all points in $\mathbb{E}^d$.

Definition 4.12 (Translation in homogeneous coordinates)
When using normalized homogeneous coordinates, the translation 4.6 simplifies to a linear transformation and, therefore, can be expressed as a vector product with a square matrix.

Let us remember our choice to use the first coordinate as homogeneous. For example, a translation of $\mathbb{E}^3$ is representable as

$$\mathbf{T}(t_x, t_y, t_z) = \begin{pmatrix} 1 & 0 & 0 & 0 \\ t_x & 1 & 0 & 0 \\ t_y & 0 & 1 & 0 \\ t_z & 0 & 0 & 1 \end{pmatrix} \tag{4.7}$$

We can see the equivalence between translation in Cartesian coordinates and in homogeneous normalized coordinates. Let $\mathbf{v} = (x, y, z)$ be a point in Euclidean space $\mathbb{E}^3$, and $\mathbf{v'} = (w = 1, x, y, z)$ the same point in $\mathbb{R}^4$:

$$\begin{pmatrix} x \\ y \\ z \end{pmatrix} + \begin{pmatrix} t_x \\ t_y \\ t_z \end{pmatrix} = \begin{pmatrix} x + t_x \\ y + t_y \\ z + t_z \end{pmatrix} \quad \text{and} \quad \begin{pmatrix} 1 & 0 & 0 & 0 \\ t_x & 1 & 0 & 0 \\ t_y & 0 & 1 & 0 \\ t_z & 0 & 0 & 1 \end{pmatrix} \begin{pmatrix} 1 \\ x \\ y \\ z \end{pmatrix} = \begin{pmatrix} 1 \\ x + t_x \\ y + t_y \\ z + t_z \end{pmatrix}$$

Notice that a translation in 3D is a shearing $\mathsf{H}(t_x, t_y, t_z)$ orthogonal to the added component in normalized homogeneous coordinates.

Coding 4.1.17 (Translation of 3D geometric object) In Plasm, we translate a geometric object of Hpc type via tensor application:

```
t_cube1 = T(1,2,3)(.5,.5,.5)(CUBE(1))
t_cube2 = T(3)(1)(CUBE(1))

VIEW(t_cube1);  VIEW(t_cube2)
```

Coding 4.1.18 (Parametric linear stairway) The step is an Hpc solid obtained as a byproduct of three line segments of given sizes. An array of n pairs [move, step] is generated and concatenated by the CAT operator. Finally, the semantics of STRUCT aggregator (see Section 4.3.3) produces the whole parametric object, shown in Figure 4.9.

```
function stairway(lx,ly,lz, n)::Hpc
    step    = QUOTE(lx) * QUOTE(ly) * QUOTE(lz)
    move    = T(1,2,3)(0, 0.8*ly, 0.8*lz)
    ramp    = STRUCT( CAT([[step, move] for k=1:n]) )
end #=
Ladder (generic function with one method) =#
stair = stairway(.8, .22, .18, 15);
VIEWCOMPLEX(LAR(stair))
```

Fig. 4.9 Simple linear stairway, demonstrating an iterative use of tensors in STRUCT. Of course, the number, size, and shape of the step model can be parametrized as arguments of a geometric function returning Hpc objects.

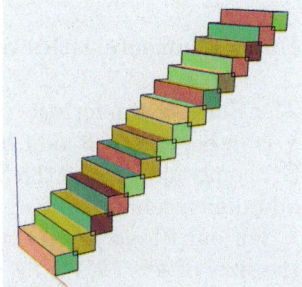

4.2 Representation Schemes

The foundations of solid modeling technology were established between the 70s and 80s. After 50 years, the time was ripe for exploring new paradigms [4, 7, 8], not based only on data structures and computational geometry algorithms, but on novel abstractions taking into account structures and operators of algebraic topology and algebra, as is happening in modern AI, which is also based on hardware tensors and linear algebra.

Ari Requicha codified the central concept in his famous 1980 paper [10], which emerged from the Project Automation Project (PAP) research guided by Herb Voelcker in Rochester (NY) during the seventies. A representation scheme of solids is a mapping $r : M \to R$ between a space of mathematical models M and a space of representations R generated by a computer grammar.

4.2.1 Boundary

The boundary scheme, denoted Brep or brep, represents a 3-solid by illustrating its 2-boundaries. The boundary is decomposed into 2-cells, each of which is denoted by the 1-cells of its boundary. Their two extreme 0-cells define the 1-cells in a PL (Piecewise-Linear) domain. The standard terminology from dimension $3 \to 0$ includes: solid cells (C), faces (F), edges (E), and vertices (V) as the main terms. Other key terms involve the "shell" (connected component) of the solid boundary and the "loop" of the face boundary. Curved objects are typically represented by combining various parametric (and, less often, algebraic) equations with the topological elements. Brep schemes are widely used in industry and academia and are frequently closely related to one or more of the schemes discussed below.

Euler Characteristics of Brep models

The Euler characteristic χ of n-dimensional spere is $\chi = 1 + (-1)^n$. Hence, χ is either 0 if n is odd or 2 if n is even. Also, the Euler characteristic of $\mathbb{E}^3$, homeomorphic to the 3-sphere (minus one point), is 0, like for any closed odd-dimensional manifold.

For our `Plasm` project, we are interested in the numeric relations between the sizes of sets CV, FV, EV, V of p-dimensional cells in the Brep datasets of 2D or 3D models. Hence, using our Brep dataset, we write the Euler characteristic of solid `Plasm` models in $\mathbb{E}^3$ as:

$$\#V - \#EV + \#FV - \#CV = 0,$$

where CV also contains the exterior unbound cell and where #CV, etc., are the cardinalities of p-cell sets. In #CV, we also consider the contributions of shells of outer unbounded cells of $\mathbb{E}^3$. For example, for a single connected model made by m 3-cells, we have $\#CV = m + 1$, where 1 takes care of the outer cell's shell. Conversely, for PL connected models, we have the well-known Euler relation for polyhedra:

$$\#V - \#EV + \#FV = 2$$

Example 4.2.1 (3D polyhedral complex) A simple two-floor building, using the Cartesian product of 1-complexes (see also Example 3.2.3).

```
X = GRID([2.4,4.5,-3,4.5,2.4])
Y = GRID([7,5])
Z = GRID([3,3]);
idea = X * Y * Z;
VIEWCOMPLEX( LAR(idea), explode = [1.2,1.2,2.0] )
VIEWCOMPLEX( LAR(idea), show=["EV"], explode=[1.2,1.2,2.0] )
```

The `GRID` operator creates 1-complex line segments of defined sizes. Negative numbers indicate empty segments. '*' serves as the Julia symbol for the Cartesian `PROD` (product) operator. □

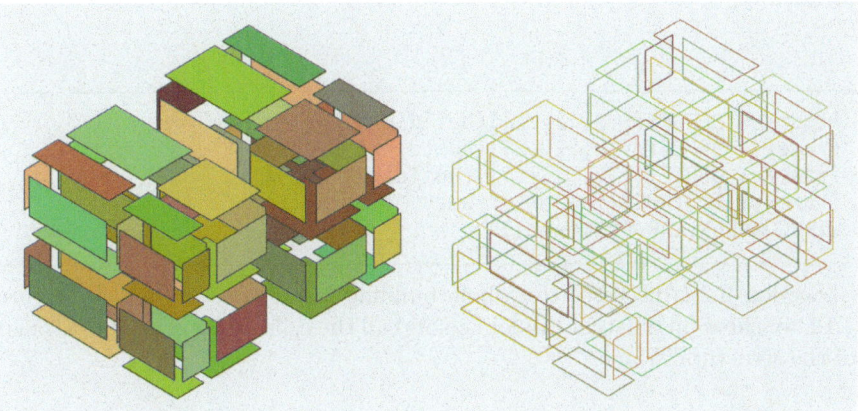

Fig. 4.10 Building idea: (a) 2-complex; (b) 1-complex of boundaries of 2-cells.

Example 4.2.2 (Chain subcomplexes) We know that `Plasm` enables the generation of complexes through the topological product of other complexes. Furthermore, we know that the '*' operator is commutative and associative. It can be used in an n-ary fashion via the `INSR` and `INSR` combinators of `FL`.

```
floors = X * Y * SK(0)(Z)
VIEWCOMPLEX(LAR(floors))
envelope = LAR(STRUCT(X*SK(0)(Y)*Z, SK(0)(X)*Y*Z))
VIEWCOMPLEX(envelope)
```

Example 4.2.1 (Building simple example) Here, we provide a simple example of an initial 3D construction project to create the Brep of the entire building and each individual unit space (3-cells for rooms).

Fig. 4.11 (a) 2-complex of horizontal floors; (b) 2-complex of vertical enclosures and partitions.

```
floors = OFFSET([.2,.2,.2])(building110);
framex = OFFSET([.2,.2,.2])(building1_011);
framey = OFFSET([.2,.2,-.4])(building1_101);
framexyz = STRUCT(framex, framey, floors);
VIEWCOMPLEX(LAR(framexyz))
```

Section 3.2.1 discusses the whole building frame construction. In Figure 4.12, we show (in different random colors) all the cells (arrangement of atoms) of the structural frame.

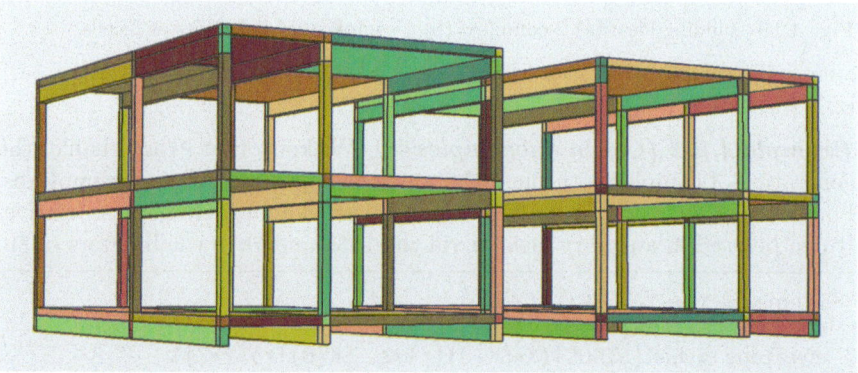

Fig. 4.12 A perspective view of the Plasm model LAR(framexyz) generated by STRUCT, SKELETON and OFFSET operators from the idea sketch in Example 4.2.1.

4.2.2 Decompositive

The solid model is often divided into three-dimensional cells with identical or similar topology, which comply with the rules for defining cellular complexes. This approach is mainly used for computer simulations of differential equations and is rarely employed for shape design. The cells are convex and represented by their vertices and other discrete boundary points called nodes, which are used for field interpolation.

In our Boolean approach, we also use a decompositive representation of geometric input $\mathcal{S}$, but with general polyhedral cells, which constitute a partition of the embedding space E^d, where each unit chain (cell) is represented by its boundary, and called atom of the arrangement $\mathcal{A}(\mathcal{S}) \equiv C_d$, such that:

$$\cup_{k=0}^d C_k = \mathbb{E}^d, \quad \text{and} \quad c^i \cap c^j = \emptyset, \quad (c^i, c^j \in C_d).$$

In our approach, the 3D Brep of an atom is represented in binary coordinates as a column of the (sparse) matrix operator.

$$\mathsf{FC} : \mathsf{C} \to \mathsf{F} \equiv \delta_{d-1} : C_{d-1} \to C_d,$$

providing its irreducible $(d-1)$-cycles, as discussed in Chapters 6 and 7.

4.2.3 Enumerative

The standard definition of this scheme relies on a grid decomposition of the space and lists the space elements belonging to the object. Examples include hierarchical decompositions of 3D domains using octrees [2] or potrees [11] to store, transmit, visualize on the web, and reconstruct the geometry from scans of solid objects, neighborhoods, and buildings with remote sensors.

In our research and development of cellular complexes, we extensively utilize multidimensional, hierarchical, and non-hierarchical grids of cuboidal and simplicial cells (see numerous examples in this book). Specifically, but not exclusively, these geometric objects serve as scaled domains of manifold curves, surfaces, and varieties of higher dimensions.

In the Plasm approach to curved geometric design, we typically map a vector function `F(u,v,w)=CONS([f(u),f(v),f(w)])` $\mapsto$ `(x,y,z)` using coordinate functions to generate the vertices of d-grid domains without altering the topology, thereby obtaining the PL approximated values of our curved geometric model. See Section 5.3.1.

Example 4.2.3 (Curved chain complexes)
First, we show in Figure 4.13 the `Plasm` generation of a cubic non-uniform spline curve depending on 10 control points and 14 knots.

```
degree = 3
ControlPoints = [[0.,0.],[-1,2],
   [1,4],[2,3],[1,1],[1,2],[2.5,1], [2.5,3],[4,4],[5,0]]) # 10
   points
knots=[0,0,0,0, 1,2,3,4,5,6, 7,7,7,7] # 14 knots
VIEW(DISPLAYNUBSPLINE(degree, knots, ControlPoints))
```

Then, we display in Figure 4.13 both the reference 3D `surface` and the generated thin solid of an Highly parameterized complex `solidMapping = THINSOLID(surface)` that computes the local normals by partial derivatives.

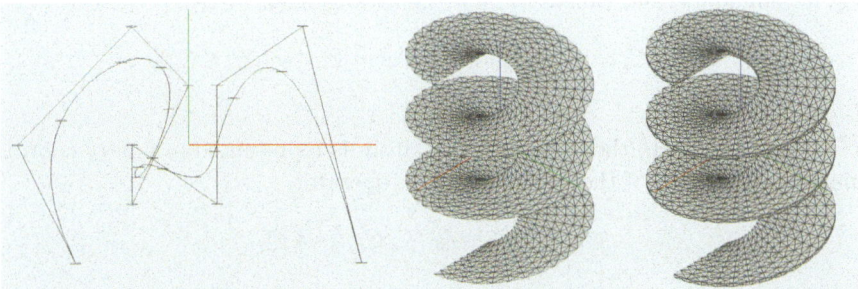

Fig. 4.13 User-defined curved 1-, 2-, and 3-manifolds: (a) Cubic non-Uniform B-spline (NURB) passing for the extreme control points, which satisfies the usual relationship $14 = 10 + 3 + 1$ of NURB splines; (b) highly parametric surface; (c) solid helicoid.

The `Plasm` code implementing the highly parameterized generator function named `SOLIDHELICOID` is provided below. The user of this generator can specify the number of turns, the external and internal radii, the resolution of cell decomposition, the surface `pitch`—the vertical distance between two spires—and the `thickness` of the thin solid. Default values are also assigned according to the Julia convention.

```
function SOLIDHELICOID(; nturns=3, R=1., r=0.5, shape=[36*nturns
    , 8], pitch=2, thickness=0.1)
    totalangle = nturns*2*pi
    grid2D = INTERVALS(36*nturns)(36*nturns) * INTERVALS(4)(8)
    Domain2D = T([2])([r])(S([1,2])([totalangle/shape[1],R-r])(
     grid2D));
    surface = p->let(u, v)=p;[v*cos(u); v*sin(u); u*(pitch/(2*pi)
     )] end
    VIEW(MAP(surface)(Domain2D))
    solidMapping = THINSOLID(surface)
    Domain3D = Domain2D * INTERVALS(thickness)(1)
    VIEW(MAP(solidMapping)(Domain3D))
end;
```

Fig. 4.14 An image of the discretized Domain2D of SOLIDHELICOID function.

Typically, Plasm grids begin with integer coordinates for their vertices and are scaled to suitable sizes depending on their use. It's important to note that these objects are very non-manifold in their interiors. Interesting generation methods and application patterns can be found in the Architecture, Engineering, and Construction (AEC) design and production field, where orthogonal geometries mostly dominate buildings.

In [6], we report by examples the interesting characteristics of combinatorics of product topological spaces. In particular, when these have dimension 1 or 0, i.e., they are made by line intervals or by discrete points on it. When X, Y, Z are 1D complexes, the space X ∗ Y ∗ Z has dimension 3; if some of product terms has dim = 0 the product dimension decreases accordingly.

Example 4.2.4 (Non-manifold cellular complex: 1/3) See Fig. 4.15a.

```
using Plasm
largrid = CUBOIDGRID([5,8,1])
VIEWCOMPLEX(largrid, explode=[1.2,1.2,2])
```

Example 4.2.5 (Non-manifold cellular complex: 2/3) See Fig. 4.15b.

```
lar = CUBOIDGRID([10,15,2]);
arrangement = ARRANGE3D(lar);
VIEWCOMPLEX(arrangement, show=["CV"], explode=[1.4,1.4,3])
```

Fig. 4.15 (a) exploded CUBOIDGRID([5,8,1]; (b) exploded arrangement generated by CUBOIDGRID([10,15,2]). The OUTER atom of space arrangement (pink) is visible, which delimits the unbounded $\mathbb{E}^3$. For details on arrangements and atoms, see Chapters 6-7.

Example 4.2.6 (Non-manifold cellular complex: 3/3) Considering the data structure of Example 4.2.5, we note that there are $(10 + 1) \times (15 + 1) \times (2 + 1) = 528$ vertices (columns) in the grid.V matrix.

```
lar.V
3×528 Matrix{Float64}:
1.0  0.0  1.0  0.0  1.0  0.0  …   6.0   7.0   8.0   9.0  10.0
1.0  1.0  0.0  0.0  0.0  0.0      15.0  15.0  15.0  15.0  15.0
0.0  0.0  0.0  0.0  1.0  1.0       2.0   2.0   2.0   2.0   2.0
```

The dictionary `lar.C` storing the topology shows that all three cells in `CF` have six boundary 2-cells; `CV` has height 0-cells for each element; all faces, `FV`, have four boundary 0-cells; and each edge, `EV`, has two 0-cells.

```
lar.C
Dict{Symbol, Vector{Vector{Int64}}} with 5 entries:
:CF => [[1, 2, 4, 9, 11, 13], [2, 3, 7, 10, 19, 125], [4, 5, 6…
:CV => [[1, 2, 3, 4, 5, 6, 7, 8], [1, 2, 7, 8, 45, 46, 47, 48]…
:FV => [[1, 2, 3, 4], [1, 2, 7, 8], [1, 2, 45, 46], [1, 3, 5, …
:EV => [[1, 2], [1, 3], [1, 7], [1, 9], [1, 45], [2, 4], [2, 8…
:FE => [[1, 2, 6, 9], [1, 3, 7, 19], [1, 5, 8, 129], [2, 3, 10…
```

Example 4.2.7 (Solid Boolean Algebra)

The `Plasm` Boolean operations create an arrangement (partition) of $\mathbb{E}^d$ space into elementary parcels called atoms. Each Boolean value in this algebra is a subset of such atoms. See Chapters 6 and 7 for details. Every Boolean expression is an enumeration of atoms.

Fig. 4.16 Boolean operations:
(a) BOOL(XOR(..)) with two atoms of $\mathbb{E}^3$ arrangement;
(b) BOOL(UNION(..)) with three atoms.

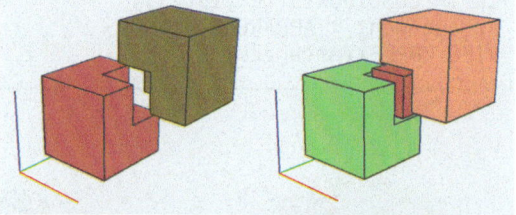

```
xor = BOOL(XOR(CUBE(1), T(1,2,3)(.5,.5,.5), CUBE(1)))
union = BOOL(UNION(CUBE(1), T(1,2,3)(.5,.5,.5), CUBE(1)))
VIEWCOMPLEX(xor, show=["CV"], explode=[1.4,1.4,1.4])
VIEWCOMPLEX(union, show=["CV"], explode=[1.4,1.4,1.4])
```

4.2.4 Constructive Solid Geometry (CSG)

This scheme was highly successful in industry because its logic emulates the design process for mechanical parts, from idea generation through refinement and production. A CSG scheme causes the concept of assembly of model parts to meet together with the elementary process operations of union, intersection, and difference.

It is often implemented as an incomplete binary tree with Boolean operations or affine transformations applied to non-leaf nodes, and with primitive objects at the leaves. The original CSG schemes, conversely, used solid objects (linear/nonlinear equalities/inequalities) as leaves and implemented the boolops as systems of polynomial inequalities. At present, a more combinable use of Brep primitives looks preferred.

As we discuss in Chapter 7, our Boolean solid algebras in `Plasm` are more general and efficient than standard CSG evaluation, succession of assessments of binary operators, where partial results are used as input to a subsequent operation. In particular, they are less subject to numerical error propagation since all numerical computations are performed in advance in 2D, and later the algorithm is only topological, i.e., symbolic.

Remark 4.10 (Plasm solid algebra is variadic) A variadic function is a function of indefinite arity, i.e., one that accepts a variable number of arguments. Our solid algebra extends CSG by including complement and ambient, which is the ambient Euclidean space. We express any Boolean formula between solid objects in Julia-based language using the primitive operators `UNION`, `INTERSECTION`, `DIFFERENCE`, and `XOR`. The Boolean solid formulas have the semantics of `STRUCT` to assemble all the geometric terms in a unique coordinate frame (see Chapter 7) and are introduced by the `BOOL` operator.

Any such formula produces the Euclidean d-space arrangement generated by its terms and the related set of atoms. By applying a logical function (equivalent to the solid formula) to the binary representation of atoms, `Plasm` selects the subset of atoms (3-chain) solution of the solid expression. The evaluation process also classifies the atoms as inner and outer chains. The last ones give the boundary between the union of atoms and the outer unbounded Euclidean space.

Our solid Boolean algebra depends directly on the arrangement created by a set of geometric objects, represented by the boundaries of their cell decomposition. This approach has enabled us to classify the entire Euclidean d-space into a collection of well-defined d-cells and the algebra of their subsets. The atoms of this set algebra are the smallest (irreducible) elements of the arrangement, represented as binary characteristic functions in 2^n, where n is the number of atoms. The quasi-disjoint union of specific subsets of these elementary components corresponds to the generating terms that form the arrangement and is isomorphic to a solid algebra.

Computing a solid shape from a Boolean expression is like establishing a logical identity between the left side, which involves operations (union, intersection, difference, xor) on solid terms, and the right side, which consists of the disjoint union of atom subsets whose point-set union encompasses the same Euclidean point set as the left side and is expressed in canonical Boolean "Sum Of Products (SOP)."

Our Boolean method generates this identity in a constructive way. While the left-hand side is expressed in Plasm by any correct combination, even hierarchical, of solid shapes and Boolean combinators, the right-hand side is algorithmically constructed step by step in canonical form[4] by the BOOL operator. A minterm, i.e., a binary tuple of bits in Computer Science, refers to a logical AND involving a set of 0/1 variables, where each bit can be itself or its complement. It is a fundamental building block in Boolean algebra that represents specific combinations of variables. Canonical sum or sum of minterms is a sum of products in which each product term is a minterm. Plasm generates the solid solution precisely as the Sum Of Minterms (SOM) or Sum of Products (SOP). Since all the variables (generators), with value 1 or 0, are present in each minterm (atom), the canonical sum is unique for a given model. Refer to Chapter 7.4.1.5 for examples and details.

4.2.5 Primitive Instancing

This scheme typically provides the user with an extendable library of geometric shapes, often characterized as generating functions and formal parameters (and multiple dispatch methods in Julia). The internal representation is usually Brep, with both default and user-definable values of semantic parameters, and with 1-, 2-, or 3-variate resolution discretization (for curves, surfaces, and solid curved primitives), depending on a suitable approximation of intrinsic shape dimensions.

The Plasm language provides a reasonably rich library (see Section 2.4) of predefined shapes and geometric constructions within the module fenvs.jl, which stands for functional environments. The many examples may help the user to make fast extensions.

Currently, there are about 75 primitive functions which produce output values of Hpc type. Such functions act as generators or transformers of geometric objects and may receive numerical and/or geometrical model values as input. They include, e.g., the following higher-level functions, among others. It is worthwhile to note that, according to Julia's powerful multiple dispatch features, many of them correspond to some different methods, and the users may add more methods and functions according to their modeling habits and needs: MAP, STRUCT, JOIN, UNION, INTERSECTION, DIFFERENCE, BOX, CUBOID, CUBE, CYLINDER, SPHERE, CONE, TRUNCONE, SQUARE, POLYLINE, SIMPLEX, PROJECT, TORUS, SIMPLEXGRID, INTERVALS, PROD, OFFSET

---

[4] In mathematics and computer science, a canonical form of a mathematical object is a standard way of presenting it as a mathematical expression. It often provides the most straightforward representation of an object and allows for its unique identification.

In all **Plasm** development, and in particular when implementing the primitive function used to generate geometric models, we have largely used the currying technique (see the start of Section 2.4) and the partial functions and methods it generates, greatly enhancing the language's descriptive power.

4.3 Assembly of geometric objects

Complex shapes are generally defined as hierarchical assemblies of geometric primitives or more intricate forms, each represented in a local coordinate system. Most graphics and modeling systems implement this semantics as a tree or hierarchical graph, where affine geometry within the nodes is defined in local coordinate frames, and arcs are linked to affine transformations that move the entire subgraph rooted at the ending node onto the coordinate system of the first node of the arc. The first node of this data structure is called the root.

4.3.1 Hierarchical graphs

Acyclic graphs/multigraphs are an abstract model of assemblies of geometric objects defined in their local coordinate system. They are also called hierarchical scene graphs because they can be associated with a tree, generated at run-time by visiting the graph with some standard traversal algorithm [3], e.g., with a depth-first-search (DFS). The ordered sequence of nodes produced by the traversal is sometimes called a linearized graph. Each node in this sequence is suitably transformed from local coordinates to world coordinates, i.e., to the coordinates of the root, by the traversal algorithm.

The main ideas concerning scene graphs can be summarized as follows. Nodes are containers of geometrical datasets stored in local coordinates. Nodes are also used and implemented as the root of subgraphs, whose data are transformed to the node coordinates by a traversal algorithm. Arcs (a, b) are associated with affine transformations, which map the data in b from their local coordinates to the coordinates of a. More than one arc may exist between the same node pair. This allows storage in memory only of one copy of each container. The composite transformations of coordinates applied to the linearized graph generated at traversal time are collectively known as the modeling transformation.

Remark 4.11 The arcs of scene graphs are normally specified implicitly in real graphical systems. For example, an arc is specified when a node is contained or referred to within another. In particular, it is possible to specify a new container node with either the matrix or the transformation parameters associated with the arc that ideally connects the new container to the current node. The `Hpc` data structure of `Plasm` attaches the affine matrix associated with the incident (symbolic) arc in front of each node.

Object Transform

Any `Plasm` geometric value of `Hpc` type can be affinely transformed by applying an affine tensor.

Coding 4.3.1 (Direct use of tensors) Let's combine with other language tensors while generating the translated 1-skeleton of a cube:

```
SK = SKELETON;
translatedcube = (SK(1) ∘ T(1)(1) ∘ CUBE)(1);
```

Coding 4.3.2 (Example.2) Then, aggregate two objects into a single object within the same coordinate frame:

```
singleframe = STRUCT(cube, tetra);
VIEW(singleframe)
```

Coding 4.3.3 (Example.3) Direct application of nodes $T_z(1)$, followed by application of $R_z(-\pi/2)$ to `tetra` value, which changes accordingly:

```
doubleframes = STRUCT(cube, (R(1,2)(-π/2) ∘ T(3)(1.0))(tetra)
VIEW(doubleframes)
```

Coding 4.3.4 (Example.4) The same effect could be obtained by the following expression because a transformation tensor is implicitly applied to every geometric value following it in the parameter sequence of a `STRUCT` combinator. Evaluation is right-to-left, according to the math composition rule:

```
doubleframe = STRUCT(cube, R(1,2)(-π/2), T(3)(1), tetra )
```

Complying with this rule, we get two objects:

$$R(1,2)(-\pi/2)(T(3)(1)(\text{tetra})) \quad and \quad \text{cube}$$

both in root coordinates, i.e., those of the first `cube` node. The `STRUCT` semantics is easier to follow end-to-start, i.e., in reverse order.   □

Assembly of components

'The aggregation of cellular complexes forms hierarchical models of complex assemblies, each defined within a local coordinate system and potentially moved through affine coordinate transformations. This process can be performed hierarchically, with subassemblies represented by aggregations of simpler components, continuing until a set of leaves containing primitive models cannot be further decomposed.

A hierarchical modeling approach has two main advantages. At every hierarchical level, each complex component and named partial assembly may be defined independently of the others using their PROPERTIES and local coordinate frames, which are suitably chosen to facilitate their definition. Furthermore, only one copy of each component is stored in memory and may be instantiated in different locations and orientations, depending on how many times it is needed.

Directed Acyclic Graph (DAG)

A hierarchical model, defined inductively as an assembly of parts, is described by an acyclic directed multigraph, commonly referred as a scene graph or hierarchical structure in computer graphics and modeling. The primary algorithm for hierarchical assemblies is the traversal function, which transforms each component of the assembly from local coordinates to global coordinates, often called world coordinates.

Definition 4.13 (Directed graph) A directed graph G is a pair (N, A), where N is a set of nodes and A is a set of directed arcs, given as ordered pairs of nodes.

Such a definition is not sufficient when more than one arc must be considered between the same pair of nodes. The notion of multigraph is hence introduced. In a multigraph, the same pair of nodes can be connected by multiple arcs.

Definition 4.14 (Directed multigraph) A directed multigraph is a triplet $G := (N, A, f)$ where N and A are sets of nodes and arcs, respectively, and $f : A \to \mathbf{N}^2$ map arcs to node pairs.

Directed graphs or multigraphs are said to be acyclic when they do not contain cycles, i.e., when no path starts and ends at the same vertex. Trees are typical examples of acyclic graphs. A tree, where each non-leaf node is the root of a subtree, is the best model of the concept of hierarchy. Nodes in a tree can be associated with their integer distance from the root, defined by the number of edges on the unique path from the node to the root. A tree can be layered by levels by putting in the same subset (level) all the nodes with equal distance from the root.

4.3.2 Hierarchical structures in Plasm

A container of geometrical objects is defined in Plasm by applying the combinator STRUCT to the contained objects sequence (or array). The value returned from this function application is of type hierarchical polyhedral complex Hpc. The coordinate system used by the returned value is associated with the first geometric object in the argument sequence.

The resulting geometrical value is often associated with a variable used as the container's name, as in

---

    obj = STRUCT(obj_1, obj_2, ..., obj_n); VIEW(obj)

The obj geometry can be pictorially described using the previously discussed graph model of hierarchical structures, as shown in Figure 4.17a. Each component object may be in turn be defined as a container of other objects, i.e., as the root of a subgraph, as shown in the following script.

---

    obj_2 = STRUCT(obj_{21}, ..., obj_{2m})

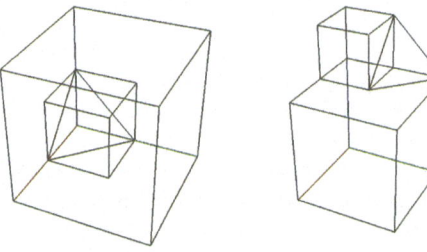

Fig. 4.17 Assembly by STRUCT: (a) without coordinate transformations all objects have the same origin; (b) with coordinate transformations and different local frames.

The same geometric result would be generated by direct nesting of STRUCT sub-expressions:

---

    obj = STRUCT(obj_1, STRUCT(obj_{21}, ..., obj_{2m}) ..., obj_n)

The sequence argument of the STRUCT operator may or may not contain affine transformations and polyhedral complexes. This fact results in generating an assembly by using the same global coordinates for the various components or using different local coordinate systems. The two cases are discussed in the following subsections.

Assembly with global vs local coordinates

Let's assume that the sequence argument of a STRUCT expression does not contain affine transformations. In other words, we assume that the evaluations of the Plasm expressions in the argument sequence only return polyhedral values. In this case, the output polyhedral complex is returned within the coordinate frame of the first element of the input sequence, and no transformations of coordinates are applied to the assembly components, which are only aggregated in the same space, as shown by the following example.

Coding 4.3.5 (STRUCT assembly (1)) The expression given below returns the object of Figure 4.17a. Note that the three-component shapes' local origin and coordinate axes coincide.

```
cube2, cube1, simplex = CUBE(2), CUBE(1), HPCSIMPLEX(3)
SK = SKELETON
obj1 = (SK(1) ∘ STRUCT)( cube2, cube1, simplex )
```

Coding 4.3.6 (STRUCT assembly (2)) The same geometry of components is used of Coding 4.3.5, while also including some transformations of coordinates to the sequence of parameters of the resulting assembly.

```
obj2 = (SK(1) ∘ STRUCT)(cube2, T(3)(2), cube1, T(2)(1), simplex)

VIEWCOMPLEX(LAR(obj1)); VIEWCOMPLEX(LAR(obj2))
```

The SK(1) operator (extraction of 1-skeleton) was @1, not available in Julia, in classic PLaSM. The resulting geometric assembly is shown in 4.17b. □

STRUCT semantics

We assume that in Plasm, the word tensor stands for affine transformation. Let's suppose that tensors T_k are contained within the sequence argument of a STRUCT expression. Each tensor in a STRUCT is applied to all polyhedral complexes that follow it. The subsequent expressions are equivalent:

```
STRUCT( pol₁, T₁, pol₂, T₂, pol₃, ... , Tₙ₋₁, polₙ ) ≡
STRUCT( pol₁, T₁(pol₂), (T₁∘T₂)(pol₃), ..., (T₁∘T₂ ⋯ Tₙ₋₁)(polₙ)
```

Looking at the internal behavior of the geometric kernel of the language, the following maps are applied to the STRUCT application at evaluation time:

```
STRUCT( pol₁, (T₁ ∘ STRUCT)( pol₂, (T₂ ∘ STRUCT)( pol₃, ...
    (Tₙ₋₂ ∘ STRUCT)( polₙ₋₁, Tₙ₋₁(polₙ) ) ... ) ) )
```

The above kind of evaluation (in DFS postorder) has inspired the design and implementation of the Hpc data structure. Looking at the geometric result shown in Figure 4.17b, we have that, according to Coding 4.3.6:

1. the output assembly is represented in the coordinate system of the first cube;
2. the second cube is translated in z direction;
3. the unit tetrahedron is translated both in z and in y directions.

Traversal algorithms

There are several ways to visit (or traverse) a graph or multigraph, ensuring that every node or edge is covered at least once. The traversal of a hierarchical structure consists of a modified Depth First Search (DFS) of its acyclic multigraph,[5] where each arc — and not each node — is traversed only once. In particular, each node is traversed several times equal to the number of paths that reach it from the root node.

A traversal algorithm aims to "linearize" a structure network by transforming all its substructures (i.e., all the subgraphs) from their local coordinates to the coordinates of the root node, assumed as world coordinates.

For this purpose, a matrix denoted as the current transformation matrix (CTM) may be maintained. Conceptually, this matrix represents the product of matrices linked to the arcs along the path from the root to the current node. For efficiency, the traversal algorithm is implemented by using a stack of CTMs. When a new arc is traversed, i.e., a reference link is encountered, the old CTM is pushed upon the stack, and a new CTM is computed by (right) multiplication of the old one times the matrix of the arc or the one heading the referred node, depending on the framework being used.

When unfolding from the forward visit of the subgraph appended to the arc,[6] The CTM is substituted by the one popped from the stack. The TRAVERSAL algorithm is specified in the pseudo-language below.

Coding 4.3.7 (Traversal of a multigraph)

> **algorithm** TRAVERSAL $((N, A, f) : multigraph)$
> $CTM :=$ identity matrix;
> TraverseNode $(root)$

---

[5] Notice that the standard DFS graph traversal (see e.g. [1]) visits all the nodes once because it recursively visits each node's children that it has not already visited.

[6] Using a pictorial image, we could say when the arc is traversed in the opposite direction.

proc TRAVERSENODE $(n : node)$
    foreach $a \in A$ outgoing from n **do**
        TraverseArc (a);
        ProcessNode (n)

proc TRAVERSEARC $(a = (n, m) : arc)$
    Stack.push (CTM);
    $CTM := CTM * a$.mat;
    TraverseNode (m);
    $CTM :=$ Stack.pop()

proc PROCESSNODE $(n : node)$
    foreach object $\in n$ **do**
        Process($CTM*$object)   □

A CTM is typically used to (left) multiply the vertices of geometric objects stored in the traversed containers. However, the reader should remember that hyperplane and normal vector equations must be multiplied conversely (right) for the inverse of the applied transformation. A double stack of matrices, which allows for pushing/popping both the CTM and its current inverse, may expedite the traversal.

4.3.3 Structure graphs in Hpc and STRUCT objects

We have to remark that the previous discussion of a generic structure graph as defined by a pair of nodes and arcs, with geometry in the nodes and transformation matrices in the arcs, is only conceptual and abstract, to make the structure concept easier to understand.

The Plasm approach is somewhat different, though it is conceptually similar to the one mentioned earlier. The main differences between the actual Plasm implementation model and the previous conceptual model are:

1. the STRUCT content is an array of Hpc objects or transformation tensors;
2. each Hpc object contains: (1) matrix object T::MatrixNd; (2) properties ::Properties field; and (3) a field childs::Union{ Vector{Hpc}, Vector {Geometry} } which may contain either a vector of Hpc objects or a vector of objects of Geometry type;
3. because of the double nature of the childs[7] array, which may contain either Hpc nodes or Geometry nodes, all the geometric datasets included in a STRUCT expression are located on the leaves of the conceptual graph structure, whereas all the non-leaf nodes are objects of Hpc type;
4. thus, any Hpc node can have any number of Hpc children nodes or any number of geometric objects stored in the same local coordinate system;

---

[7] Historically born with this flaw name.

5. the role of arc as a carrier of a transformation matrix for the subgraph rooted in its second node is played by the `T::MatrixNd` matrix that transforms all the objects within the `childs` vector.

Remark 4.12 (About the Hpc type) The `Plasm` type `Hpc` is itself a hierarchical graph where all leaves are `Geometry` objects and where all non-leaves are `Hpc` objects.

Remark 4.13 (Multigraph implementation) Of course, a multigraph saves memory space since nodes duplicated many times[8] are stored in actual memory only once. Of course, this saving is achieved in `Plasm` by giving separate definitions (named constants or variables) for semantically meaningful objects and by inserting their instances, with proper transformation matrices, into named structures of a higher level. For a noteworthy example, see the `refectory` example starting at Coding 4.3.8.

Remark 4.14 (Traversal is structure linearization) Of course, a traversal algorithm has a structure multigraph as input and, by definition, produces a linear stream of transformed objects, each one spatially relocated in its unique position in space.

Assembly examples

The two examples in this section are finalized to introduce some powerful `Plasm` higher-level functions and combinators. In particular, we use `STRUCT` and `MAP`, `CAT`, `N`, `PROPERTIES`, `IF`, and `THINSOLID`. The geometric primitives `CUBE`, `CYLINDER`, and `SIMPLEX` are also used.

Refectory room model

Some coordinated coding examples are provided here to illustrate the bottom-up development of the geometric model of a refectory room, featuring its main appliances, tables, and user chairs, as shown in Figure 4.18.

Coding 4.3.8 (Table) Both the `tableTop` and `tablelegs`, with 4 `tableLeg` instances, are generated from a 3D cube with a side length of 1:

```
cube = T(1,2)(-.5,-.5)(CUBE(1));
tableTop  = STRUCT( T(3)(.85), S(1,2,3)(1,1,.05), cube );
```

---

[8] Think about a window, a door, or any repeated object within a building

```
tableLeg  = STRUCT( T(1,2)(-.475,-.475), S(1,2,3)(.1,.1,.89),
    cube );
tablelegs = STRUCT( NN(4)([tableLeg, R(1,2)(π/2)]) );
table = STRUCT( tableTop, tablelegs );
VIEW( table )
```

The operator NN(n)(array) in Plasm repeats n times its array argument and catenates the result. Of course, R, T, and S are geometric transformation tensors.   □

Coding 4.3.9 (Chair)   The chair object is made by a chairTop and 4 cylindrical chairLeg with radius 0.06 and height 0.5:

```
cylinder  = CYLINDER([.06, .5])(8)
chairTop  = STRUCT( T(3)(0.5), S(1,2,3)(0.5,0.5,0.04), cube );
chairLeg  = STRUCT( T(1,2)(-.22,-.22), S(1,2)(.5,.5), R(1,2)(π
    /8), cylinder );
chairlegs = STRUCT( NN(4)([chairLeg, R(1,2)(π/2)]) );
chair = STRUCT( chairTop, chairlegs );
VIEW( chair )
```

Coding 4.3.10 (Four sits)   Four positions are created by alternating chairs with local rotations in object fourChairs:

```
theChair   = STRUCT( T(1)(-.8), chair )
fourChairs = STRUCT( NN(10)([R(1,2)(π/2), theChair]) );
fourSit    = STRUCT( fourChairs, table );
VIEW( fourSit )
```

Coding 4.3.11 (Single row of tables and chairs)   A 4-sit's row is generated by alternating ten instances of fourSit and T(2)(2.5) tensors:

```
singleRow = STRUCT( NN(10)([fourSit, T(2)(2.5)]) );
VIEW( singleRow )
```

Coding 4.3.12 (Whole refectory)   Analogously, the refectory room is furnished by juxtaposing ten instances of singleRow and x-translations:

```
refectory = STRUCT( NN(10)([singleRow, T(1)(3)]) );
VIEW( refectory )
```

Fig. 4.18 A hierarchical assembly generated by multilevel nested **STRUCT** expressions, starting from a single **table** model.

4.4 Attach properties to geometry

In this initial stage of developing the **Plasm** language as a Julia package, the properties associated with a geometric value of type **Hpc**, and thus with any collection of other **Hpc** values, mainly concern the object's display and visual interaction. Anyway, the language is designed to easily support design platforms for specialized domains.

Another key use for the **properties** field in the **Hpc** data structure is to store symbols that represent a Boolean operation to be executed among the children of the current node (see Section 7.5.2). In future versions, nearly any property type can be defined either by the Plasm computational infrastructure or by the user when creating a **Plasm** library.

In particular, the **properties** to attach to any **Hpc** object value are stored as **Dict{Any}{Any}** within the object's storage. Just remember the **Hpc** type definition:

```julia
mutable struct Hpc
    T::MatrixNd
    childs::Union{Vector{Hpc},Vector{Geometry}}
    properties::Properties
    # constructor
    function Hpc(T::MatrixNd=MatrixNd(0), childs::Union{Vector{
    Hpc},Vector{Geometry}}=[],
        properties=Properties())
      self = new()
      self.childs = childs
      self.properties = properties
      if length(childs) > 0
        Tdim = maximum([dim(child) for child in childs]) + 1
        self.T = embed(T, Tdim)
      else
        self.T = T
      end
      return self
    end
end
```

and the `PROPERTIES` function

```
function PROPERTIES(hpc :: Hpc, properties :: Properties)
   ret = STRUCT([hpc])
   ret.properties = copy(hpc.properties)
   for (key, value) in properties
      ret.properties[key] = value
   end
   return ret
end
```

and the `Properties` exported definition from `Viewer.jl` module.

```
Properties = Dict{String,Any}
```

References

1. Aho, A.V., Hopcroft, J.E.: The Design and Analysis of Computer Algorithms, 1st edn. Addison-Wesley Longman Publishing Co., Inc., USA (1974) pages 152
2. Ayala, D., Brunet, P., Juan, R., Navazo, I.: Object representation by means of nonminimal division quadtrees and octrees. ACM Trans. Graph. **4**(1), 41–59 (1985). DOI 10.1145/3973.3975. URL http://doi.acm.org/10.1145/3973.3975 pages 141
3. Cormen, T.H., Leiserson, C.E., Rivest, R.L., Stein, C.: Introduction to Algorithms, Third Edition, 3rd edn. The MIT Press (2009). URL https://mitpress.mit.edu/9780262533058/introduction-to-algorithms/ pages 147
4. DiCarlo, A., Paoluzzi, A., Shapiro, V.: Linear algebraic representation for topological structures. Computer-Aided Design **46**, 269–274 (2014). DOI 10.1016/j.cad.2013.08.044. URL https://doi.org/10.1016/j.cad.2013.08.044 pages 47, 137, 184, 186, 218
5. Paoluzzi, A.: Geometric Programming for Computer Aided Design. John Wiley Sons, Chichester, UK (2003). URL https://onlinelibrary.wiley.com/doi/book/10.1002/0470013885 pages 37, 38, 48, 73, 127, 130, 201, 206, 209, 244
6. Paoluzzi, A., Scorzelli, G.: Computational topology, boolean algebras, and solid modeling. Computer-Aided Design **181**, 103,839 (2025). DOI https://doi.org/10.1016/j.cad.2025.103839. URL https://www.sciencedirect.com/science/article/pii/S0010448525000016 pages 37, 47, 65, 143, 218, 221, 239
7. Paoluzzi, A., Shapiro, V., DiCarlo, A., Furiani, F., Martella, G., Scorzelli, G.: Topological computing of arrangements with (co)chains. ACM Trans. Spatial Algorithms Syst. **7**(1) (2020). DOI 10.1145/3401988. URL https://doi.org/10.1145/3401988 pages 47, 94, 137, 216, 218, 221, 224, 227, 228, 231, 233, 234, 235, 251, 258, 260, 278
8. Paoluzzi, A., Shapiro, V., DiCarlo, A., Scorzelli, G., Onofri, E.: Finite algebras for solid modeling using julias sparse arrays. Computer-Aided Design **155**, 103,436 (2023). DOI https://doi.org/10.1016/j.cad.2022.103436. URL https://www.sciencedirect.com/science/article/pii/S0010448522001695 pages 47, 65, 97, 137, 184, 218, 221, 254, 260, 270, 277
9. Permutohedron: Permutohedron — Wikipedia, the free encyclopedia (2023). URL https://en.wikipedia.org/wiki/Permutohedron. [Online; accessed 31-May-2024] pages 130

10. Requicha, A.: Representations for rigid solids: Theory, methods and systems. ACM
    Computing Surveys **12**(4), 437–464 (1980). URL `https://doi.org/10.1109/TC.`
    `1980.1675470` pages 64, 74, 138, 218, 244
11. Schutz, M., Ohrhallinger, S., Wimmer, M.: Fast out-of-core octree generation
    for massive point clouds. Computer Graphics Forum **39**(7), 13 (2020). DOI
    10.1111/cgf.14134. URL `https://www.cg.tuwien.ac.at/research/publications/`
    `2020/SCHUETZ-2020-MPC/` pages 141, 237

Chapter 5
Symbolic modeling with Julia **Plasm.jl**

The Julia `Plasm` package provides a robust functional framework for geometric and solid modeling, integrating tensor calculus with Boolean solid algebra. It introduces high-level, dimension-independent geometric types and specialized maps: `Hpc` for hierarchical assemblies and visualization, and `Lar` for representing chain complexes, spatial decompositions, and Boolean operations. Together, these components make `Plasm` a versatile and efficient tool for a wide range of modeling tasks.

This chapter focuses on the symbolic modeling of geometric tools and primitives in Julia `Plasm`, used to define parametric families of named solids. It presents topological operators for analyzing space partitions in cell and chain complexes, as well as incidence or adjacency relations, along with methods for constructing parametric curves, surfaces, and solids at any scale.

`Plasm` combines speed with usability by leveraging Julia's advanced functional programming features. At its foundation are paradigmatic primitives, generators, and transformers for solid modeling, supported by homological operators implemented as sparse matrices—crucial for computational topology and geometry. Its symbolic approach uses explicit names to capture and communicate geometric semantics, providing both clarity and precision in model definition of complex and composite shapes.

Parametric generators and topological operators demonstrate how to create flexible, reusable models using symbolic definitions and affine transformations. Geometric mapping of complexes and grids illustrates effective techniques for representing and manipulating structured geometric data. Linearized approximations of curved manifolds reveal how curves, surfaces, and solids can be generated with uniform piecewise linear (PL) methods across 1D, 2D, and 3D domains.

Through practical examples and detailed explanations, this chapter aims to equip you with the concepts and tools needed to exploit `Plasm`'s full capabilities for advanced geometric modeling.

© The Author(s), under exclusive license to Springer Nature Switzerland AG 2026   159
A. Paoluzzi and G. Scorzelli, *BIM Geometry with Julia Plasm—Functional Language for CAD Programming*, Digital Innovations in Architecture,
Engineering and Construction, https://doi.org/10.1007/978-3-031-90244-4_5

5.1 Parametric primitives

This section introduces a large subset of `Plasm` tensors (higher-level functions) along with several examples that create geometric objects by applying the functions to actual parameter values. As the user will see in the following pages, it is important to note that the `FL`-based style of Julia `Plasm` programming significantly enhances the flexibility of primitive functions, allowing them to be easily combined in various ways.

Example 5.1.1 (Geoidal coordinate model)   The surface of the Earth, at sea level, is a 2-sphere (also called geoid) where every point (except the two poles) is defined by latitude and longitude angles. In Example 5.1.2, we show how to calculate the (x, y, z) coordinates of each point p on this surface.

   We are going to generate the coordinate functions $x(u, v), y(u, v), z(u, v)$ where u, v are latitude, longitude and use them to build a polyhedral approximated solid model of the geosphere with any desired approximation.

   The `Plasm` function generates 2-spheres at at every radius r and resolution (`m,n`); see Figure 5.1.2:

```
SPHERE(r :: Number)([m :: Int, n :: Int]) :: Hpc
```

---

where the real r is the sphere radius, and `m,n` are the integer numbers that tell the function how to approximate the surface with planar faces; it will be generated with $m \times n$ planar facets. Note that such surfaces are not regular at the poles, where the partial derivatives are not linearly independent.   □

Example 5.1.2 (Spherical to Cartesian coordinate system)   Let us start with point $(r, 0, 0) \in \mathbb{E}^3$ at position vector $\mathbf{r}$ on the x-axis. Then, generate an arc of the circle by rotating it about the y-axis with angle $-\pi/2 \leq u \leq \pi/2$. The vector $\mathbf{r}$ is rotated as a function of angle u and moves to position $\mathbf{r}(u)$. The parametric matrix equation represents all points of this curve:

$$\mathbf{r}(u) = R_y(u)\,\mathbf{r}, \quad -\pi/2 \leq u \leq \pi/2 \tag{5.1}$$

Then rotate this curve about the z axis of angle v, with $-\pi \leq v \leq \pi$:

$$\mathbf{r}(u, v) = R_z(v)\,R_y(u)\,\mathbf{r}, \quad -\pi/2 \leq u \leq \pi/2, \tag{5.2}$$

Finally, provide the explicit parametric equations for the generated surface, specifically the three coordinate functions of a generic point on the sphere's surface in $\mathbb{E}^3$.

$$\mathbf{r}(u, v) = \begin{pmatrix} x(u, v) \\ y(u, v) \\ z(u, v) \end{pmatrix} = \begin{pmatrix} \cos v & -\sin v & 0 \\ \sin v & \cos v & 0 \\ 0 & 0 & 1 \end{pmatrix} \begin{pmatrix} \cos u & 0 & \sin u \\ 0 & 1 & 0 \\ -\sin u & 0 & \cos u \end{pmatrix} \begin{pmatrix} r \\ 0 \\ 0 \end{pmatrix}. \tag{5.3}$$

These will be used to MAP the discrete points (p is the generic) of the cellular complex domain via three scalar-valued functions fx -> x(p[1], p[2]), fy -> y(p[1], p[2]), fz -> z(p[1], p[2]), where p = (u,v).

$$
\begin{pmatrix} x(u,v) \\ y(u,v) \\ z(u,v) \end{pmatrix} = \begin{pmatrix} \cos v & -\sin v & 0 \\ \sin v & \cos v & 0 \\ 0 & 0 & 1 \end{pmatrix} \begin{pmatrix} r\cos u \\ 0 \\ -r\sin u \end{pmatrix} = \begin{pmatrix} r\cos u\cos v \\ -r\cos u\sin v \\ -r\sin u \end{pmatrix}
$$

The Plasm tensor SPHERE is a direct higher-function implementation:

```
function SPHERE(radius=1.0::Number)
   function SPHERE0(subds=[16, 32]) # subdivision
      N, M = subds
      domain = T(1, 2)(-π/2, -pi)(INTERVALS(π)(N) * INTERVALS(2π
   )(M))
      fx = p ->  radius * cos(p[1]) * cos(p[2])
      fy = p -> -radius * cos(p[1]) * sin(p[2])
      fz = p -> -radius * sin(p[1])
      return MAP([fx, fy, fz])(domain)
   end; return SPHERE0
end
```

Three productions of faceted surfaces and viewing examples follow:

```
VIEW( SPHERE(1)([8, 16]) )
VIEW( SPHERE(1)([16,32]) )
VIEW( SPHERE(1)([32,64]) )
```

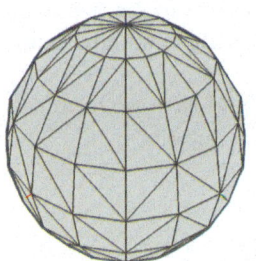  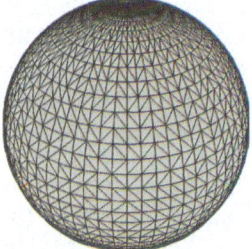

Fig. 5.1 Three geoidal SPHERE models generated with increasing resolution.

Coding 5.1.1 (The ICOSPHERE generator)

The obj object below, with the value generated by the function ICOSPHERE() without arguments, is shown in Figure 5.2a as Hpc object and in Figure 5.2b as a Lar object.

```
using Plasm
obj = ICOSPHERE()
obj = ICOSPHERE(obj)
VIEW(obj) # Hpc value
```

VIEW is used for interactive visualization of objects of Hpc type, VIEWCOMPLEX is used for objects of Lar types. See Figure 5.2.

```
VIEWCOMPLEX(LAR(obj)) # Lar value
VIEWCOMPLEX(LAR(obj), explode = [1.2,1.2,1.2])
```

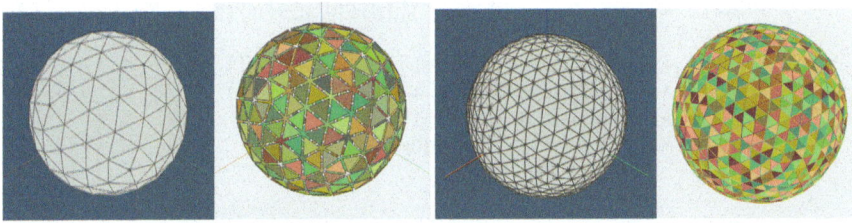

Fig. 5.2 Icospheres at different resolutions: (a,c) values of Hpc type (convex hull of points); (b) exploded Brep of Lar type (chain complex); (d) better resolution.

Of course, using an ICOSPHERE instead of a geoidal SPHERE is preferable when the surface is to be combined with other shapes in Boolean operations.

Example 5.1.3 (Parametric Boolean expressions) Here, we create a hollow solid sphere and demonstrate its thickness by subtracting a portion using a Boolean difference with a solid object, specifically a standard unit cube.

First, we create a standard second-order icosphere, and then we formulate the Boolean expression representing the hollow sphere's interior:

```
sphere = ICOSPHERE(ICOSPHERE())
solid(r) = BOOL(DIFFERENCE(
    DIFFERENCE(sphere, S(1,2,3)(r,r,r), sphere),
    CUBE(1)
)); VIEWCOMPLEX(solid(0.8))
```

In this example, you can see a Plasm function taking itself as a shape parameter (!) and a solid object with a Float64 parameter r.

Fig. 5.3 The partial hollow spheres generated by
the parametric function
solid(r), with (a) solid(0.2),
and (b) solid(0.8)

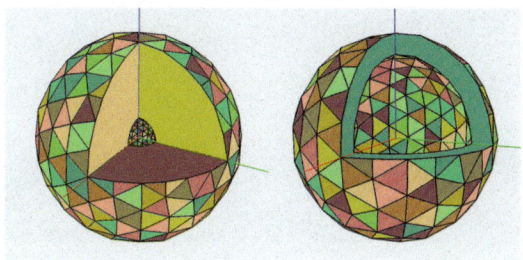

5.1.1 Generators of solid classes

This section introduces various classes of parametric solids and discusses the
writing style used to define parametric function primitives and generators of
3D named solid models. A Plasm script is often multidimensional, producing
geometric objects of different dimensions

Remember that objects are considered solids when their intrinsic and embedding dimensions match, meaning the number of coordinates equals the
number of minimal generator parameters. However, in many cases, using a
Brep (closed surface) also qualifies as a solid. Therefore, many definitions
may produce 2D and 3D geometric objects or be used by users to represent
boundary patches.

Coding 5.1.2 (Multidimensional CUBOID*)* The code fragment below is
fascinating because it combines features of functional and object-oriented
programming styles, both allowed by Julia, and produces a dimension-
independent generator function of Hpc values, also working for dimensions
greater than three, which are currently non-visualizable. ;-)

Let us note the semantics of the Scale operator. Just try the code evaluation, which produces hyper-parallelopipeds of any size.

```julia
function Scale(self :: Hpc, vs :: PointNd)
    return Hpc(scale(vs), [self])
end
function CUBOID(vs)
    return Scale(Cube(length(vs)), [Float64(it) for it in vs])
end
function CUBE(size)
    return CUBOID([Float64(size), Float64(size), Float64(size)])
end
SQUARE(d) = CUBOID([d, d])    # short Julia function syntax
```

The code above is Julia's implementation. The user's style is simply to write
CUBOID([1,1,1,1,1]) to obtain the unit five-dimensional hypercube, a convex Hpc object generated by 32 vertices.

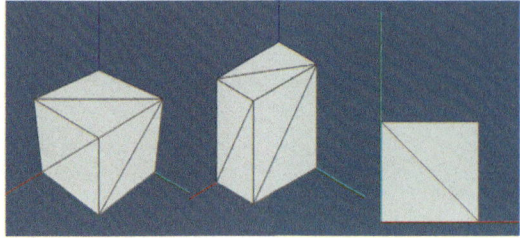

Fig. 5.4 (a) VIEW(CUBE(0.5)) perspective projection; (b) VIEW(CUBOID([.5,.3,.7]))
parallel projection of multidimensional primitive; (c) VIEW(CUBOID([0.5,.5])) 2D view.

*Coding 5.1.3 (Dimension-independent **CROSSPOLYTOPE**)* The Platonic solid OCTAHEDRON is the three-dimensional example of the standard n-
dimensional CROSSPOLYTOPE. Therefore, the only parameter of this definition
is the integer dim = D. Similarly, the standard OCTAHEDRON is generated
around the origin and the Cartesian frame of $\mathbb{E}^3$. As always, for the class of
basic generator functions, the output object is of Hpc multidimensional type.

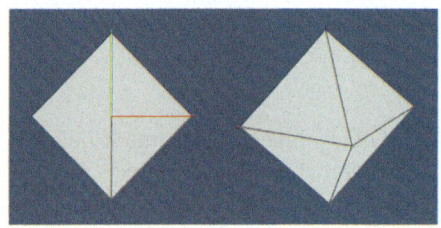

Fig. 5.5 (a) VIEW(CROSSPOLYTOPE(2)); (b) VIEW(CROSSPOLYTOPE(3)).

```
function CROSSPOLYTOPE(D)
   points = Vector{PointNd}()
   for i in 1:D
      point_pos = [0 for x in 1:D]; point_pos[i] = +1
      point_neg = [0 for x in 1:D]; point_neg[i] = -1
      push!(points, point_pos, point_neg)
   end
   cells = [collect(1:D*2)]
   pols = [[1]]
   return MKPOL(points, cells, pols)
end
function OCTAHEDRON()
   return CROSSPOLYTOPE(3)
end
```

***Coding 5.1.4** (CONE **and** FINITECONE **generator**)* The `CONE` 3D primitive is quite standard, with given `radius` and `height` and generation with the basis on the $z = 0$ plane and axis along the z direction. More interesting is the multidimensional `FINITECONE` tensor, which generates the polyhedral cone from its origin to any `Hpc` value (multidimensional) object. □

```
function CONE(args); radius, height = args
   function CONE0(N)
      basis = CIRCLE(radius)([N, 1])
      apex = T(3)(height)(HPCSIMPLEX(0))
      return JOIN([basis, apex])
   end; return CONE0
end
function FINITECONE(pol)
   point = [0.0 for i in 1:RN(pol)]
   return JOIN([pol, MK(point)])
end
```

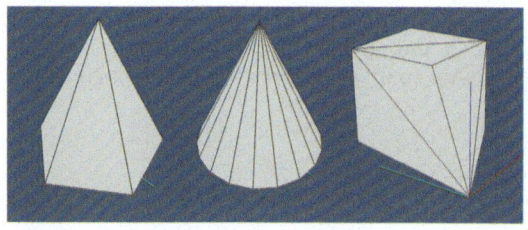

Fig. 5.6 (a) VIEW(CONE([1,2])(5)); (b) VIEW(CONE([1,2])(24)); (c) VIEW(FINITE-CONE(T(1,2)(.3,.6,.9)(CUBE(1)))).

As you may see in the `CONE` definition and many other definitions, many `Plasm` functions maintain the classic PLaSM syntax, which requires application over a single parameter to increase the compositional flexibility of language expressions.

***Coding 5.1.5** (TRUNCONE **and** CONVEXHULL **primitives**)* The `TRUNCONE` parametric generator function is a `MAP` (see Section 2.5.3) of three coordinate functions on a simplicial 2D domain $u \times v$. The first two coordinate functions $x(u,v), y(u,v)$ combine the coordinates of domain points with cos and sin Julia functions. The third coordinate map $z(u,v)$ scales their $y \in [0,1]$ component by the H parameter. Of course, `MAP` is applied only to domain vertices.

```
TRUNCONE([1,0.5,2])(16)
UKPOL(TRUNCONE([1,0.5,2])(16))[1]
VIEW(CONVEXHULL(UKPOL(TRUNCONE([1,0.5,2])(16))[1]))
VIEWCOMPLEX(LAR(PERMUTAHEDRON(3)))
```

The previous UKPOL (UnmaKe POLyhedron) is the inverse map of the tensor (MaKe POLyhedron—MKPOL) in the next script. The arguments of MKPOL are three lists $< points >$, $< convexcells >$, $< polyhedralcells >$, of integer indices on the previous entity.

```
function CONVEXHULL(points)
   return MKPOL(points, [collect(1:length(points))], [[1]])
end
```

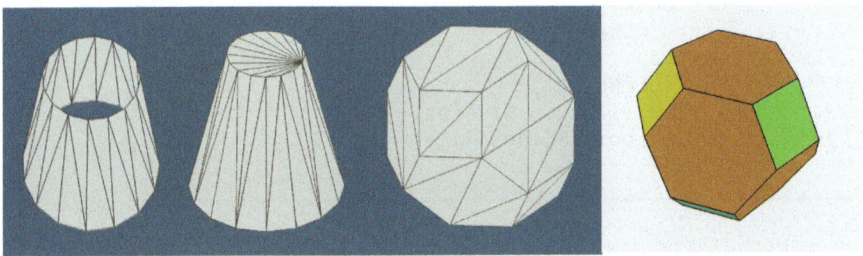

Fig. 5.7 (a) open TRUNCONE surface; (b) closed Brep; (c) The PERMUTAHEDRON function is multidimensional: it combines PERMUTATIONS, MEANPOINT, MKPOL, Translate, and PROJECT tensors; (d) VIEWCOMPLEX(LAR(PERMUTAHEDRON(3))).

2D surfaces and curves

In the following pages, we will examine the construction and use of some two-dimensional objects and discuss design choices adopted in developing such Plasm scripts. We begin with the generator function for CIRCLE objects, which has a relatively standard construction for many curved model generators. Nesting a function once to separate the formal parameters into two sets requires a double application to the actual parameter values.

Coding 5.1.6 (CIRCLE 2D solid) First, we observe the structure of a higher-level function by nesting one function inside another, as is often done in Plasm. When CIRCLE is applied to an actual radius value — say, CIRCLE (1.5) — returns the function circle0, which is now specialized to generate only circles with radius R=1.5. Then, circle0 requires the parameter subs decomposed into N, M integers to specify the piecewise-linear object decomposition. Of course, the generated objects are "solid". The subs parameter stands for "subdivision" array.

Let's notice that the fun map produces the array of the two coordinate functions $x(u,v) = v \cos u$, $y(u,v) = v \sin u$. The R parameter is used in generating the proper size of the domain object.

```
function CIRCLE(R::Number)
    function CIRCLE0(subs); subs=N,M
        domain = INTERVALS(2π)(N) * INTERVALS(R)(M)
        fun = p -> [p[2] * cos(p[1]), p[2] * sin(p[1])]
        return MAP(fun)(domain)
    end
    return CIRCLE0
end
```

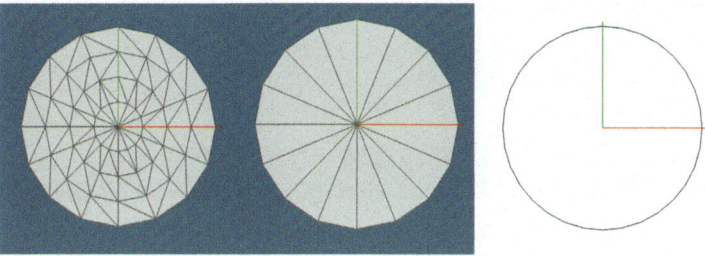

Fig. 5.8 (a,b) Inner structure of two instances of the 2-complex CIRCLE; (c) CIRCUM-FERENCE of radius 1 and N = 32 sides.

Below are the NGON definition and three language expressions generating the images in Figure 5.8. Consider that NGON is very simple because CIRCUMFERENCE provides the actual implementation.

```
function NGON(N)
    return CIRCUMFERENCE(1)(N)
end
view = Dict("background_color"=>WHITE)
VIEW( CIRCLE(1)([16,4]) )
VIEW( CIRCLE(1)([16,1]) )
VIEW( CIRCUMFERENCE(1)(32), properties=Dict("background_color"=>
    WHITE) )
```

The reader may have noticed that one functional language's best feature is its concise, easy-to-write-and-read code. This example illustrates a case where this valuable trait was not maintained. "assets=" would be better.

Coding 5.1.7 (A well defined CIRCUMFERENCE*)* Specifically, for instance, the notation CIRCUMFERENCE(R)(N) as POLYLINE(points) (with points generated on the unit circle) was concise, clear, correct, and performed very quickly. However, it contained a bug: one duplicated point, corresponding to angle values u = 0 and u = 2π. As a result, the generated 1-complex was a 1-chain but not a 1-cycle. In other words, its boundary was not empty, which is required for topological n-cycles. Two minor adjustments regarding u and cell values generation using modular arithmetic resolved the issue.

```
function CIRCUMFERENCE(R)
   function CIRCUMFERENCE1(N)
      points = [[cos(u), sin(u)] for u=0:2π/N:(2π-2π/N)]
      cells = [[i,(i%N)+1] for i=1:N]
      return MKPOL(points.* R, cells, [[N]])
   end
end
```

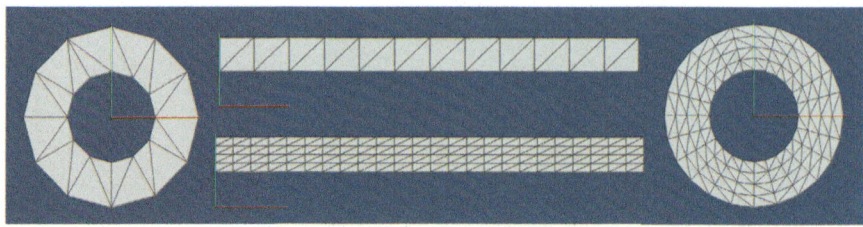

Fig. 5.9 RING construction. The parameter radius is a vector with two components R1, R2, as well the subdomain subds = [M,N]. (a) RING([0.5,1])([12,1]); (b) the two domain values; (c) RING([0.5,1])([24,4]).

Coding 5.1.8 (Other solid primitives)

We start by explaining the MAP:fun:domain mechanism. The MAP tensor applies its argument fun (function) to all vertices of the domain. In turn, fun is built using the coordinates of the domain point, which acts as the parameter of an anonymous vector function that defines the coordinate functions of the curved output manifold. We have already seen this mechanism using the p vector components p[1], p[2] for brevity. Here, we explicitly use two symbols u,v for function parameters via direct assignment (u,v) = p.

```
function RING(radius::PointNd); R1,R2 = radius
   function RING0(subds); N,M = subds
      domain = T(2)(R1)(INTERVALS(2*π)(N)*INTERVALS(R2-R1)(M))
      fun = p -> begin (u,v) = p; [v*cos(u), v*sin(u)] end
      return MAP(fun)(domain)
   end; return RING0
end
```

```
function TUBE(args::PointNd); r1, r2, height = args
   function TUBE0(N)
      return Power(RING([r1, r2])([N, 1]), QUOTE([height]))
   end return TUBE0
end
```

Let us note two Julia general issues for coding parametric `Plasm` functions. The problems concern the syntax of anonymous functions used within a `MAP`: (a) the necessity of a single symbol as the argument, and (b) the use of the compound expression `begin...end` to denote the value expression when this requires more than one instruction.

Example 5.1.4 (Variations of CYLINDER concept) Of course, there are several methods (algorithms) to generate the same solid representation of a curved manifold, i.e., the same Brep.

For example, a cylindrical surface could be constructed as a `MAP` of parametric equations or as a Cartesian product of either the `CIRCUMFERENCE` by `INTERVALS`, resulting in an empty lateral surface, or as a product of `CIRCLE` times `INTERVALS`, yielding a solid `CYLINDER` with internal 3-cells. A further variation of the resulting object topology is given by `RING` times `INTERVALS`.

```
cyl1 = CIRCUMFERENCE(1)(16) * INTERVALS(2.)(4);
cyl2 = CIRCLE(1)([16,3]) * INTERVALS(2.)(4);
cyl3 = RING([0.3,1])([16,3]) * INTERVALS(2.)(4);
```

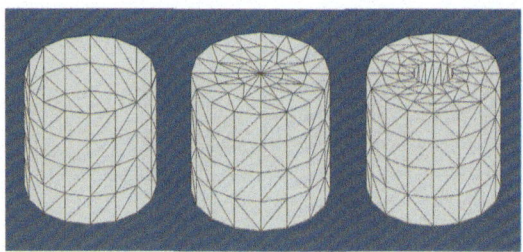

Fig. 5.10 (a) Cylindrical surface; (b) solid cylinder; (c) hollow solid cylinder.

Using basic functional abstractions, we can connect different types of cylinder objects to their corresponding primitive generator tensors by appropriately selecting parameters, possibly using default values. Additionally, we can pass a single geometric parameter to a more general primitive constructor, `CYLINDER`, which is an excellent example of programming at the Function Level (`FL`) with `Plasm` in Julia.

```
function CYLINDER(section::Hpc, measure::Hpc)
    return section * measure
end
cyl = CYLINDER(RING([0.6,1])([32,2]), INTERVALS(6.)(20))
VIEW( R([1,3])(π/2)(cyl) )
```

Example 5.1.5 (From CYLINDER concept to AEC idea)

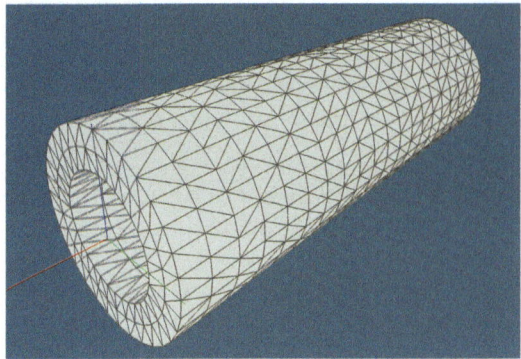

Fig. 5.11 The cyl::Hpc object, starting from vertical configuration, was rotated of $\pi/2$ angle about the y axis, so getting its axis coinciding with the x axis (the red one).

In this example, we present a straightforward shape-ideation process, starting with the hollow-cylinder concept and progressing to the development of vertical enclosures for a futuristic building.

```
SK = SKELETON
obj1 = LAR(CIRCLE(1)([6,2]) * INTERVALS(2)(4));
obj2 = LAR(CIRCLE(1)([6,2]) * SK(0)(INTERVALS(2)(4)));
VIEWCOMPLEX(obj1, explode=[1.3,1.3,2], show=["FV"])
VIEWCOMPLEX(obj1, explode=[1.3,1.3,2], show=["EV"])
VIEWCOMPLEX(obj2, explode=[1.3,1.3,2], show=["FV"])
```

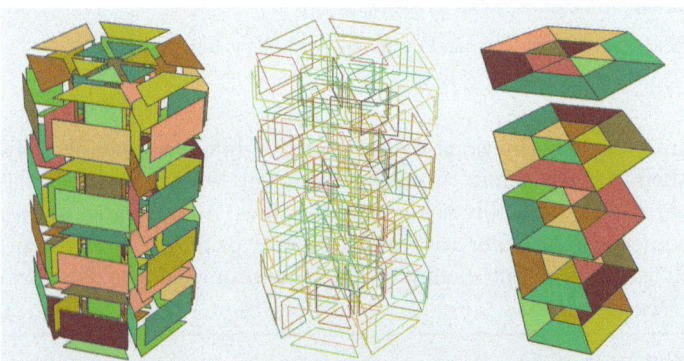

Fig. 5.12 Some elaboration about the cylindrical tower idea: (a) exploded 2-cells; (b) exploded boundary polygons; (c) horizontal partitions.

The torus surface can be generated as the product of two 1-circles, but the resulting manifold is a 2-surface (1+1) embedded in 4D (2+2). This surface instance can be ported to three coordinates by a proper projection. A more straightforward generation is obtained through parametric equations, where a 1-circle within a coordinate Cartesian plane is rotated about the z axis.

Coding 5.1.9 (Two hollow solids) A 3D torus is a surface of revolution generated by rotating with angle 2π a 1-circle (circumference) in $\mathbb{E}^3$ about an axis coplanar with the circle. It is obtained here by coordinate equations that combine scaling and translation of the generating 1-circle and its rotation about a coordinate axis.

```
function TORUS(radii=[1.0, 2]::Vector); r1, r2 = radii
   function TORUS0(subds=[32, 16]::Cell); N, M = subds
      a = 0.5 * (r2 - r1)
      c = 0.5 * (r1 + r2)
      domain = INTERVALS(2 * pi)(N) * INTERVALS(2 * pi)(M)
      fx = p -> (c + a * cos(p[2])) * cos(p[1])
      fy = p -> (c + a * cos(p[2])) * sin(p[1])
      fz = p -> a * sin(p[2])
      return MAP([fx, fy, fz])(domain)
   end; return TORUS0
end
```

Suppose we want to generalize the torus surface concerning the revolution angle. In that case, we might introduce one more level in TORUS tensor below, with parameter the bounds vector storing the extreme angles B1, B2 with default [0, 2π] and changing accordingly the domain definition with proper interval and translation.

```
function TORUS(bounds=[0, 2π]::Vector); B1, B2 = bounds
   function TORUS0(radii=[1.0, 2]::Vector); r1, r2 = radii
      function TORUS1(subds=[32, 16]::Cell); N, M = subds
         a = 0.5 * (r2 - r1)
         c = 0.5 * (r1 + r2)
         domain = T(1)(B1)(INTERVALS(B2-B1)(N) * INTERVALS(2pi)(M))
         VIEW(domain)
         fx = p -> (c + a * cos(p[2])) * cos(p[1])
         fy = p -> (c + a * cos(p[2])) * sin(p[1])
         fz = p -> a * sin(p[2])
         return MAP([fx, fy, fz])(domain)
      end; return TORUS1
   end; return TORUS0
end;
```

To introduce a user-defined thickness in the output model, we could add a third radius value, r0, and provide a suitably transformed 3-dimensional domain. The reader could try this, considering that the VIEW(domain) inserted within the function body gives good feedback and is expected to be removed when the code is fixed. Try VIEW(TORUS()()) and note the default parameterization.

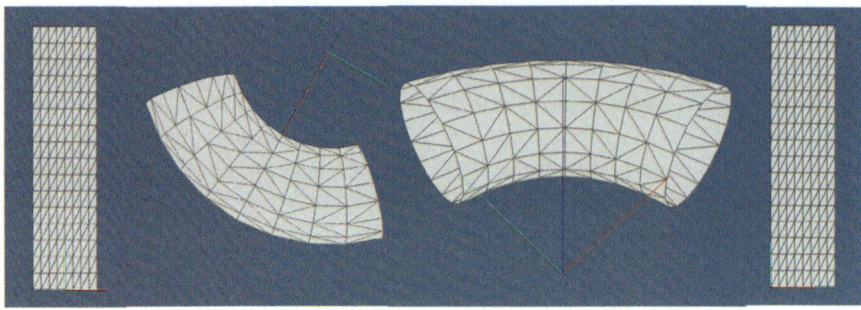

Fig. 5.13 (a, b) VIEW(TORUS($[-\pi/4, \pi/4]$)( )([8, 16])); (c, d) VIEW(TORUS($[0, \pi/2]$)( )([8, 16])). Radiuses are the default. Note the width and translation of domain.

Plasm design of a turbo pump

This section discusses the stepwise generation of a highly parameterized family of geometric objects. First, a code template, SOLIDHELICOID, is produced starting from the one-dimensional SPIRAL curve, and then a definitive model, TURBOPUMP. This is difficult to generate or even impossible to build with any GUI-based CAD system.

Coding 5.1.10 (Solid helicoid) A parametric helicoid surface and solid follow. The source code represents a whole family of ∞^6 different shapes:

```julia
function SOLIDHELICOID(; nturns=3,R=1.,r=0.0,shape=[36*nturns,
    8], pitch=2, thickness=0.1)
  totalangle = nturns*2*pi
  grid2D = INTERVALS(36*nturns)(36*nturns)*INTERVALS(4)(8)
  Domain2D=T(2)(r)(S(1,2)([totalangle/shape[1],R-r])(grid2D))
  surface = p->((u,v)=p;[v*cos(u);v*sin(u);u*(pitch/(2*pi))])
  solidMapping = THINSOLID(surface)
  Domain3D = Domain2D * INTERVALS(thickness)(1);
  view = Dict("background_color"=>WHITE)
  VIEW(MAP(surface)(Domain2D), properties=view)
  VIEW(MAP(solidMapping)(Domain3D), properties=view)
end;
```

When executing the expression below, the code template visualizes both a surface and a thin solid shown in Figure 5.14.

```julia
julia> SOLIDHELICOID( r=0.4, thickness=0.2 )
```

This atypical method—inserting a VIEW expression within a function—was used in the SOLIDHELICOID template to VIEW both the first step (a parametric surface) and the definitive solid template. This procedure is typical of development coding and testing with `Plasm` as a generator of geometric models.

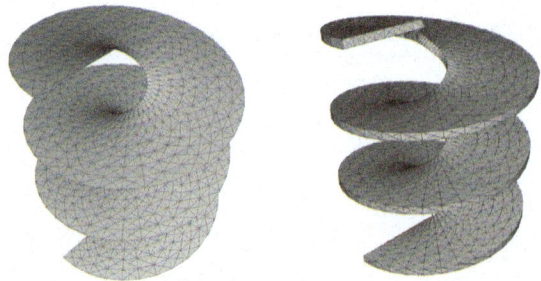

Fig. 5.14 Parametrized helicoid values: (a) surface; (b) thin solid.

Fig. 5.15 Two prospectives of the turbo pump component: (a) from top; (b) from side.

Coding 5.1.11 (Turbo pump) The main difference with `SOLIDHELICOID` template is the piecewise-linear `Dom2D` mapped from the rectangular `grid2D`.

```
function TURBOPUMP(; nturns=3, R=1., r=0.0, shape=[36*nturns,8],
    pitch=2, thickness=0.1)
    totalangle = nturns*2*pi;   xM = 36*nturns
    (x0,xm1,xm2,xM) = (0.0, xM/2nturns, 6xM/2nturns, xM)
    G = (x-> x<xm1 ? x/xm1 : (x)xm2 ? 1+(xm2-x)/(xM-xm2) : 1))
    grid2D = INTERVALS(xM)(xM) * INTERVALS(1)(8)
    dom = MAP( p->begin (u,v)=p; [u, (G(u)+0.1)*v] end )(grid2D)
    Dom2D= T([2])([r])(S([1,2])([totalangle/shape[1],R-r])(dom))
    surface = p->((u,v)=p;[v*cos(u); v*sin(u); u*(pitch/(2*pi))])
    solidMapping = THINSOLID(surface)
    Domain3D = Dom2D * INTERVALS(thickness)(1)
    return MAP(solidMapping)(Domain3D)
end
```

The piecewise linear `dom` shape is produced by the Julia function assigned to symbol G, implemented as a nested conditional triple statement. The three piecewise G interval subdomains are defined by consecutive number pairs in (x0,xm1,xm2,xM).

Coding 5.1.12 (Turbo pump's object viewing)

```
obj = TURBOPUMP( pitch=0.15, thickness=0.015, nturns=5, r=1/2 );
view = Dict("background_color"=)[1,1,1]); VIEW(obj, view)
```

The other parameters have default values, which are given in the definition head.   □

The profile of the sectioned mechanical object in Figure 5.1.1 is defined here as piecewise-linear, but it would not be difficult to define it as any type of curve or spline by introducing a helper function.

5.1.2 Primitive design devices

This section covers basic methods and tools for merging geometric information to create geometric objects with correct features and properties. Some design tools were already used in previous sections, but will be discussed in more detail here.

Higher order, partial, and multi-dimensional functions

As we saw in Chapter efchapt:2, Julia `Plasm` is more advanced because it supports functions that accept other functions as arguments and/or may return a function as a result. All functions are objects of the Julia `Function` type. As objects (containing a reference to the function code), they can be assigned to a name (identifier).

Definition 5.1 (Function order) The order of an object of `Function` type is the ordinal number of applications to actual parameters needed to return the final actual value, not a partial function value (which requires an application to other actual parameter values).

In other words, the order of a Julia function is the number of times it must be applied to actual parameters to return the final function value. Most Julia functions are first-order. Conversely, many `Plasm` functions are higher-order.

***Coding 5.1.13 (Primitive function* INTERVALS(size::Number)(n::Int))**
The second-order function INTERVALS()(), which requires two applications to actual values, generates a 1-complex made by n adjacent line segments for a total given `size` length. The length of each 1-cell is `size`/n.

```
segments = INTERVALS(10 :: Number)(4 :: Int)        #=
Hpc(MatrixNd(2), Geometry([[0.0], [2.5], [5.0], [7.5], [10.0]],
    hulls=[[1, 2], [2, 3], [3, 4], [4, 5]]))       =#
```

Note that the segments value is 1D since its five vertices have exactly one coordinate.   □

Coding 5.1.14 (Primitive QUOTE(measures::ArrayNumber))
QUOTE generates a 1D complex as a sequence of segments with specified lengths, which may be either solid or empty. The formal parameter is an array of signed numbers. Positive numbers represent solid intervals of a specified size, while negative numbers indicate hollow space, meaning the displacement of subsequent segments.

```
julia> two_aligned_segments = QUOTE([1,-2.5,-2.5,1])   #=
Hpc(MatrixNd(2), Geometry([[0.0], [1.0], [6.0], [7.0]], hulls
    =[[1, 2], [3, 4]]))   =#
```

Successive negative numbers are allowed.   □

Coding 5.1.15 (Primitive Q(measure::Number)) The formal parameter is a signed number.

```
segment = Q(10)    #=
Hpc(MatrixNd(2), Geometry([[0.0], [10.0]], hulls=[[1, 2]])) =#
```

The Q operator produces a single 1D segment of given size with one vertex on the origin, and the other with a coordinate equal to the parameter value. □

Single convex cell

Julia Plasm features an extensive library of generator functions, which includes many elementary objects created from a single convex cell, specified solely by the set of vertices. Two examples are provided below. Users can extract or generate additional examples by examining the file Plasm/src/fenvs.jl, which includes the five Platonic solids. The multidimensional *d*-permutahedron was generated in Coding 4.1.11.

*Coding 5.1.16 (*CUBOID(size))
The function CUBOID, generator of multi-dimensional hyper-parallelepipeds, was already met at Coding 5.1.2. It has a parameter of type Vector{Number} that specifies the size of the sides. We see the textual Plasm output for dimensions 1,2,3,4. Looking at some geometric datasets may be helpful.

Note the single value in `hulls`, with vertex indices from 2^1 to 2^4 of the convex d-cells named `hulls`.

```
CUBOID([1])           #=
Hpc(MatrixNd(2), Hpc(MatrixNd(2), Geometry([[0.0], [1.0]], hulls
    =[[1, 2]]))) =#
CUBOID([1,2])         #=
Hpc(MatrixNd([[1.0, 0.0, 0.0], [0.0, 1.0, 0.0], [0.0, 0.0,
    2.0]]), Hpc(MatrixNd(3), Geometry([[0.0, 0.0], [1.0, 0.0],
    [1.0, 1.0], [0.0, 1.0]], hulls=[[1, 2, 3, 4]]))) =#
CUBOID([1,2,3])          #=
Hpc(MatrixNd([[1.0, 0.0, 0.0, 0.0], [0.0, 1.0, 0.0, 0.0], [0.0,
    0.0, 2.0, 0.0], [0.0, 0.0, 0.0, 3.0]]), Hpc(MatrixNd(4),
    Geometry([[0.0, 0.0, 0.0], [1.0, 0.0, 0.0], [0.0, 1.0, 0.0],
    [1.0, 1.0, 0.0], [0.0, 0.0, 1.0], [1.0, 0.0, 1.0], [0.0,
    1.0, 1.0], [1.0, 1.0, 1.0]], hulls=[[1, 2, 3, 4, 5, 6, 7,
    8]]))) =#
```

The four-dimensional hyperparallelepiped with edges of size $1, 2, 3, 4$ follows:

```
CUBOID([1,2,3,4])        #=
Hpc(MatrixNd([[1.0, 0.0, 0.0, 0.0, 0.0], [0.0, 1.0, 0.0, 0.0,
    0.0], [0.0, 0.0, 2.0, 0.0, 0.0], [0.0, 0.0, 0.0, 3.0, 0.0],
    [0.0, 0.0, 0.0, 0.0, 4.0]]), Hpc(MatrixNd(5), Geometry
    ([[0.0, 0.0, 0.0, 0.0], [1.0, 0.0, 0.0, 0.0], [0.0, 1.0,
    0.0, 0.0], [1.0, 1.0, 0.0, 0.0], [0.0, 0.0, 1.0, 0.0], [1.0,
    0.0, 1.0, 0.0], [0.0, 1.0, 1.0, 0.0], [1.0, 1.0, 1.0, 0.0],
    [0.0, 0.0, 0.0, 1.0], [1.0, 0.0, 0.0, 1.0], [0.0, 1.0, 0.0,
    1.0], [1.0, 1.0, 0.0, 1.0], [0.0, 0.0, 1.0, 1.0], [1.0,
    0.0, 1.0, 1.0], [0.0, 1.0, 1.0, 1.0], [1.0, 1.0, 1.0, 1.0]],
    hulls=[[1, 2, 3, 4, 5, 6, 7, 8, 9, 10, 11, 12, 13, 14, 15,
    16]]))) =#
```

Of course, the unit hypercube in $\mathbb{E}^6$ is generated by `CUBOID([1,1,1,1,1,1])`

.   □

The `Plasm` coding of the "icosphere," which stands for the polyhedral approximation of the 2-sphere obtained by subdivision of `ICOSAHEDRON()` surface, is given below, starting from the Platonic solid. The generation method is simple. We get the vertices at step $i+1$ by adding to the vertices at step i obtained by subdivision of edges. The implementation uses the `Hpc` and the `Lar` data structures.

Coding 5.1.17 (The ICOSPHERE(seed::Hpc)::Hpc *generation)*
First, we take the input `obj` cell complex using the `LAR` combinator, which transforms a `Hpc` value to a `Lar` value. Then, for each edge, we compute the mean point and aggregate the new ones, scaled by the factor `r1/s1` built with the distance from the `[0,0,0]` center of both models to the old vertices.

Finally, the `[W; V*(r1/s1)]::Vector{Vector{Float64}}` created from old vertices and new scaled ones is provided to the `CONVEXHULL` operator, which transforms this collection of points into their geometric exititconvex hull and returns an `Hpc` value.

```
function ICOSPHERE(obj::Hpc)::Hpc
  W  = LAR(obj).V
  EV = LAR(obj).C[:EV]
  W  = [W[:,k] for k=1:size(W,2)]
  V  = [(W[v1]+W[v2])./2 for (v1,v2) in EV]
  r1 = sqrt(sum(W[1].^2))
  s1 = sqrt(sum(V[1].^2))
  CONVEXHULL([W; V*(r1/s1)]);
end
```

Remember that such polyhedra are convex sets; hence, they have a single (convex) cell.

```
out0 = ICOSAHEDRON(); VIEW(out0)
out1 = ICOSPHERE(out0); VIEW(out1)
out2 = ICOSPHERE(out1); VIEW(out2)
out3 = ICOSPHERE(out2); VIEW(out3)
...
```

Successive approximations of the icosphere have 12, 42, 162, 600, and so on, vertices. Let's note the remarkable simplicity of such polyhedral generations. To get a good approximation of the unit sphere, just scale `out3`:

```
R = 2/SIZE(1)(out3) # => 0.7110806012572247
SPHERE = S(1,2,3)(R,R,R)(out3) # icosphere with r = 1
```

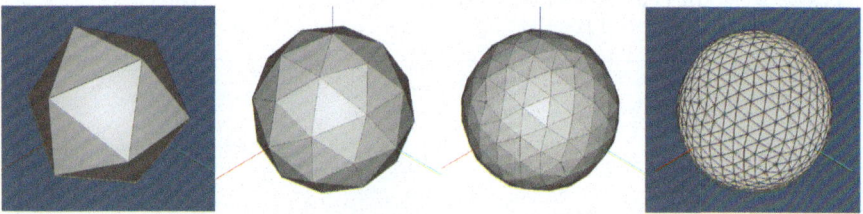

Fig. 5.16 (a) icosahedron; (b, c, d) icospheres with increasing numbers of vertices.

Patterns of cell objects

The functions `INTERVALS` or `QUOTE` may be used to create many types and patterns of grid geometries. We show non-trivial examples in the following.

Coding 5.1.18 (Building frame)

First, we provide the primary dataset for a building frame by "quoting" the side measurements, the 2D design plan, and section.

```
# Longitudinal trusses
Y = QUOTE([0.3, -6, 0.3, -6, 0.3])
# transverse beams
X = QUOTE([0.3, -3, 0.3, -4.2, 0.3, -3, 0.3])
# vertical measurements
Z = QUOTE([3,0.3])
```

Then, an alternate set of `INTERVALS` vector parameters is generated by Julia broadcast `.*` of the scalar `-1` to invert all the signs.

```
X1 = QUOTE([0.3, -3, 0.3, -4.2, 0.3, -3, 0.3] * -1)
Y1 = QUOTE([0.3, -6, 0.3, -6, 0.3] * -1)
Z1 = QUOTE([3,-0.3] * -1) # broadcast non-needed
```

Below, the 3D `frame` subsystems are generated. They are the C_3 basis of a local cellular complex. Note the variation pattern at `trusses1`.

```
# Cartesian product
pillars = COLOR(Plasm.COLORS[1])(X*Y*Z);
trusses = COLOR(Plasm.COLORS[2])(X*Y1*Z1);
trusses1 = COLOR(Plasm.COLORS[2])(X1*Y*QUOTE([-2.7,0.6]));
floorslab = COLOR(Plasm.COLORS[3])(X1*Y1*Z1);
```

Finally, the sub-complexes of 3D cells are aggregated in a single `Plasm` complex using the `STRUCT` combinator discussed in the next session. All the aggregated models live in the same reference frame in this example.

```
frame = STRUCT(pillars, trusses, trusses1, floorslab);
VIEW(frame, properties=Dict("background_color"=>WHITE))
```

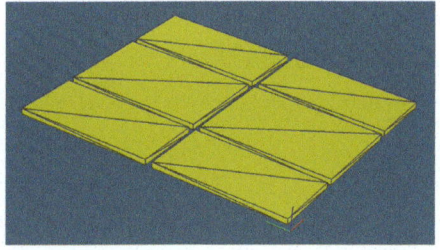

Fig. 5.17 The negative space associated with a Cartesian product of negative values in **QUOTE** combination: see **floorslab** = ... and notice the suitable complementation of solid intervals in **X1,Y1,Z1**.

In geometric modeling of complex assemblies, the programmer frequently uses the hierarchical scene graphs or hierarchical structures, where subassemblies are defined in local coordinates and are transformed into the coordinate

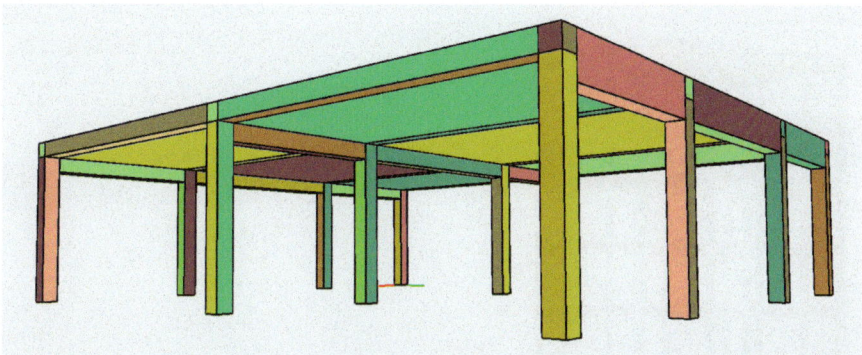

Fig. 5.18 frame = STRUCT(pillars, trusses, trusses1, floor slab); Note the different heights between beams and curbs of reinforced concrete and the reference frame position at the corner.

frame of the output assembly by explicitly applying the proper coordinate transformations. The measurement unit used here is the meter.

It is worth noting that such "modeling transformations"—as they are referred to in graphical systems—are entirely the programmer's responsibility.

Coding 5.1.19 (Building skeleton)
We assemble a building skeleton model by creating a STRUCT assembly with n =7 instances of the Julia Vector, generated from the Hpc value frame and the MatrixNd value T(3)(3.3), which produces a translation in the z direction.

```
skeleton = STRUCT(NN(7)([frame, T(3)(3.3)]));
VIEW(skeleton, Dict("background_color"=>WHITE))
```

Here the STRUCT semantics is unambiguous: the evaluated subexpression NN (7)([frame, T(3)(3.3)]) is an array of type Vector{Union{Hpc, MatrixNd }} and contain 14 items with alternating types. When STRUCT is evaluated on this vector, an Hpc node is generated, whose MatrixNd field contains a 4×4 identity matrix, and the childs vector contains the reference to the first frame.

As shown in the next section, Plasm features unique combinations of coordinate shapes that alleviate the programmer of this responsibility. These are especially beneficial when the aggregation is complex.

Alignment aggregators TOP, BOTTOM, LEFT, RIGTH

In the present section, we discuss some simplified assembly operators where the language computes coordinate transformations automatically.

Fig. 5.19 Building structural VIEW(STRUCT(NN(7)([frame, T(3)(3.3)]))).

Plasm has some predefined binary functions used to locate two complexes with respect to each other. In particular, the second argument of such functions will be positioned as either TOP the first one, or ABOVE, or LEFT, or RIGHT, or UP, or DOWN, respectively, according to the alignment semantics given below.

Such relative positioning allows for the easy construction of complex assemblies without considering the local coordinate frames where the sub-assemblies are defined. This way, the user avoids applying affine transformations, which are conversely needed in hierarchical scene graphs.

Coding 5.1.20 (alternate method for vertical aggregation)

```
multifloor(model,n) = STRUCT(INSR(TOP)(N(n+1)(model))) #=
multifloor (generic function with one method)           =#
VIEW(multifloor(frame,7))
```

The object generated here equals the **skeleton** model of Figure 5.19.   □

TOP is a binary function of two Hpc models that create their vertical aggregation. Any binary function is transformed into an n-ary function by the second-order operator INSR (for insert right). INSR is applied first to the binary operator argument to make it n-ary, then to the list of objects the operator applies to.

N(n)(value :: Any) is a Plasm function called # in classic PLaSM, impossible to port in Julia, where denotes comments, that returns a list of n instances of value object or expression, including suitable transformations when needed.

Definition 5.2 (Primitive ALIGN function.) Any pair of polyhedral complexes can be aligned along any chosen set of coordinates using the primitive binary tensor ALIGN. This second-order operator must be applied to a sequence of triples that define specific behavior for each affected coordinate, which are designated as 1, 2, or 3.

Each alignment directive along a coordinate is a triple that must belong to the set:

$$\{1,2,3\} \times \{\texttt{MIN},\texttt{MED},\texttt{MAX},\texttt{K}(\alpha)\} \times \{\texttt{MIN},\texttt{MED},\texttt{MAX},\texttt{K}(\alpha)\}$$

where $n \in \{1,2,3\}$ is a coordinate index and $\alpha \in \{1,2,3\}$, is used to align on a given coordinate. The resulting specialized operator is then applied to a pair of Hpc values and returns a single Hpc. Use examples of the ALIGN combinator for other operator definitions as follows:

```
TOP = ALIGN([[3, MAX, MIN], [1, MED, MED], [2, MED, MED]])
BOTTOM = ALIGN([[3, MIN, MAX], [1, MED, MED], [2, MED, MED]])
LEFT = ALIGN([[1, MIN, MAX], [3, MIN, MIN]])
RIGHT = ALIGN([[1, MAX, MIN], [3, MIN, MIN]])
UP = ALIGN([[2, MAX, MIN], [3, MIN, MIN]])
DOWN = ALIGN([[2, MIN, MAX], [3, MIN, MIN]])
# 294 (generic function with one method)
```

Coding 5.1.21 A very simplified view of a column model is built here to show the use of the INSR(TOP) and INSR(RIGHT) functions applied to a list of geometric objects, each given in its local coordinate system:

```
function Column(r,h)
  basis = CUBOID([ 2*r*1.2, 2*r*1.2, h/12.0 ])
  trunk = CYLINDER([ r, (10.0/12.0)*h ])(12)
  capital = CUBOID([ 2*r*1.2, 2*r*1.2, h/12.0 ])
  beam = S(1)(3)(capital)
  return INSR(TOP)([basis,trunk,capital,beam])
end
# Column (generic function with two methods)
```

```
function ColRow(N,r,h)
  col = Column(r,h)
  columnlist = [col for k in 1:N+1]
  return INSR(RIGHT)(columnlist)
end
# ColRow (generic function with two methods)
```

```
VIEW(ColRow(4,1.,12.), properties=Dict("background_color"=>WHITE
    ))
```

Fig. 5.20 This figure exemplifies two fellow tensor
devices used together in
Plasm model design by the
parametric function Col-
Row(N,r,h).

The generated model is displayed in Figure 5.20.   □

The `ColRow(N,r,h)` function not only defines a single parametric column
via the nested expression `Column(r,h)` but also instantiates it a number `N` of
times, combining the `INSR` and `RIGHT` operators. Remember, from Chapter
2, that the `FL` primitive `INSR` transforms a function of two arguments into a
function of n arguments. Also, the `RIGHT` operator puts its two arguments in
the desired relative positions.

Internally, it creates a `STRUCT` recursive expression with a proper `Hpc` graph
structure enclosing column instances and transformations. Of course, the
`Plasm` coding of the parametric `ColRow(N,r,h)` function is much easier for
laypeople.

Fig. 5.21 The fully parameterized Temple() function features some param-
eters that are intercon-
nected through algebraic
equations. Let's note the
number of side and front
columns and their relationship to column height and
gable width. The entire
function code consists of
about 15 lines. Addition-
ally, observe the mixture
of 3D and 2D objects,
where **PROJECT** the **BOX**
of **TEMPLE** generates the
ground rectangle of the
right size.

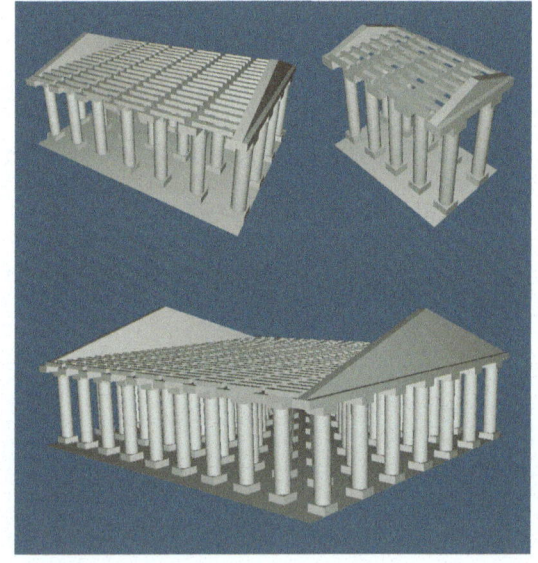

5.2 Topological operators

For simplicity's sake, we begin our presentation of topological operators by discussing one of the simplest example models, the numbered `Lar` version of the 3-cube with a unit side (see Figure 5.22). The `Plasm` itself provides the vector font of the text on the image when needed, along with its integrated display of the evaluated expression. This integration of code with text has been invaluable in algorithm development.

First, in this model we have a 3D chain complex (see section 3.4) with one 3-cell, 6 faces (2-cells `FV`), 12 edges (1-cells `EV`), and 8 vertices (0-cells `V`).

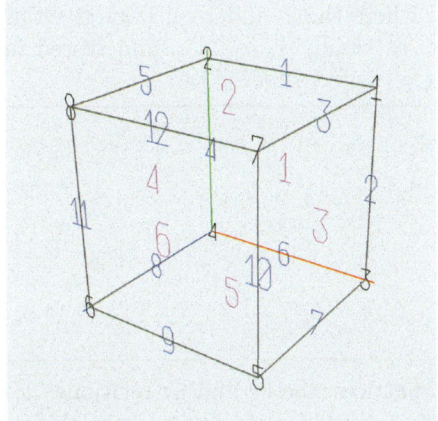

Fig. 5.22 The numbered unit 3-cube cellular complex. The ordinal numbers for vertices, edges, and faces are set in the middle of each 0-, 1-, and 2-cell. Note that the indexes have lexicographic and not geometric order (say, clockwise or counterclockwise), but this is not a constraint.

The reader is invited to check for a comparison of the following dataset with Figure 5.22. In general, the cells and cell indices are unordered.

```
LAR(CUBE(1))        #=
Lar(3, 3, 8, [1.0 0.0 … 1.0 0.0; 1.0 1.0 … 1.0 1.0; 0.0 0.0 …
      1.0 1.0], Dict{Symbol, AbstractArray}(:CV => [[1, 2, 3, 4,
      5, 6, 7, 8]], :FV => [[1, 2, 3, 4], [1, 2, 7, 8], [1, 3, 5,
      7], [2, 4, 6, 8], [3, 4, 5, 6], [5, 6, 7, 8]], :EV => [[1,
      2], [1, 3], [1, 7], [2, 4], [2, 8], [3, 4], [3, 5], [4, 6],
      [5, 6], [5, 7], [6, 8], [7, 8]])) =#
```

Consider the three entities `V`, `E`, `F` representing the boundaries of solid objects (`B-rep`). As a result, we have three binary adjacency relations: `VV`, `EE`, and `FF`, along with six binary incidence relations: `VE`, `VF`, `EV`, `EF`, `FV`, and `FE`. The latter set is well-known among topologists, while software engineers mainly use the former to design efficient data structures for solid modeling.

We introduce here a new entity, C, to denote solid cells in 3D, and two new incidence relations, CF, FC, used by our Lar (Linear Algebraic Representation) [1, 4] to build the B-reps itself and the Boolean Algebras of solids. The relations CV, VC are instead used by the Hpc representation. All the following datasets and matrices refer to the topology of CUBE(1) shown above, as readers may check.

5.2.1 Incidence operators

The "unevaluated" Lar representation [4] is based on FV and EV (other than on the V matrix of vertex coordinates) where the 2- and 1-cells are given as unordered lists of 0-cells's indices, i.e., as chains of vertices, and stored in Julia by using the type Chains = Vector{Vector{Int64}}.

```
EV = LAR(CUBE(1)).C[:EV]          #=
12-element Vector{Vector{Int64}}:
[[1, 2], [1, 3], [1, 7], [2, 4], [2, 8], [3, 4], [3, 5], [4, 6],
     [5, 6], [5, 7], [6, 8], [7, 8]]        =#
FV = LAR(CUBE(1)).C[:FV]          #=
6-element Vector{Vector{Int64}}:
[[1, 2, 4, 6], [1, 3, 5, 12], [2, 3, 7, 10], [4, 5, 8, 11], [6,
     7, 8, 9], [9, 10, 11, 12]]        =#
```

When represented as sparse binary matrices, the FV and EV relations, defined as subsets of the Cartesian products F×V and E×V, facilitate a straightforward computation of the FE matrix, as seen below. This also applies to their transposed operator matrices [VF]=[FV]', [VE]=[EV]', and [EF]=[FE]'.

The sparse matrix KEV is of Julia type ChainOp[1] and is generated by the lar2cop function, where cop refers to ChainOp, and thus lar2cop means "from Lar to Chain Operator". The same applies to KFV.

```
KEV = lar2cop(EV)             #=
12×8 SparseArrays.SparseMatrixCSC{Int8, Int64} with 24 stored
     entries:
 1  1  ·  ·  ·  ·  ·  ·
 1  ·  1  ·  ·  ·  ·  ·
 1  ·  ·  ·  ·  ·  1  ·
 ·  1  ·  1  ·  ·  ·  ·
 ·  1  ·  ·  ·  ·  ·  1
 ·  ·  1  1  ·  ·  ·  ·
 ·  ·  1  ·  1  ·  ·  ·
 ·  ·  ·  1  ·  1  ·  ·
 ·  ·  ·  ·  1  1  ·  ·
 ·  ·  ·  ·  1  ·  1  ·
```

---

[1] ChainOp is defined in Julia as SparseArrays.SparseMatrixCSC{Int8, Int64}, where CSC stands for Compressed Sparse Column.

```
.  .  .  .  .  .  1  .  1
.  .  .  .  .  .  1  1   .   =#
```

```
KFV = lar2cop(FV)              #=
6×8 SparseArrays.SparseMatrixCSC{Int8, Int64} with 24 stored
    entries:
 1  1  1  1  .  .  .  .
 1  1  .  .  .  .  1  1
 1  .  1  .  1  .  1  .
 .  1  .  1  .  1  .  1
 .  .  1  1  1  1  .  .
 .  .  .  .  1  1  1  1   =#
```

Remark 5.1 (Meaning of sparse matrices)
KEV: V→E and KFV: V→F are operators between the linear subspaces V, E, and F of 0-, 1-, and 2-chains, respectively. EV and FV represent the elementary bases of the chain subspaces E and F in terms of the V basis.

To clarify, it might be helpful to revise the mathematical notation for "chain complex" (see 3.4), and we restate it here with our data structure notation for the user's convenience. To simplify, we often leave out the K from the names and use EV, FE, and CF, with meanings that change depending on the context. For example, FE represents the δ_1 operator; KFE is its sparse matrix.

$$C_\bullet = (C_p, \partial_p) := C_3 \underset{\partial_3}{\overset{\delta_2}{\leftrightarrows}} C_2 \underset{\partial_2}{\overset{\delta_1}{\leftrightarrows}} C_1 \underset{\partial_1}{\overset{\delta_0}{\leftrightarrows}} C_0 \quad \equiv \quad CV \underset{FC}{\overset{CF}{\leftrightarrows}} FV \underset{EF}{\overset{FE}{\leftrightarrows}} EV \underset{VE}{\overset{EV}{\leftrightarrows}} V.$$

Coding 5.2.1 (Algebraic computation of FE $\equiv \delta_1$*)*
The operator FE: E→F is unknown and must be computed and added to the Lar dataset when needed. We may compute it algebraically as FV * VE = EF. By multiplying the two sparse matrices, we get:

```
KFV * KEV'
6×12 SparseArrays.SparseMatrixCSC{Int8, Int64} with 48 stored
    entries:
 2  2  1  2  1  2  1  1  .  .  .  .
 2  1  2  1  2  .  .  .  .  1  1  2
 1  2  2  .  .  1  2  .  1  2  .  1
 1  .  .  2  2  1  .  2  1  .  2  1
 .  1  .  1  .  2  2  2  2  1  1  .
 .  .  1  .  1  .  1  1  2  2  2  2
```

The second term was transposed (') to get multiplication compatibility (size(KFV,2)==size(KEV',1)). Note that each product term (x_{ij}) represents the number of vertices shared between row i and column j of the two matrices (see Figure 5.22).

This matrix must be filtered to the remainder of ÷ `Int8(2)` (integer division by 2 in 8 bits). We get algebraically the binary matrix of the incidence relation `FE`:

```
KFE = KFV * KEV' .÷ Int8(2)                          #=
6×12 SparseArrays.SparseMatrixCSC{Int8, Int64} with 24 stored
  entries:
  1  1  ·  1  ·  1  ·  ·  ·  ·  ·  ·
  1  ·  1  ·  1  ·  ·  ·  ·  ·  ·  1
  ·  1  1  ·  ·  ·  1  ·  ·  1  ·  ·
  ·  ·  ·  1  1  ·  ·  1  ·  ·  1  ·
  ·  ·  ·  ·  ·  1  1  1  1  ·  ·  ·
  ·  ·  ·  ·  ·  ·  ·  ·  1  1  1  1        =#
```

Remark 5.2 We note that setting up the incidence data structures using this algebraic approach has the same space and time complexity as using standard data structures for solid modeling [1]. Conversely, the aggregated queries across several entity items are much faster, since they can be computed by a single Matrix-Vector or Matrix-Matrix product.

Remark 5.3 Let us notice that the operator matrix `FE` for the 3D object `CUBE` has two nonzeros in each column and four nonzeros in each row, so we correctly implemented the incidence relation between faces and edges.

5.2.2 Adiacency operators

Adjacency relations are the sparse symmetric matrices `FV × VF`, `EV × VE`, and `EV × VE`. We have, using `KFV` and `KEV` with transposition for compatibility:

```
A = KFV * KFV';
KFF = (A - Diagonal(A)) .÷ 2     #=
6×6 SparseMatrixCSC{Int64, Int64} with 24 stored entries:
  ·  1  1  1  1  ·
  1  ·  1  1  ·  1
  1  1  ·  ·  1  1
  1  1  ·  ·  1  1
  1  ·  1  1  ·  1
  ·  1  1  1  1  ·        =#
```

Note that `Diagonal` comes from the **LinearAlgebra** package of the Julia ecosystem. Our matrices `KFF`, `KEE`, and `KVV` indicate that in a cube, each face is adjacent to four faces, each edge to four edges, and each vertex to three vertices.

```
B = KEV * KEV'; KEE = (B - Diagonal(B))      #=
12×12 SparseMatrixCSC{Int8, Int64} with 48 stored entries:
 ·  1  1  1  1  ·  ·  ·  ·  ·  ·  ·
 1  ·  1  ·  ·  1  1  ·  ·  ·  ·  ·
 1  1  ·  ·  ·  ·  ·  ·  ·  1  ·  1
 1  ·  ·  ·  1  1  ·  1  ·  ·  ·  ·
 1  ·  ·  1  ·  ·  ·  ·  ·  ·  1  1
 ·  1  ·  1  ·  ·  1  1  ·  ·  ·  ·
 ·  1  ·  ·  ·  1  ·  ·  1  1  ·  ·
 ·  ·  ·  1  ·  1  ·  ·  1  ·  1  ·
 ·  ·  ·  ·  ·  ·  1  1  ·  1  1  ·
 ·  ·  1  ·  ·  ·  1  ·  1  ·  ·  1
 ·  ·  ·  ·  1  ·  ·  1  1  ·  ·  1
 ·  ·  1  ·  1  ·  ·  ·  ·  1  1  ·      =#
```

```
C = KEV' * KEV; KVV = (C - Diagonal(C))      #=
8×8 SparseMatrixCSC{Int8, Int64} with 24 stored entries:
 ·  1  1  ·  ·  ·  1  ·
 1  ·  ·  1  ·  ·  ·  1
 1  ·  ·  1  1  ·  ·  ·
 ·  1  1  ·  ·  1  ·  ·
 ·  ·  1  ·  ·  1  1  ·
 ·  ·  ·  1  1  ·  ·  1
 1  ·  ·  1  ·  ·  ·  1
 ·  1  ·  ·  ·  1  1  ·      =#
```

5.2.3 Atomic decomposition

Here, we briefly recall the main aspects of algebraic topological computation with linear spaces of chains of cells. We start by remarking the implementation as sparse matrices of the two coboundary FE, EV, and two boundary operators EF, VE of the (co)chain complex between three linear spaces of chains and cochains (with identified standard bases) C_2, C_1, C_0, namely, FE, EV, and V.

To be able to perform quite any kind of 3D geometric computation in the broadest area of Solid and Geometric Modeling, in last years enlarged by design and simulation of multi-material patterns, we need a fourth 3-chain space C_3, i.e., a linear space of 3D solid chains, which is not known *a priori*.

Definition 5.3 (Elementary cells) Set of elementary cells of a cellular d-complex is the arrangement $\mathcal{A} = \{\mathcal{A}_p\}$ $(0 \leq p \leq d)$ of open connected p-cells generated by a given collection $\mathcal{S}$ of closed $(d-1)$-manifolds by partitioning the Euclidean d-space, i.e., such that $\cup_p \mathcal{A}_p = \mathbb{E}^d, \mathcal{A}_p \cap \mathcal{A}_q = \emptyset$.

Definition 5.4 (Elementary chain bases) The chain p-basis of an arrangement $\mathcal{A}(\mathcal{S})$ of E^d is the set of singleton chains corresponding one-to-one to elementary p-cells of $\mathcal{A}$.

Definition 5.5 (Chain spaces and subspaces) A chain subspace C_p is the powerset 2^{C_p} of the set C_p. The chain space C is the direct sum of the chain subspaces C_p ($0 \leq p \leq d$). Each chain may be uniquely formed by the sum (mod 2) of the basis elements or the product of a chain by a scalar in $\{0,1\}$.

Using chain concepts, we have discussed in this book how to produce 3D solid models based on a **generalization** of Constructive Solid Geometry (CSG), which is a modeling method used in computer graphics, CAD, and BIM systems. CSG allows the development of complex 3D models by solid operations such as regularized [2] union, intersection, and difference, either of simple primitive shapes or by mixing closed boundary models (Breps).

Therefore, in the following chapters, we discuss a more powerful modeling method in which n-ary Boolean operations are performed without the need for a binary tree of subexpressions, by computing over Breps of elementary solids using a full-featured chain complex generated from the data.

Logic of algebraic geometric computing

We aim here to show the "why" of this algebraic approach. In reading the examples, let us consider the actual minimality of the exemplifications. Our queries on real object arrays, even without GPU, are fast.

In fact, each p-chain subspace C_p, with cardinality $n = \# 2^{C_p}$ is a linear subspace over the field $\mathbb{Z}_2$. Therefore, we can parameterize it using coordinates with scalar numbers in $\{0,1\}$ and perform addition (mod 2). In particular, the coordinates of every $x_p \in C_p$ form a binary vector of length n, which we represent in Julia algorithms as a sparse array of type `Chain`.

Example 5.2.1 (Chain-based computing – 1) We use our `cube` prototype model of Figure 5.22, for its simplicity and compactness. Let's consider the chain $= f_1 + f_6 \in C_2$ given in coordinates as $[1,0,0,0,1]$ and compute the coordinates of its boundary β:

```
β = (KFE' * [1, 0, 0, 0, 0, 1])'   #=
1×12 adjoint(::Vector{Int64}) with eltype Int64:
 1  1  0  1  0  1  0  0  1  1  1  1     =#
```

---

[2] Regularization simplifies the solution of a problem. It is often used to address ill-posed problems or prevent overfitting. Specifically, a point set is regularized when it coincides with the closure of its boundary. The union, intersection, and difference set operations are considered regularized when the resulting set is regularized. In short, it is made regular by removing non-regular subsets (appendices) where boundary points have neighborhoods containing only boundary points.

The result of the sparse matrix-vector computation is $\beta = e_1 + e_2 + e_4 + e_6 + e_9 + e_{10} + e_{11} + e_{12} \in C_1$, transposed here for typographic reasons. □

Example 5.2.2 (Chain-based computing – 2)
Similarly, let's compute the faces incident of edge $e_5 \in C_1$:

```
KFE * [0,0,0,0,1,0,0,0,0,0,0,0])'       #=
1×6 adjoint(::Vector{Int64}) with eltype Int64:
  0  1  0  1  0  0    =#
```

The above matrix-vector product gives $\gamma = f_2 + f_4 \in C_2$. □

Example 5.2.3 (Chain-based computing – 3)
Let's finally compute the boundary 1-cycle[3] of the chain $\alpha = f_1 + f_2 \in C_2$ of two adjacent faces:

```
A = KFE' .* [1, 1, 0, 0, 0, 0]        #=
6×12 SparseMatrixCSC{Int64, Int64} with 8 stored entries:
  1  1  1  1  .  .  .  .  .  .  .  .
  1  .  .  .  1  1  1  .  .  .  .  .
  .  .  .  .  .  .  .  .  .  .  .  .
  .  .  .  .  .  .  .  .  .  .  .  .
  .  .  .  .  .  .  .  .  .  .  .  .
  .  .  .  .  .  .  .  .  .  .  .  =#
α = sum(A, dims=1) .% 2      #= sum by rows mod 2
1×12 Matrix{Int64}:
  0  1  1  1  1  1  1  0  0  0  0  0    =#
```

The result is $\alpha = e_2 + e_3 + e_4 + e_5 + e_6 + e_7 \in C_1$. □

Remark 5.4 (Storage of sparse matrices) Of course, Julia displays the small sparse matrices (`ìsize(m,n)` using the dot symbols, but stores the sparse matrices in efficient $O(nnz)$, i.e., $O(\#\ non\ zeros)$ space.

5.3 Parametric manifold mapping

This section presents concepts essential for understanding computer-generated curves and surfaces, which will be explored in detail. Specifically, we introduce the concepts of curve and surfaceas point-valued functions of one or two variables, respectively. We provide basic mathematics and numerous examples of generating piecewise-linear polyhedral approximations of curves, surfaces, and parametric solids.

---

[3] A chain of whatever dimension is called "cycle" when its boundary is zero (i.e., empty).

Definition 5.6 (Curve) A curve in $\mathbb{E}^d$ is a point-valued mapping $\mathbf{c}(u)$ defined by summing a vector-valued function $\alpha : \mathbb{R} \to \mathbb{R}^d$ of one parameter to the origin of a Cartesian system so that curve points are generated as:

$$\mathbf{c}(u) = \mathbf{o} + \alpha(u), \qquad u \in [a, b] \subset \mathbb{R}.$$

One-dimensional manifolds include lines and circles, but not self-crossing curves, since the intersection point does not satisfy the manifold condition.

The important part of the curve definition is the vector-valued function α, so that sometimes we use the word "curve" for it. From a notational viewpoint, we normally use a bold Latin letter, say $\mathbf{a}$, $\mathbf{b}$ or $\mathbf{c}$, to indicate a map $\mathbb{R} \to \mathbb{E}^d$, and a bold Greek letter, say α, β or γ, to indicate a map $\mathbb{R} \to \mathbb{R}^d$.

Definition 5.7 (Image and domain) The image of a curve is the set $\mathbf{c}[a, b]$ of its $\mathbb{E}^d$ points. The domain of the curve is the parameter interval $[a, b]$, i.e. the set $\{u \in \mathbb{R} : a \leq u \leq b\}$, often normalized to the standard unit interval $[0, 1] \subset \mathbb{R}$.

Definition 5.8 (Surface) A surface in $\mathbb{E}^d$ is a point-valued mapping $\mathbf{S}(u, v)$ summing a vector-valued function $\beta : \mathbb{R}^2 \to \mathbb{R}^d$ of two parameters to the origin so that a point of the surface is generated as:

$$\mathbf{S}(u, v) = \mathbf{o} + \beta(u, v), \qquad (u, v) \in [a, b] \times [c, d] \subset \mathbb{R}^2.$$

A d-manifold is a topological space in which every point has a neighborhood homeomorphic to the interior of a d-sphere in Euclidean space of the same dimension. A 2-manifold is a collection of points forming a topologically closed or open surface. The surface is said to be closed when it has no boundary. Surfaces often arise as the closed boundaries of three-dimensional solids and are used as Brep patches of the solid boundary.

Example 5.3.1 (3D curve) A curve $\mathbf{c}$ in three-dimensional space has 3 coordinate functions, and is often denoted as

$$\mathbf{c}(u) = (x(u), y(u), z(u))^T, \qquad u \in [a, b]$$

where $x(u) = \alpha(u) \cdot \mathbf{e}_1$, $y(u) = \alpha(u) \cdot \mathbf{e}_2$ and $z(u) = \alpha(u) \cdot \mathbf{e}_3$. The tangent vector function $\mathbf{t}$ is defined by

$$\mathbf{t}(u) = D_u \mathbf{c}(u) = D_u \alpha(u) = (x'(u), y'(u), z'(u))^T, \qquad u \in [a, b].$$

Coding 5.3.1 (Toolbox) Some fairly simple generalizations of the Plasm function INTERVAL are given here. In particular, we show a simpler and more terse method to define higher-level functions without recurring to nesting a different function for each level of parameters.

```
INTERVALS(a,b) = n -> T(1)(a)(INTERVALS(b-a)(n))

linearinterval = INTERVALS(2,5)(10);
white = Dict("background_color"=>WHITE)
VIEW(linearinterval, properties=white)

INTERVALS2D(m1,m2,M1,M2) = (n1,n2) ->
    T(1,2)(m1,m2)(INTERVALS(M1-m1)(n1) * INTERVALS(M2-m2)(n2))
intervals2D = INTERVALS2D(.4,.2,1,.8)(5,5);
VIEW(intervals2D, properties=white)

INTERVALS3D(m1,m2,m3,M1,M2,M3) = (n1,n2,n3) ->
    T(1,2,3)(m1,m2,m3)(INTERVALS(M1-m1)(n1) *INTERVALS(M2-m2)(n2)
        * INTERVALS(M3-m3)(n3))
intervals3D = INTERVALS3D(-.4,-.4,0,.6,1,1)(10,10,5);
VIEW(intervals3D, properties=white)
```

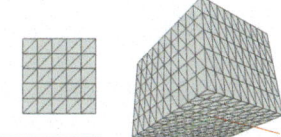

Fig. 5.23 Examples of cuboidal domain generators INTERVALS, INTERVALS2D, and INTERVALS3D.

The mapping machinery

Accordingly, with the functional approach of the Plasm language and its derivation from FL, we often denote a 3D curve, as well as its derivative curves, by using a variable-free notation:

$$\mathbf{c} = (x, y, z)^T$$

with $x = \alpha \cdot \mathbf{e}_1$, $y = \alpha \cdot \mathbf{e}_2$, and $z = \alpha \cdot \mathbf{e}_3$.

It should be clearly understood here that x, y, z are maps $\mathbb{R} \to \mathbb{R}$ and that each $\mathbf{e}_i$ has the constant maps $\underline{0} : \mathbb{R} \to 0$ and $\underline{1} : \mathbb{R} \to 1$ as components.

Analogously, we write a curve $\mathbf{c} : \mathbb{R} \to \mathbb{E}^d$, as $\mathbf{c} = \mathbf{o} + \alpha$, with $\alpha = (\alpha_i)$, where $\alpha_i : \mathbb{R} \to \mathbb{R}$, for all i. The variable-free notation, where functions are directly added and multiplied exactly like numbers, is very useful for quickly implementing curves and surfaces in Plasm, as seen in the following. To have a pretty good understanding of variable-free functions is a prerequisite for user implementation of new curves, surfaces, and parametric solids classes.

Useful maps

Some special maps are needed to perform such variable-free calculus with functions. As the reader knows (see Section 2.3), they are coded in **Plasm**:

identity $\mathrm{id} : \mathbb{R} \to \mathbb{R}; \quad x \mapsto x$
constant $\underline{c} : \mathbb{R} \to \mathbb{R}; \quad x \mapsto c$
selection $\sigma : \{1, \ldots, d\} \times \mathbb{R}^d \to \mathbb{R}; \quad (i, (x_1, \ldots, x_d)) \mapsto x_i$

A computer scientist would probably prefer the following specification, just to point out that function σ (**SEL**) is often used as a partial function (**S1 = SEL (1)**, etc.), i.e. a function which may be applied to a subset of its arguments:

selection $\sigma : \{1, \ldots, d\} \to (\mathbb{R}^d \to \mathbb{R}); \quad i \mapsto ((x_1, \ldots, x_d) \mapsto x_i)$

Actually, the **FL** primitives ID, K and SEL used by the **Plasm** language for identity, constant, and selection have no domain restrictions, and can be applied to any data objects.

Algebraic operations

We must also recall how to perform algebraic operations in the linear algebra of maps $\mathbb{R} \to \mathbb{R}$. For each map $\alpha, \beta : \mathbb{R} \to \mathbb{R}$ and each scalar $a \in \mathbb{R}$

$$\alpha + \beta : u \mapsto \alpha(u) + \beta(u), \quad \alpha\beta : u \mapsto \alpha(u)\beta(u), \quad a\,\beta : u \mapsto a\,\beta(u).$$

Consequently, we have that:

$$\alpha - \beta : u \mapsto \alpha(u) - \beta(u), \quad \alpha/\beta : u \mapsto \alpha(u)/\beta(u).$$

Coding 5.3.2 Therefore, we have extended several infix Julia functions in the Julia **Base** package to work with **Function** arguments:

```
import Base.+, Base.-, Base./, Base.*, Base.+, Base.^, Base.sqrt
-(f::Function, g::Function) = (x...) -> f(x...) - g(x...)
+(f::Function, g::Function) = (x...) -> f(x...) + g(x...)
/(f::Function, g::Function) = (x...) -> f(x...) / g(x...)
*(f::Function, g::Function) = (x...) -> f(x...) * g(x...)
^(f::Function, g::Function) = (x,y) -> f(x)^g(y)
*(pol1::Hpc, pol2::Hpc) = Power(pol1, pol2)
SQRT(f::Function) = x -> f(x)^(1/2)
```

In particular, the infix product * may also work with geometric type **Hpc**, as we have seen in several examples of the last two chapters.   □

Coordinate representation

Finally, let's recall that the coordinate functions of a curve $\alpha = (\alpha_i)$ are maps $\mathbb{R} \to \mathbb{R}$. The variable-free vector notation allows the linear combination of coordinate functions with the basis vectors $\mathbf{e}_i$ of the target space:

$$\alpha = (\alpha_1, \cdots, \alpha_d)^T : \mathbb{R} \to \mathbb{R}^d; \quad u \mapsto \sum_{i=1}^{d} \alpha_i \mathbf{e}_i.$$

Example 5.3.2 (Circular arc) Some different curves are given here. They have the same image in $\mathbb{E}^2$ but different coordinate representation in the space of functions $\mathbb{R} \to \mathbb{R}$. All such curves generate a circular arc of unit radius centered at the origin.

1. trigonometric representation:

$$\alpha(u) = \left(\cos\left(\frac{\pi}{2}u\right), \sin\left(\frac{\pi}{2}u\right) \right)^T \qquad u \in [0,1]$$

2. rational representation:

$$\beta(u) = \left(\frac{1-u^2}{1+u^2}, \frac{2u}{1+u^2} \right)^T \qquad u \in [0,1]$$

3. Cartesian representation:

$$\gamma(u) = \left(u, \sqrt{1-u^2} \right)^T \qquad u \in [0,1]$$

It is possible to verify that the image sets of such curves formally coincide: $\alpha[0,1] = \beta[0,1] = \gamma[0,1]$, but parameterizations are different.

Coding 5.3.3 ((Variable-free circular arc) It may be useful to give the variable-free representation of the three maps on the $[0,1]$ interval shown in previous Example, that is exactly the representation we need to give a Plasm implementation of such maps:

According to the semantics of the MAP function, the curve mapping is applied to all vertices of a simplicial decomposition of the polyhedral domain. But all vertices are represented as sequence (Vector) of coordinates, say $[u]$ for a curve, so that in order to act on u the mapping must necessarily select it from the sequence through the SEL function or its friends S1, S2, etc.

```
α = [cos ∘ (K(π/2) * ID), sin ∘ (K(π/2) * ID)];
β = [(K(1) - SQR)/(K(1) + SQR), (K(2) * ID)/(K(1) + SQR)];
γ = [ID,sqrt ∘ (K(1) - (ID * ID))];

VIEW( MAP(CONS(α)∘S1)(INTERVALS(1)(32)) )
VIEW( MAP(CONS(β)∘S1)(INTERVALS(1)(32)) )
```

```
VIEW( MAP(CONS(γ)∘S1)(INTERVALS(1)(32)) )
```

The circle segment representations of Example 5.3.2 are directly used in the `Plasm` implementation of `FL` curves in this script.   □

To understand the implementation, notice that we generate a polyhedral complex by mapping the curve `Vector` function (either α, β or γ of Example 5.3.2) on the polyhedral representation of the $[0, 1] \subset \mathbb{R}$ domain.

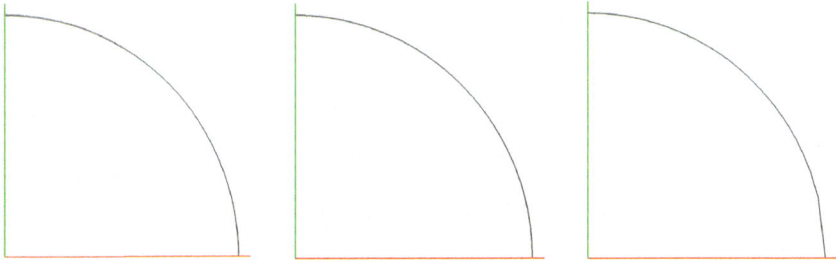

Fig. 5.24 Circular arc segment implementation: (a) trigonometric `alpha` curve; (b) rational `beta` curve; (c) Cartesian `gamma` curve. Note the bad approximation for small angles in the Cartesian representation.

In particular, let's explain the `alpha` curve visualization. `INTERVALS(1)` `(32)` is an `Hpc` object used as domain by the `MAP`. This one applies the vector function `CONS(alpha)∘S1` to all the domain vertex points. Each domain vertex is a coordinate vector (just one for a curve). Consequently, the selector `S1` extracts the numeric value from the point array and passes it to `CONS(alpha)`. According to the `FL` semantics of vector functions, it provides the coordinate as an argument to all the `alpha` component functions. Finally, the output of `MAP` returns the geometric curve as an object of `Hpc` type.

`Plasm` uses this semantic mechanism for all its implementations of curves and surfaces, resulting in a natural and straightforward coding solution for the higher-level `MAP` that generates both standard and transfinite d-manifolds.

A few other general utility functions are used in the remainder of this chapter. In particular, the `SQR` function that returns its squared input.

Remark 5.5 (Coordinate functions) Let us finally note that, e.g., `alpha` is a sequence of coordinate functions. Conversely, `CONS(alpha)` is the correct implementation of the vector-valued function α, which only can be composed with other functions, say `S1`. Notice also that `SQR = ID*ID` (square) is a `Plasm` implementation of the id^2 function and that the language explicitly requires the operator `*` to denote also the product of functions.

Example 5.3.3 (Full 1-Circle) The following `MAP` expression generates a polygonal approximation with 36 line segments of the unit circle centered in the origin, i.e. of the set $\mathbf{c}[0, 2\pi]$, with $\mathbf{c}(u) = (0, 0) + (\cos u, \sin u)$.

```
domain = INTERVALS(2π)(36)
F = CONS([cos, sin]) ∘ S1
# FL vector function: F = [f1,f2]; F(x) = f1(x),f2(x)
circle = MAP(F)(domain) #=
Hpc(MatrixNd(3), Hpc(MatrixNd(3), Geometry([[1.0, 0.0],
    [0.984807753012208, ... =#
VIEW( circle, properties=Dict("background_color"=>WHITE) )
```

A manifold curve with different radiuses may be obtained via uniform scaling transformation of the `Hpc` value returned by the `MAP` functional.

```
circle = MAP(F)(domain)
```

This example illustrates the powerful functional mechanism behind the `Plasm` generation of curves and surfaces, as well as the roles of `CONS` and `SEL(d)` ≡ `Sd` in implementing variable-free manifold functions. It serves as the model for `Plasm`'s implementation of surfaces using curves and trivariate manifolds utilizing surfaces, and so forth.   □

5.3.1 Curve generation methods

This section discusses several geometric programming methods for implementing curve, surface, and solid generating functions. Among the various classes of mathematical techniques, we discuss the parametric ones. They are both more used by industrial CAD platforms and more adaptable to being implemented by a functional language like `Plasm`.

We make extensive use, in quite all the discussed examples, of the `MAP` combinator (function of the second order), to be applied first to a vector-valued function coding the parametric expression of one, two, or three coordinate function $\mathbb{R} \to \mathbb{R}$ to apply to domain vertices, then to a polyhedral complex of `Hpc` type to transform it correspondingly. The `MAP` output will always be generated using the `Hpc` type.

Naturally, the various curved shapes created by `Plasm`, similar to those produced by any platform for geometric or graphical computing or visualization, are a linear approximation of the original curved shape. The piecewise-linear approximation is achieved using 1-, 2-, or 3-simplices —triangles with hardware optimization.

Remark 5.6 (Transfinite implementation) In some cases, the name of a class of geometry-generating curves, say `Bezier` or `Hermite`, for instance, will be followed by a selector function (e.g., `S1` or `S2` or `S3`) since some manifold-generating functions are implemented in a "transfinite" way and may accept as input either control points or vector-valued control functions.

Definition 5.9 (Curve definition) A curve $\mathbf{c} : dom \to \mathbb{E}^d$ is a vector-valued function of a parameter $u \in \mathsf{dom} \subset \mathbb{R}$, where it is often $\mathsf{dom} = [0,1]$. The points supported by the curve image may belong to any $\mathbb{E}^d$, with $d \geq 1$. In every case, $\mathbf{c}$ will contain d coordinate functions.

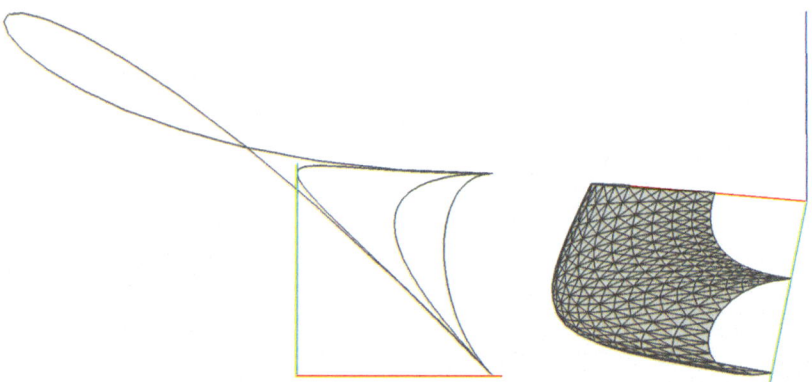

Fig. 5.25 Curve examples: (a) **Hermite** curves with same extreme point, and scaled extreme tangents (see Example 5.3.5); (b) **transfinite Hermite** interpolation of two curves $c1$ and $c2$ with two fields of constant derivatives $[1,1,1]$ and $[-1,-1,-1]$ (see Example 5.3.6).

Of course, the primary **Plasm** generator of curves is the second order **MAP** (**curve :: Function**)(**domain :: Hpc**) combinator, that we already know.

Remark 5.7 (Semantics of MAP operator) The first function argument of the **MAP** combinator extracts the **1D** points from the polyhedral complex decomposition of the curve **domain**, which is the second argument of the **MAP** operator. The mapped points are returned embedded as a transformed **domain** object, changing the domain shape and the dimension of the embedding Euclidean space but not its topology, including connectedness.

Remark 5.8 (Control curve of a robot) It is interesting to note that the control curve of an industrial robot with six degrees of freedom (for example, joint angles) is a **MAP** of a vector function applied to 1D points, and returns a 1D manifold in $\mathbb{E}^6$, where points have six coordinates. The reader is warmly invited to produce such a curve, produced by a sequence of 6D control points (see belove, Bézier curves).

Example 5.3.4 ((Hermite cubic curve definition) In the case of Hermite cubic curves, four constraints impose the passage through assigned initial and final points $(\mathsf{P_1},\mathsf{P_2})$, with assigned initial and final tangents $(\mathsf{T_1},\mathsf{T_2})$. Four constraints, hence degree 3. The **Plasm** native code is given below.

```
function HERMITE(args)
   P₁,P₂, T₁,T₂ = args
   return CUBICHERMITE(S1)([P₁,P₂, T₁,T₂])
end
```

Example 5.3.5 (Hermite cubic curves) (See Figure 5.25a) Here, we observe four CUBICHERMITE curves with increasing values of extreme tangents. At some point, the curve self-intersects, which negatively impacts geometric design. The low-degree Hermite curve is useful in splines, where long sequences of points and tangents are interpolated.

```
domain = INTERVALS(1.0)(20)
VIEW( STRUCT([    # aggregation of four curves
MAP(CUBICHERMITE(S1)([[1,0],[1,1],[ -1, 1],[ 1,0]]))(domain),
MAP(CUBICHERMITE(S1)([[1,0],[1,1],[ -2, 2],[ 2,0]]))(domain),
MAP(CUBICHERMITE(S1)([[1,0],[1,1],[ -4, 4],[ 4,0]]))(domain),
MAP(CUBICHERMITE(S1)([[1,0],[1,1],[-10,10],[10,0]]))(domain) ]))
```

Example 5.3.6 ((Transfinite Hermite surface) (See Figure 5.25b)

The transfinite Hermite implementation of Julia Plasm is used here. Let us note that c1 and c2 are Plasm curves since their expressions contain the S1 selector. Conversely, the cubic object is a surface since it contains S2. The two extreme curves are interpolated here. Of course, the domain is 2D, obtained as the Cartesian product of two polyhedral complexes of dimension one.

```
c1 = CUBICHERMITE(S1)([[1  ,0,0],[0  ,1,0],[0,3,0],[-3,0,0]])
c2 = CUBICHERMITE(S1)([[0.5,0,0],[0,0.5,0],[0,1,0],[-1,0,0]])
cubic = CUBICHERMITE(S2)([c1,c2,[1,1,1],[-1,-1,-1]])
# cubic : domain ⊂ E²
domain = INTERVALS(1.0)(14) * INTERVALS(1.0)(14)
VIEW( MAP(cubic)(domain)), properties=white)
```

The implementation of Julia Plasm doesn't always require a CONS combinator, often implicit in Julia code.   □

Definition 5.10 (Bézier curve) Bézier curves of degree n in $\mathbb{E}^d$ are fully defined by an ordered sequence of $n+1$ control d-points.

These curves have several valuable properties, including (a) interpolating the first and last control points and (b) being contained within the convex hull of their control points.

With two control points, we obtain a linear curve; with three, we achieve a quadratic curve; with four, we create a cubic curve; and so forth. A cubic

Bézier curve may have one flex (bend) point, a quartic curve has two, and an n-degree curve has $n-2$ such points, making the shape flexible. For numerical reasons, it is advisable to keep the curve degree low.

Example 5.3.7 (Bézier quartic curve)
This script shows the first simple example of the important class of Bézier curves. The curve `domain` is partitioned into 32 segments of the interval $[0,1]$. The degree of the curve is 4 because there are five control points. See Figure 5.26a.

```
controlpoints = [[0,0],[1,0],[0.5,2],[2,1],[2,2]]
curve   = BEZIER(S1)(controlpoints)
domain = INTERVALS(1.0)(32)
VIEW(STRUCT(MAP(curve)(domain), POLYLINE(controlpoints), FRAME2)
    , properties=white))
```

Definition 5.11 (Spline curve) A spline is a continuous curve constructed to pass through a given set of points with a certain number of continuous derivatives. Splines may either interpolate or approximate the control points. Each control point's coordinate generates a coordinate function of the curve.

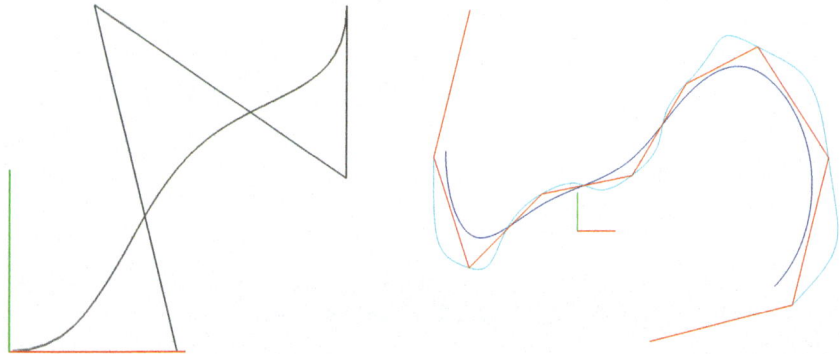

Fig. 5.26 Curve examples: (a) **Bézier** curve of the fourth degree, depending on five control points (see Example 5.3.7); (b) Cardinal (green) and **uniform B-spline** (blue) cubic curves, together with the polyline (red) of their control points (see Example 5.3.8).

Example 5.3.8 ((Cubic spline curves) (See Figure 5.26b) Here, we show two examples of 2D cubic splines, piecewise curves defined by joining several segments of a low-degree curve. Each consecutive quadruple of points in cubic splines represents a different curve segment. In Figure 5.26b, we see together the `polyline` of control points, a cardinal interpolating curve (green), and a uniform B-spline approximating curve (blue). Let us notice that the two extreme control `points` are only given to set the direction and size of extreme tangent vectors to the first and last spline segments.

```
domain = INTERVALS(1.0)(20)
points = [[-3.0, 6.0], [-4.0, 2.0], [-3.0,-1.0], [-1.0, 1.0],
    [1.5, 1.5], [3.0, 4.0], [5.0, 5.0], [7.0, 2.0], [6.0,-2.0],
    [2.0,-3.0]]
polyline = COLOR(RED)(POLYLINE(points))
cardinal_spline = COLOR(CYAN)(SPLINE(CUBICCARDINAL(domain))(
    points));
uniform_Bspline = COLOR(BLUE)(SPLINE(CUBICUBSPLINE(domain))(
    points));
VIEW( STRUCT( polyline, cardinal_spline, uniform_Bspline ) )
```

Example 5.3.9 (Exact* ELLIPSE *definition) (See Figure 5.27a)
The exact ellipse curve, whose quarter is generated as an exact quadratic rational Bézier curve, is shown here. The example shows one of two ellipses degenerating into a rational circle. There are three control points because each quarter is a cubic Bezier curve. Note the mirroring of partial objects.

```
function ELLIPSE(args :: Vector{Float64})
    A, B = args
    function ELLIPSE0(N :: Int)
        C = sqrt(2)/2
        curves = RATIONALBEZIER([[A, 0.0, 1.0], [A*C, B*C, C],
        [0.0, B, 1.0]])
        quarter = MAP(curves)((INTERVALS(1.0)(N)))
        half = STRUCT([quarter, S(2)(-1.0)(quarter)])
        return STRUCT([half, S(1)(-1.0)(half)])
    end
    return ELLIPSE0
end
```

```
VIEW( STRUCT( ELLIPSE([1.0,2.0])(15), ELLIPSE([2.0,2.0])(15)),
    properties=white  )
```

The aggregation of two exact `RATIONALBEZIER` ellipses with different parameter values. □

Example 5.3.10 (Non-uniform cubic B-spline) (Figure 5.27a)
This class of splines has consecutive curve segments defined between pairs of consecutive parameter values, called knots. To interpolate the extreme control points, the extreme knots must have multiplicity equal to $degree + 1$.

In this example, for the number of control points, we have $n = m - (degree + 1) = 14 - 4 = 10$, where $m = n + degree + 1$ is the number of knots. The number of cubic curve segments is $n - degree = 7$. The knots are not

necessarily integer. The knots (in parameter space) may be subject to affine
maps (translation and scaling).   □

```
degree = 3
ControlPoints = [[0.,0],[-
    1,2],[1,4],[2,3],[1,1],[1,2],[2.5,1],[2.5,3],[4,4],[5,0]]
# 10 control points
knots=[0,0,0,0, 1,2,3,4,5,6, 7,7,7,7] # 14 knots
VIEW(DISPLAYNUBSPLINE(degree, knots, ControlPoints), properties=
    white)
```

Example 5.3.11 (Procedure to fix a spline parameter dataset)
As a rule of thumb, start fixing the n control points and, hence, the approximate spline shape by viewing their POLYLINE. Choose the degree and,
consequently, the number of curve segments and knots ($m = n + degree + 1$).
Finally, write the knot sequence starting from zero (if you want to interpolate
the first and last control points) and repeat $n + 1$ times, then append a knot
at a time until it reaches the number of segments, and repeat the previous
knot (Example 5.3.10). Other examples of NUBSPLINE (nonrational NURBS)
with ten control points and different examples of degrees and segments are
given below: □

$$n = 10; degree = 2; segments = 8; m = 13; knots = 0, 0, 0, 1, 2, 3, 4, 5, 6, 7, 8, 8, 8$$
$$n = 10; degree = 4; segments = 6; m = 15; knots = 0, 0, 0, 0, 0, 1, 2, 3, 4, 5, 6, 6, 6, 6, 6$$

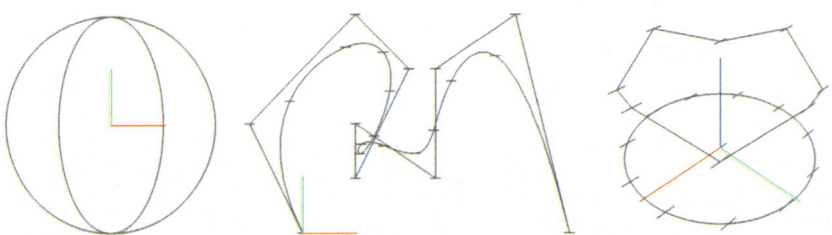

Fig. 5.27 Non-uniform B-spline examples: (a) **exact ellipse** and circle curves via NUBS
(nonrational) curve spline with two parameters (see Example 5.3.9); (b) (c) exact **NURBS**
(rational) circle of unitary radius (see Example 5.3.12); NUB-spline curve defined in
Example 5.3.10.

Example 5.3.12 ((Quadratic NURBS circle) See Figure 5.27a. There
are several ways to construct an exact 1D circle with NURBS curves of
degrees two, three, and four [2]. In the case implemented in this example
(see also Figure 5.27c), we have $n = 9(2$ coincident$), degree = 2, m = 12$
satisfying our parameter relation formulas:

```
knots = [0,0,0,1,1,2,2,3,3,4,4,4]
_p = sqrt(2)/2
controlpoints = [[-1,0,1], [-_p,_p,_p], [0,1,1], [_p,_p,_p],
    [1,0,1], [_p,-_p,_p], [0,-1,1], [-_p,-_p,_p], [-1,0,1]]
VIEW(DISPLAYNURBSPLINE([2,knots,controlpoints]))
```

5.3.2 Surface generation methods

Parametric curves can be combined to create parametric surfaces, solids, or higher-dimensional manifolds. In this section, we introduce generative methods for parametric surfaces as point-valued functions of two real parameters while presenting extensions to certain three-variate solids.

Practical classes of surfaces are discussed, such as the profile product surfaces, which include rotational surfaces; the ruled surfaces, encompassing generalized cylinders and cones; and the surfaces generated by tensor product of curves or splines. The straightforward combinatorial semantics of the Plasm language may provide valuable insights to the reader regarding tensor operations and transfinite combinations.

To generate a 2D surface, i.e., a two-dimensional manifold embedded in $\mathbb{E}^3$, we have to produce a vector-valued function with three coordinate functions $S(u, v) = [x(u, v), y(u, v), z(u, v)]$.

Definition 5.12 (Transfinite maps) Transfinite maps are interpolation or approximation operators that combine not only to a discrete set of points but also to a non-numerable set of points, such as functions of curves or functions that generate maps. The concept was introduced by Gordon and Coons in the 1960s, during the dawn of CAD systems and research. Embedded in classic PLaSM (1995), they were mainly used in [3].

Example 5.3.13 ((Parametric sphere) (See Figure 5.1.2) A named sphere with unit radius centered at the origin is presented again. The default domain represents a cell decomposition of the $[0, \pi] \times [0, 2\pi]$ interval into 16×32 quads. The SPHERE generating function in 3D is quickly formulated as Hpc object with empty triangles at the poles and double vertices and edges at the v frontier of the u, v discretized domain.

To extract from this generator the exact discrete topology, a conversion to Lar data structure is needed, as done by sphere = LAR(SPHERE(1.0) ([16,32])).

```
function SPHERE(radius=1.0::Number)
   function SPHERE0(subds=[16,32]::Vector{Int})
      N, M = subds
```

```
        dom = INTERVALS(π)(N) * INTERVALS(2π)(M)
        domain = T(1,2)(-π/2,-π)(dom)
        fx = p -> radius * (-cos(p[1])) * sin(p[2])
        fy = p -> radius * cos(p[1]) * cos(p[2])
        fz = p -> radius * sin(p[1])
        return MAP([fx, fy, fz])(domain)
    end
    return SPHERE0
end

VIEW( SPHERE(1.0)([16,32]) )
```

This text discusses a unit radius sphere surface, which is piecewise-linearly approximated by 16×32 quads (default). To curate the `Hpc` surface from empty triangles and double vertices and edges, the sphere representation must be ported to the `Lar` data structure.   □

Remark 5.9 (Curation of inexactly mapped manifolds) When generating a curved manifold through the `MAP` of a 2D grid domain, the resulting curved object is often not topologically closed (i.e., a 2-cycle) because by changing the topology some 2-cells may become void while some 1-cells may become duplicated. The outcome may be visually precise and quickly produced, yet it can be topologically incorrect. Plasm provides an easy curation method: convert the `Hpc` value into a `Lar`, topologically exact value. In this case, we have: V - E + F = 482 - 992 + 512 = 2

Example 5.3.14 (Parametric torus) (See Figure 5.3.2a) The parametric equation of the torus surface in 3D can be computed by rotating a parametric semicircle in the plane x=0 about the z-axis. The reader may construct it for exercise. The three coordinate functions are denoted as `fx`, `fy`, `fz`, and the `domain` is $[0, 2\pi] \times [0, 2\pi]$. The surface depends on the minor and major radii (`r1`, `r2`).

An implementation of the `TORUS` function of the second order is given below. In case of need for exact topology, convert to `Lar` via `LAR()`.

```
function TORUS(radii=[1.0,2.0]::Vector{Number})
    r1, r2 = radii
    function TORUS0(subds=[16,32]::Vector{Int})
        N, M = subds
        a = 0.5 * (r2-r1)
        c = 0.5 * (r1+r2)
        domain = INTERVALS(2*pi)(N) * INTERVALS(2*pi)(M)
        fx = p -> (c + a*cos(p[2])) * cos(p[1])
        fy = p -> (c + a*cos(p[2])) * sin(p[1])
        fz = p -> a * sin(p[2])
        return MAP([fx, fy, fz])(domain)
    end
    return TORUS0
end
```

An application example of the TORUS function to actual parameters follows. [1.0,2.0] are the smaller and bigger radii.

```
VIEW( TORUS([1.0,2.0])([36,12]) )
```

Fig. 5.28 Closed and open surface examples: (a) TORUS()() surface ; (b) RULEDSURFACE (bilinear, in this case).

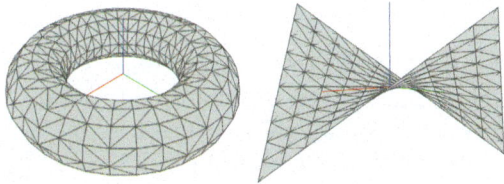

Example 5.3.15 ((Cubic Bézier surface generated by Bézier curves)
(See Figure 5.29a) A transfinite surface generation is presented here. The standard (bi)cubic Bézier surface requires a 4×4 tensor of point parameters; in contrast, the cubic Bézier manifold in this example is parameterized by the four Bézier curves C0, C1, C2, C3 of various degrees listed below.

The presence of S1 selectors in parametric curves and of S2 in the surface generator expression is worth mentioning.

```
C0 = BEZIER(S1)([[0,0,0],[10,0,0]])
C1 = BEZIER(S1)([[0,2,0],[8,3,0],[9,2,0]])
C2 = BEZIER(S1)([[0,4,1],[7,5,-1],[8,5,1],[12,4,0]])
C3 = BEZIER(S1)([[0,6,0],[9,6,3],[10,6,-1]])
domain = INTERVALS(1.0)(10) * INTERVALS(1.0)(10)
VIEW( MAP(BEZIER(S2)([C0,C1,C2,C3]))(domain) )
VIEW( STRUCT([C0,C1,C2,C3]))(domain) )
```

Fig. 5.29 Transfinite surfaces: (a) Bezier surface defined by four different degrees boundary curves; (b) Coons patch defined by four generic boundary curves (S1, S2 are used to select the **domain** points) of first/second coordinate.

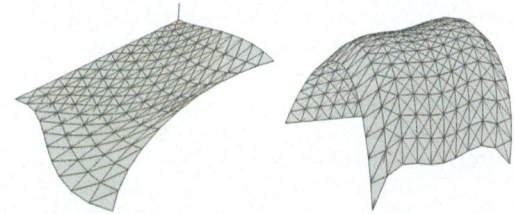

Example 5.3.16 ((Coons' patch) (See Figure 5.29b)
The Coons patch provides a method for constructing a surface supported on a given contour when the contour is composed of 4 arcs of curves: Su0,Su1, Sv0,Sv1. They are created using the first and second coordinates of control points, respectively.

```
Su0 = BEZIER(S1)([[0,0,0],[2,0,0]])
Su1 = BEZIER(S1)([[0,2,0],[2.5,2,3],[1,2,-2],[2.5,2,3],[2,2,0]])
Sv0 = BEZIER(S2)([[0,0,0],[0,0,1.5],[0,2,1],[0,1.5,0]])
Sv1 = BEZIER(S2)([[1,0,0],[1.8,1.5,3],[1,2,0]])
domain = INTERVALS(1.0)(10) * INTERVALS(1.0)(20)
VIEW( MAP(COONSPATCH([Su0,Su1,Sv0,Sv1]))(domain) )
```

Example 5.3.17 ((Ruled surface) (See Figure 5.3.2d)

The RULEDSURFACE function builds a ruled 2-manifold between two given curves. Of course, the domain must be 2D and embedded in an Euclidean space of at least dimension $d = 2$.

```
alpha = point -> [point[1], point[1], 0 ]
beta = point -> [ -1, +1, point[1] ]
domain = T(1,2)(-1,-1)(INTERVALS(2.0)(10) * INTERVALS(2.0)(10))
VIEW( MAP(RULEDSURFACE([alpha,beta]))(domain) )
```

The two parametric functions alpha and beta define a patch of hyperbolic paraboloid surface, with two parametric maps $\mathbb{E}^2 \to \mathbb{E}^3$ such that $(u,v) \mapsto [u,u,0]$ and $(u,v) \mapsto [-1,1,u]$ over the domain patch

$$T(1,2)(-1.-1)([0,2] \times [0,2]) = ([-1,1] \times [-1,1]) \subset \mathbb{E}^2$$

. A hyperbolic paraboloid is a saddle surface (see Figure 5.3.2), as its Gauss curvature is negative at every point. Therefore, although it is a ruled surface, it is not developable. □

Example 5.3.18 ((Rotational surface and solid) (See Figure 5.3.2b) The

primitive function ROTATIONALSURFACE is a profile product where the section function is already known. The control points of the profile curve are set to start with a horizontal tangent vector at $u = 0$. Of course, the profile curve rules entirely the surface shape. Figures 5.3.2 and 5.3.2 show the rotational surface and the thin solid derived from it.

```
profile = BEZIER(S1)([[0,0,0],[3,0,0],[4,0,1],[4,0,4]])
surface = ROTATIONALSURFACE(profile)
domain = INTERVALS(1.0)(10) * INTERVALS(2*PI)(30)
VIEW( MAP(surface)(domain), properties=white )
VIEW( MAP(THINSOLID(surface))(domain*INTERVALS(0.25)(1)) )
```

Example 5.3.19 ((Cylindrical surface and solid) (See Figure 5.3.2d)

A cylindrical surface is implemented here by the CYLINDRICALSURFACE primitive by the tensor product of a Bézier curve alpha in $z = 0$ and a vertical vector [0,0,1].

```
alpha = BEZIER(S1)([[1,1,0],[-1,1,0],[1,-1,0],[-1,-1,0]])
Udomain = INTERVALS(1.0)(20)
Vdomain = INTERVALS(1.0)(2)
domain = Udomain * Vdomain
fn = CYLINDRICALSURFACE([alpha,[0,0,1]])
VIEW(MAP(fn)(domain))
VIEW(MAP(THINSOLID(fn))(domain * Q(0.1)) )
```

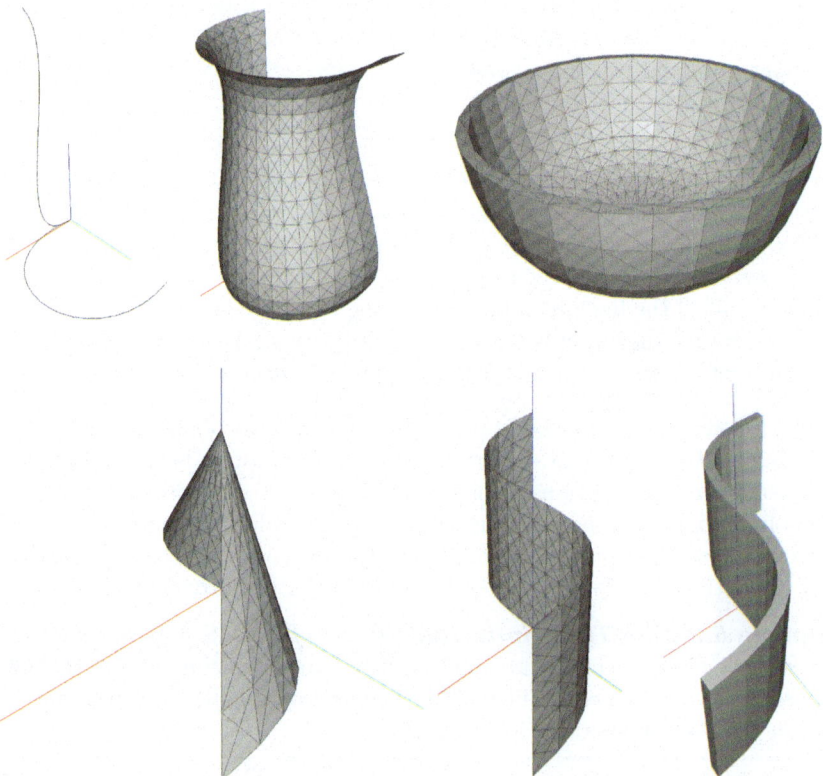

Fig. 5.30 Surface examples: (a) profile-product patch, a transfinite combination of two curves; (b) thin solid associated to a rotational surface (see Example 5.3.18); (c) generalized conical surface patch generated by a Bézier curve and an apex point (see Example 5.3.20); (d) generalized cylindrical surface and solid, generated by a cubic Bézier curve and by **CYLINDRICALSURFACE** and **THINSOLID** (see Example 5.3.19).

Example 5.3.20 ((Conical surface) (See Figure 5.3.2c) A conical surface is a ruled surface whose straight lines join a base curve, the beta Bézier
in this example, and a fixed point, in this case [0,0,1] out of the curve.

```
domain = Power(INTERVALS(1.0)(20),INTERVALS(1.0)(6))
beta = BEZIER(S1)([ [1,1,0],[-1,1,0],[1,-1,0],[-1,-1,0] ])
VIEW( MAP(CONICALSURFACE([[0,0,1],beta]))(domain) )
```

Definition 5.13 (Tensor product surface) The parametric surfaces known
as tensor product surfaces serve as a bivariate generalization of parametric
curves. For example, polynomial surfaces represent a linear combination of
either points or vectors using an appropriate basis of bivariate polynomial
functions obtained by tensor product of two univariate bases.

Remark 5.10 (Tensor product surfaces)
Let's remember that the matrix of the tensor product of two function bases
(say, univariate Hermite o Bézier) can be obtained as the $m \times n$ product
matrix of the first used as $m \times 1$ column-vector times the second used as
$1 \times n$ row-vector (or vice-versa).
    The result of the linear combination of the bivariate basis with a $m \times n$ set
of geometric handles (control point or vectors) is a point-valued function of
two parameters, whose image set is the surface considered as a point locus.

    The following examples show more than one generation method in action,
encouraging the reader to reason about and distinguish between bivariate
parametric surface functions, coordinate functions, transfinite combinations,
and polynomial tensor product functions, which are directly mapped on $m \times n$
matrices of control points. In some cases, the interested reader may directly
refer [3] at the open-source implementation of the primitive functions.

Example 5.3.21 ((Transfinite Hermite surface) (See Figure 5.25b) In
this case, we present a cubic transfinite Hermite interpolating two CUBICHERMITE
curves, and with two constant (u) fields (initial and final), of partial derivatives in the v directions (see Figure 5.25b).

```
c1 = CUBICHERMITE(S1)([[1,0,0],[0,1,0],[0,3,0],[-3,0,0]])
c2 = CUBICHERMITE(S1)([[0.5,0,0],[0,0.5,0],[0,1,0],[-1,0,0]])
sur3 = CUBICHERMITE(S2)([c1,c2,[1,1,1],[-1,-1,-1]])
domain = INTERVALS(1.0)(14) * INTERVALS(1.0)(14)
VIEW( MAP(sur3)(domain) )
```

Example 5.3.22 ((Bilinear surface) (See Figure 5.3.2b) Tensor product
surface function generated by the Plasm operator TENSORPRODUCT with arguments two arrays of 3D controlpoints, each containing two points, the first

and the last of a linear curve in one parametric direction. The same applies to the transposed tensor in the other parametric coordinate direction. The ruled surface generated by four given points interpolates the (four) extreme points of domain $[0, 1] \times [0, 1]$.

```
controlpoints = [
    [[0.0, 0.0, 0.0], [2.0, -4.0, 2.0]],
    [[0.0, 3.0, 1.0], [4.0, 0.0, 0.0]]
]
domain = Power(INTERVALS(1.0)(10), INTERVALS(1.0)(10))
mapping = BILINEARSURFACE(controlpoints)
VIEW(MAP(mapping)(domain), title="TestBilinarSurface")
```

The Plasm implementation of the BILINEARSURFACE follows. BERNSTEINBASIS of first degree is just the convex combinator $[1 - u, u]$ of two control points:

```
function BILINEARSURFACE(controlpoints)
    return TENSORPRODSURFACE([BERNSTEINBASIS(S1)(1),
        BERNSTEINBASIS(S1)(1)])(controlpoints)
end
```

Example 5.3.23 ((Biquadratic surface) (See Figure 5.3.2b) The operator BIQUADRATICSURFACE is generated by TENSORPRODSURFACE of two instances of quadratic Bézier, given as $basis = [(1 - u)^2, \ 2(1 - u)u, \ u^2] \ (0 \le u \le 1)$. Note the tensor of controlpoints arranged as three arrays of three 3D points.

```
controlpoints = [
    [[0,0,0],[2,0,1],[3,1,1]],
    [[1,3,-1],[3,2,0],[4,2,0]],
    [[0,9,0],[2,5,1],[3,3,2]]]
domain = INTERVALS(1.0)(10) * INTERVALS(1.0)(10)
surfaces = BIQUADRATICSURFACE(controlpoints)
VIEW(MAP(surfaces)(domain) )
```

Example 5.3.24 ((Bicubic Hermite surface) (See Figure 5.3.2a) This surface is an extension of the Hermite curve. It is defined by 16 three-dimensional parameters (4 endpoints, eight tangent vectors at endpoints, and four twist vectors). You have (a) position vectors, the four corner points; (b) tangent vectors, the tangent vectors in u,w directions at the four position vectors of the surface; (c) twist vectors, the twist vectors at all four position vectors of the surface. The elements of the controlpoints tensor can be put in correspondence with the function of the basis tensor:

$$
\mathcal{G}_{hh} = \begin{pmatrix}
\mathbf{S}(0,0) & \mathbf{S}(0,1) & \mathbf{S}^u(0,0) & \mathbf{S}^u(0,1) \\
\mathbf{S}(1,0) & \mathbf{S}(1,1) & \mathbf{S}^u(1,0) & \mathbf{S}^u(1,1) \\
\\
\mathbf{S}^v(0,0) & \mathbf{S}^v(0,1) & \mathbf{S}^{uv}(0,0) & \mathbf{S}^{uv}(0,1) \\
\mathbf{S}^v(1,0) & \mathbf{S}^v(1,1) & \mathbf{S}^{uv}(1,0) & \mathbf{S}^{uv}(1,1)
\end{pmatrix}
$$

It is easy to see that the `controlpoints` geometric tensor $\mathcal{G}_{hh}$ can be decomposed into four submatrices corresponding to the surface $\mathbf{S}$, its first partial derivatives $\mathbf{S}^u$, $\mathbf{S}^v$ and its mixed second partial derivative $\mathbf{S}^{uv}$, called twist vectors, respectively evaluated in the four corners of $[0,1]^2$ domain:

```
controlpoints = [
    [[0,0,0 ],[2,0,1], [3,1,1],[4,1,1]],
    [[1,3,-1],[3,2,0], [4,2,0],[4,2,0]],

    [[0,4,0 ],[2,4,1], [3,3,2],[5,3,2]],
    [[0,6,0 ],[2,5,1], [3,4,1],[4,4,0]]
]
domain = INTERVALS(1.0)(10) * INTERVALS(1.0)(10)
surfaces = HERMITESURFACE(controlpoints)
VIEW( MAP(surfaces)(domain), properties=Dict("background_color"
    =>WHITE) )
```

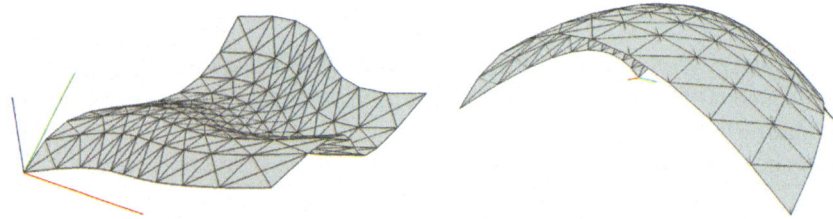

Fig. 5.31 Bicubic surfaces: (a) bicubic HERMITESURFACE; (b) bicubic BEZIERSURFACE.

Example 5.3.25 ((Bicubic Beziér surface) See Figure 5.3.2b.

Conversely, the generating `controlpoints` of a cubic Bézier surface in 3D are a 4×4 matrix of 3D point coordinates. This kind of surface interpolates the four extreme points, the two cubic curves generated by the first and last row, and the two cubic curves generated by the first and last column of the tensor of points.

```
controlpoints = [
    [[ 0,0,0],[0 ,3  ,4],[0, 6, 3],[0,10,0]],
    [[ 3,0,2],[2 ,2.5,5],[3, 6, 5],[4,8,2]],
    [[ 6,0,2],[8 ,3 , 5],[7,6,4.5],[6,10,2.5]],
    [[10,0,0],[11,3  ,4],[11,6, 3],[10,9,0]]
]
domain = INTERVALS(1.0)(10) * INTERVALS(1.0)(10)
surfaces = BEZIERSURFACE(controlpoints)
VIEW( MAP(surfaces)(domain) )
```

5.3.3 Solid generation methods

This section explores some well-known and unusual techniques for creating geometric models of solid objects. They range from constrained Delaunay triangulation to Minkowski's addition of convex polygons and some generalization of the Cartesian product. Regarding the output, we note that the Hpc type represents objects as sets of convex hulls of vertices. Conversely, the Lar type stores the entire cellular complex.

From polyline 2D to polygon

Definition 5.14 (Polyline and polygon) A polyline (1-chain) is a series of points connected by line segments drawn between consecutive points. A polyline is described as closed when it has no boundary points. A polygon (2-chain) is a collection of 2D points whose boundary forms a closed polyline.

In general, producing a 2D polygon object from its 1D boundary–made by one or more closed polylines–is a nontrivial task when starting from a nonintersecting set of 1D boundary polylines. Conversely, in Plasm, this task is relatively easy, using a fairly simple algorithm [3], as shown below.

Example 5.3.26 (From polyline 2D to polygon)
Let us start by defining a "unit test," a simple step-by-step method of implementing the coding solution. The input dataset is a triple of closed cycles of 2D points, i.e., a sequence of consecutive points closed by repeating the first point. This trick is commonly used in Computer Graphics to generate closed polylines. In applying the POLYLINE primitive, you get an array of three closed polylines (1-cycles). STRUCT transforms their array into a single Hpc value. See Figure 5.3.5

```
polylines = [
    [[0,0],[4,2],[2.5,3],[4,5],[2,5],[0,3],[-3,3],[0,0]],
    [[0,3],[0,1],[2,2],[2,4],[0,3]],
    [[2,2],[1,3],[1,2],[2,2]] ]
pol = STRUCT(AA(POLYLINE)(polylines))
VIEWCOMPLEX(LAR(pol))    # 1-complex from (closed) points lists
```

Coding 5.3.4 (SOLIDIFY implementation) The operator function is defined by connecting any 1-cell (edge in EV) with its image projected in the direction of a coordinate axis. It then performs the XOR operation on the generated stripes, meaning the union minus the intersection of the non-degenerating 2-cells that have been previously filtered, to ensure that the vertical edges remain unprocessed.

A vertical quadrilateral stripe with four vertices is generated for each non-stripped input edge via the CONVEXHULL operator. Finally, the XOR of all the stripes is returned.

```
function SOLIDIFY(pol)
    V,EV,_ = UKPOL(pol)
    stripes = Hpc[]
    Y = 2 * maximum(hcat(V...), dims=2)[1] # far away y value
    for (v1,v2) in EV
        if V[v1][1] != V[v2][1]
            stripe = CONVEXHULL([V[v1], V[v2],
                [V[v2][1],Y], [V[v1][1],Y]])
            push!(stripes, stripe)
        end
    end
    return BOOL(XOR(stripes...))
end
```

The UKPOL (UnmaKe POLhyedron) returns the essential input of the MKPOL primitive, i.e., a triple of vertices, convex cells, and polyhedral cells.

```
VIEWCOMPLEX(SOLIDIFY(pol)) # solid polygon's atoms
```

Coding 5.3.5 (From polylines to polygons) In particular, we get fourteen 0-cell points in V, fourteen 1-cell (edges) in EV, and one polyhedral cell decomposed into six cells (chain atoms) in CV. The last image shows the 2D-generated polygon. Hence: $\chi = 14 - 14$

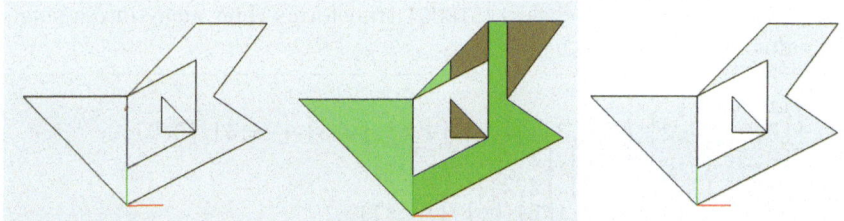

Fig. 5.32 From polylines to polygon: (a) Three 1D polylines (wire-frame); (b) the 2-chain generated by XOR of 2D non-degenerated trapezoids (stripes); (c) noncontractible 2-chain (nonmanifold polygon) generated by UNION().

A final example shows the result of our computation, where SOLIDIFY maps Hpc objects to Lar objects and BOUNDARY from Lar to Hpc.

```
lar = SOLIDIFY(STRUCT(AA(POLYLINE)([
    [[0,0],[4,2],[2.5,3],[4,5],[2,5],[0,3],[-3,3],[0,0]],
    [[0,3],[0,1],[2,2],[2,4],[0,3]],
```

```
   [[2,2],[1,3],[1,2],[2,2]]])) )
VIEWCOMPLEX(lar)
```

```
function BOUNDARY2D(lar::Lar)
    EV = lar.C[:EV]
    edgeuses = sum(lar2cop(lar.C[:FE]),dims=1)
    newedges = [lar.C[:EV][k] for k=1:LEN(edgeuses) if edgeuses[k
       ]!=2]
    return MKPOL(lar.V, newedges)
end
VIEW(BOUNDARY2D(lar))
```

Triangulation

The Constrained Delaunay triangulation (CDT) [5] is a specialization of the Delaunay triangulation where some edges (usually the boundary of a shape) are forced to belong to the triangulation itself. So, it becomes unnecessary to output the convex hull of a set of points, and, in the case of interior cycles, those are triangulated too and can be eliminated a posteriori. In this way, 2D polygons with internal holes can be generated.

Triangulation methods depend on the object's dimensionality. Even if not natively triangulated, the 2-cells (faces) of Plasm polyhedra need, in two cases, to be associated with their CDT. For this purpose, Plasm uses the 2D Triangulate.jl package, a Julia wrapper around Jonathan Richard Shewchuk's original C library Triangle [6, 7].

The two uses of triangulation concern the visualization of modal algorithms and the TGW algorithm. In fact, the dataset must be passed to the graphics library and/or to the GPU, which may only consume triangles. The other use is TGW (see Section 6.3.5), where we must correctly process the non-convex boundary faces. Whenever face triangulation is needed, the 3D geometry is ported to 2D using the Newell method [9], a numerically robust way to compute the plane equation of an arbitrary 3D polygon.

For solid triangulation, Julia uses the TetGen.jl package [8], which is a Julia wrapper for the C++ project TetGen. This wrapper enables TetGen-based tetrahedral meshing and (constrained) 3D Delaunay and Voronoi tessellation.

Boolean operations and Cartesian product

We already know the Cartesian product of cellular complexes as cellular and simplicial grids (see sections 3.2 and 3.2.2). They also work explicitly between any pair of Hpc or Lar datasets and implicitly between any pair of lists (Array) of such types through the use of the Plasm combinator INSR (see Section 2.3). Remember that the result of the Cartesian product has intrinsic dimension d+g and embedding dimension m+n if the two arguments were endowed with intrinsic and embedding dimensions (d,m) and (g,n), respectively.

The Plasm Boolean methods for computer-generated solid models will be thoroughly discussed in Chapter 6 and Chapter 7. We understand that the scope of the language's representation scheme ranges from the category of Chain Complexes to the set of valid values of Julia user types Hpc and Lar, which are appropriately utilized by the language's operators according to their semantics and specific applications and objectives.

Minkowsky methods

The Minkowski sum (or difference) of two convex polyhedra is the convex hull of the points obtained by replicating the vertices of one polyhedron around every vertex of the other one. Plasm was born as multidimensional from the very beginning. Hence, the object representations and the affine transformations, particularly the shearing and the Cartesian products, are dimension-independent. Therefore, exciting and powerful methods for increasing the intrinsic dimension of geometrical objects can be defined by bringing them into higher dimensions, then transforming them affinely, and finally projecting back into the original embedding space.

Definition 5.15 (EXTRUDE operation) The standard extrude operation creates a 3D solid from a solid 2D Hpc value or a 1D Hpc (open or closed) embedded in 2D. Objects are extruded orthogonally from the plane of the source object, in a specified direction, or along a selected path.

Example 5.3.27 (Extrude example) The Plasm function EXTRUDE at present accepts a vector triple [_N, Pol, H] as a single argument, where _N is a dummy argument, Pol is a Hpc of intrinsic dimension p embedded in d-space (i.e., having d coordinates), and N is a Number. The function returns a $p + 1$ dimensional value in $(d + 1)$-space.   □

Definition 5.16 (SWEEP operation) An elementary SWEEP operation takes an Hpc p-object in d-space and returns a $(p + 1)$-object in $\mathbb{E}^d$.

Remark 5.11 (Difference between EXTRUDE and SWEEP operations) Both operations grow the intrinsic dimension of the input by one unit, but SWEEP returns its output in the same space, whereas EXTRUDE returns it in the dimensionally-grown space.

Cellular grid generation

We have already encountered this topic when we introduced the multidimensional primitives CUBOIDGRID (see Section 3.2) and SIMPLEXGRID (see Section 3.2.2). In both cases, the only argument is an integer vector shape whose length defines the spatial dimension, i.e., the number of coordinates of generated grid points. In contrast, the vector content specifies the number of grid elements in each space dimension.

For example, CUBOIDGRID([2,3,1]) produces a 3-dimensional object made by a cellular grid (a 3-chain) of 3-cubes with $2 \times 3 \times 1 = 6$ elements and consistent 2-, 1-, and 0-skeletons. The analogous organization is possessed by SIMPLEXGRID, where each 3-cube is further decomposed into two 3-chains of tetrahedra generated by two triangles of basis. Of course, both such grids satisfy the compatibility constraints of cellular complexes—such compatibility is conserved in dimensions two, three, four, etc.

It is also remarkable to remember here the association of such cellular complexes with the Plasm data structures Hpc and Lar. The first stores the spatial organization of the grid elements through the underlying graph of nodes, the head matrix associated with each node, and the cube geometry as the (implicit) convex hull of 2^d references to E^d points.

References

1. DiCarlo, A., Paoluzzi, A., Shapiro, V.: Linear algebraic representation for topological structures. Computer-Aided Design **46**, 269–274 (2014). DOI 10.1016/j.cad.2013.08.044. URL https://doi.org/10.1016/j.cad.2013.08.044 pages 47, 137, 184, 186, 218

2. Eberly, D.: Representing a circle or a sphere with nurbs. Geometric Tools (2020). URL https://www.geometrictools.com/Documentation/NURBSCircleSphere.pdf pages 200

3. Paoluzzi, A.: Geometric Programming for Computer Aided Design. John Wiley Sons, Chichester, UK (2003). URL https://onlinelibrary.wiley.com/doi/book/10.1002/0470013885 pages 37, 38, 48, 73, 127, 130, 201, 206, 209, 244

4. Paoluzzi, A., Shapiro, V., DiCarlo, A., Scorzelli, G., Onofri, E.: Finite algebras for solid modeling using julias sparse arrays. Computer-Aided Design **155**, 103,436 (2023). DOI https://doi.org/10.1016/j.cad.2022.103436. URL https://www.sciencedirect.com/science/article/pii/S0010448522001695 pages 47, 65, 97, 137, 184, 218, 221, 254, 260, 270, 277

5. Paul Chew, L.: Constrained delaunay triangulations. Algorithmica **4**, 97–108 (1989). URL https://doi.org/10.1007/BF01553881 pages 211

6. Shewchuk, J.R.: Triangle: Engineering a 2D Quality Mesh Generator and Delaunay Triangulator. In: M.C. Lin, D. Manocha (eds.) Applied Computational Geometry: Towards Geometric Engineering, Lecture Notes in Computer Science, vol. 1148, pp. 203–222. Springer-Verlag (1996) pages 211

7. Shewchuk, J.R.: Delaunay refinement algorithms for triangular mesh generation. Computational Geometry **22**(1), 21 – 74 (2002). DOI https://doi.org/10.1016/S0925-7721(01)00047-5. URL http://www.sciencedirect.com/science/article/pii/S0925772101000475 pages 211

8. Si, H.: Tetgen, a delaunay-based quality tetrahedral mesh generator. ACM Trans. Math. Softw. **41**(2) (2015). DOI 10.1145/2629697. URL https://doi.org/10.1145/2629697 pages 211

9. Tampieri, F.: V.5 - Newell's method for computing the plane equation of a polygon. In: D. Kirk (ed.) Graphics Gems III (IBM Version), pp. 231–232. Morgan Kaufmann, San Francisco (1992). DOI https://doi.org/10.1016/B978-0-08-050755-2.50052-X. URL https://www.sciencedirect.com/science/article/pii/B978008050755250052X pages 211

Chapter 6
Space arrangement pipeline

This chapter introduces our approach to the Boolean algebra of solid models, grounded in the algebraic topology of piecewise-linear geometry. It presents computational topology algorithms for determining the two- and three-dimensional space partitions (arrangements) generated by collections of one- and two-dimensional geometric objects, respectively.

Sparse arrays, equipped with their standard algebraic operations, serve as the essential data structures of our computational framework. This approach enables the representation and manipulation of general cellular complexes homeomorphic to d-polyhedra, modeling triangulable spaces that may be non-convex and multiply connected.

Our presentation includes collections of geometric complexes, linear spaces of chains and cochains, the chain complex of linear operators between pairs of chain spaces, and their compositions. We also discuss the computational aspects—particularly parallelization—of the algorithms developed to compute a boundary representation (Brep) of all d-atoms in an $\mathbb{E}^d$ arrangement and to perform local fragmentation of primitive input datasets. As explained in the next chapter, this arrangement computation is a preliminary and necessary step toward constructing a Boolean solid algebra and resolving any Boolean expression within it.

The discussion focuses on piecewise-linear topology and on spaces of dimension three or lower. The homology pipeline, described in Section 6.3, constitutes the computational core of Plasm—even more fundamental than its Boolean solid operations, for which our research has proposed a novel method, based on Boolean binary algebra, discussed in the next chapter.

© The Author(s), under exclusive license to Springer Nature Switzerland AG 2026   215
A. Paoluzzi and G. Scorzelli, *BIM Geometry with Julia Plasm—Functional Language for CAD Programming*, Digital Innovations in Architecture, Engineering and Construction, https://doi.org/10.1007/978-3-031-90244-4_6

6.1 Cellular and boundary models

This section demonstrates that the `Plasm` representation scheme can be seen as a hybrid of cellular and boundary schemes, summarized and discussed below. Specifically, `Plasm` encapsulates a unique blend of hierarchical cellular schemes based on convex cell nodes defined by their vertices alongside a flat scheme that represents the entire topology of such assemblies. Chain complex representations avoid the issues and complications associated with geometric non-manifoldness, which need not be considered.

6.1.1 Cellular models

Most computational models used for physical simulations, which utilize differential equations to study the electrical, thermal, and mechanical behavior of the components of human-made artifacts, are cellular models. In this case, the object under consideration is partitioned into quasi-disjoint spatial parcels of equal or similar topology, interacting through their boundaries.

The Hierarchical Polyhedral Complex (HPC) representation scheme is implemented in Julia `Plasm` through the `Hpc` type and the related `Geometry` type.

The datasets of their concrete values are represented in computer memory as a directed acyclic graph, where non-leaf nodes have an array `childs` of memory references pointing to their children. Leaf nodes contain a `Geometry` value embedding a multidimensional `points` array and a `hulls` array of arrays of integer indices of points. Any `Hpc` d-value (node) is equipped with a multidimensional matrix of type `MatrixNd` of $(d+1) \times (d+1)$ dimensions, containing an affine transformation in homogeneous coordinates.

Remark 6.1 (Position of homogeneous coordinate) Unlike affine matrices in computer graphics, the `Plasm` homogeneous coordinates occupy the first row and column of the `Hpc` matrix. An analogous `MatrixNd` value is also present in objects of the `Geometry` type, intended to encode the coordinate transformation associated with the (virtual) referencing arc.

In conclusion, the geometry of the cells in `Plasm` models coincides with the leaves of their graph representation in memory. By utilizing the directed acyclic graph (DAG) structure of `Hpc` type, `Plasm` can represent both individual convex cells and assemblies of `Hpc` objects, potentially identified by Julia names or referenced by elements of `childs` arrays.

Another type of cellular decomposition is performed by `Plasm` and is the focus of this chapter: the computation of the atoms of the space partition (arrangement) [10] induced by any collection of geometric objects embedded in Euclidean d-space. If one or more geometric objects do not overlap, their set of atoms corresponds to the highest-dimensional cells of their boundary representation, such as the 3-cells in 3D space. If the objects in the input collection overlap, the boundaries of their $(d-1)$-cells intersect mutually, and new atoms are generated as unitary parts of the induced space arrangement, initially unknown to Plasm.

6.1.2 Boundary models

Boundary models of solid objects, known as Breps, are created using representation schemes where the shape of a solid object is depicted through the assembly of its closed boundary surfaces (shells). Each open patch of these surfaces (faces) is represented by its boundary cycles (loops) of oriented edges, which are subsequently represented (at least in the piecewise-linear case) by their ending vertices. When representing the $(d-1)$-boundaries of d-cells within a cellular d-complex, the $(d-1)$-cells are associated with at most two different d-cells. Storing all boundaries of every d-cell in a raw manner would be incorrect, as their $(d-1)$-cells would be mistakenly stored twice.

Representations of rigid solids using Breps are the most common in academia and industry, especially in the libraries developed over the past forty years, where local non-manifoldness is explicitly stored. This feature demands highly complex data structures and equally sophisticated handling algorithms. Usually, an intermediate layer for conceptual and software implementation is required to manage the data structure complexity, which often involves multiple linked lists for adjacency, incidence, and ordering—typically stored in duplicate with inverted orderings. This approach relies on the so-called Euler operators baumgart:1972,mantyla:1988, which modify the connection graph by adding or removing mesh details while preserving its topology. They enable developers to create complex algorithms with a relatively small set of functions or procedures that correctly alter internal data structures while preserving topological and numerical consistency.

Remark 6.2 An important consequence is that only a few industrial-made libraries remain in the market and are licensed by most companies in the CAD and BIM sectors. Another consequence is the rigidity and lack of innovation caused by this approach, affecting both supply and demand. After many years of making only minor updates to existing software tools, no competitor was encouraged to invest in research and the in-house development of new, even simpler, innovative solutions, which would require years and significant capital. This lack of vision has significantly hindered the development of new geometric kernels over the past two decades.

In recent years, something has changed with the advent of 3D printing technologies, and even more changes are anticipated as AI-assisted design research continues. We hope for even more with the printing of this book ;-)

Therefore, the book aims to demonstrate that an unconventional new mathematical approach to geometric data structures and algorithms, supported by a simple yet powerful functional language and capable of covering all the design operations required by CAD and even more by BIM, could be essential for promoting innovation in AI-driven design and production.

6.1.3 Julia Plasm representation scheme

The foundations of solid modeling technology were established between the 70s and 80s. After 50 years, the time was ripe for exploring new paradigms [5, 10, 11], not based only on data structures and computational geometry algorithms, but on novel abstractions taking into account structures and operators of algebraic topology and algebra [9], as is happening in modern AI, which is also based on hardware tensors and linear algebra.

Remark 6.3 Ari Requicha formalized the core concept in his foundational 1980 paper [12], which arose from the PAP (Project Automation Project) research directed by Herb Voelcker in Rochester, NY, during the seventies. A representation scheme of solids is a mapping $r : M \to R$ between a set of mathematical models M and a set of representations R produced by a computer grammar.

Definition 6.1 (Plasm representation scheme) In our approach, the set of mathematical models is the category of chain complexes, and the set of representations is the class of functions that map tuples of numeric/geometric/topological parameters to a few typed Julia objects. Their syntax and type are introduced as FL+Julia language computer grammar.

This book and the associated software language, Plasm, combine concepts and methods from functional languages, algebraic topology, affine geometry, and computer programming. In particular, we aim to clarify the concepts introduced to BIM and CAD developers and advanced users.

The syntax-by-examples approach in this book, which demonstrates Plasm expressions evaluating to Julia values, is illustrated with straightforward representative cases, typically visualized in the book's figures. These cases would be better experienced through interactive evaluation in a computational environment running a Julia Plasm notebook. [1] For this purpose, check the language site on GitHub, which also covers the use of Plasm in a Jupiter notebook, where the reader or designer can integrate Plasm code with explanations, comments, static images, as well as 3D renderings and user interaction.

---

[1] A remote computational environment is provided by PlasmRemote on JuliaHub.com.

Plasm **Types and Objects**

Julia `Plasm` is primarily based on three geometric types, with constructors and conversion methods:

Hpc: Hierarchical Polyhedral Complex, for hierarchical assemblies and interactive visualization, even on notebooks.

Lar: Linear Algebraic Representation, for instancing cellular complexes and chain operators.

Geo: Geometry format, for enumeration of 0- and d-chains in hierarchical space domains, such as octrees and potrees, and in nD triangulations.

Example 6.1.1 (Values of Hpc and Lar types)

```
using Plasm
obj = ICOSPHERE()
VIEW(obj) # Hpc value
```

`VIEW` is used for interactive visualization of objects of `Hpc` type; `VIEWCOMPLEX` is used for geometric objects of `Lar` types. See Figure 6.1.

```
VIEWCOMPLEX(LAR(obj)) # Lar value
VIEWCOMPLEX(LAR(obj), explode = [1.2,1.2,1.2])
```

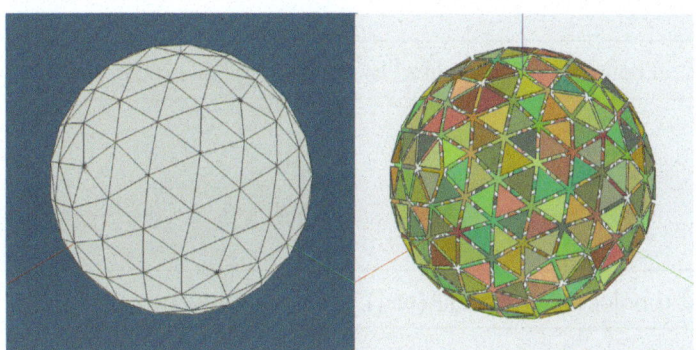

Fig. 6.1 Icosphere: (a) of `Hpc` type (convex hull of points); (b) exploded Brep of `Lar` type (chain complex).

In Examples 6.1.2 and 6.1.3 below, we show the Julia coding of `Plasm` types `Hpc` and `Lar`, as well as the storage of `Hpc` and `Lar` objects respectively.

Example 6.1.2 (Julia Plasm Data Types)

```
mutable struct Hpc #outer constructors
  T :: MatrixNd
  childs :: Union{Vector{Hpc},Vector{Geometry}}
  properties :: Dict{Any, Any}
end
```

```
mutable struct Lar
  V :: Points     # object geometry
  # object topology (C for cells)
  C :: Dict{Symbol, AbstractArray}
  # inner constructor
  Lar(V :: Points=Matrix{Float64}(undef, 0, 0), C :: Dict=Dict{
    Symbol,AbstractArray}()) = new(V, C}())
end
```

Example 6.1.3 (Storage of Hpc and Lar types)

As we know, the storage of values of Julia types Hpc and Lar is quite different, i.e., a recursive graph of Hpc nodes and a pair made by the matrix .V and the dictionary .C:

```
obj = ICOSPHERE()  #=
Hpc(MatrixNd(4), Geometry([[0.0, -0.36967054873218,
    -0.59813951248849], … [0.0, 0.36967054873218,
    0.70315516858082877, 0.0]], hulls=[[1, 2, 3, 4, 5, 6,  …
    157, 158, 159, 160, 161, 162]])) =#
```

The representation conversion is executed by the LAR(:: Hpc) :: Lar function:

```
LAR(obj).V  #=
3×162 Matrix{Float64}:
  0.217287  -0.217287   0.0        …  -0.217287  -0.568864
 -0.568864  -0.568864  -0.703155       0.568864   0.351578
 -0.351578  -0.351578   0.0           -0.351578   0.217287
```

Geometry V and topology C of the LAR(obj) converted object:

```
LAR(obj).C  #=
Dict{Symbol, Vector{Vector{Int64}}} with 5 entries:
  :CF => [[1, 2, 3, 4, 5, 6, 7, 8, 9, 10  …  71, 72, 73, 74,…
  :CV => [[1, 2, 3, 4, 5, 6, 7, 8, 9, 10  …  33, 34, 35, 36,…
  :FV => [[1, 2, 3], [1, 2, 19], [1, 3, 36], [1, 17, 19], [1…
  :EV => [[1, 2], [1, 3], [1, 17], [1, 19], [1, 29], [1, 36]…
  :FE => [[1, 2, 7], [1, 4, 9], [2, 6, 15], [3, 4, 72], [3, …
```

where we see some distinct maps from 0-cells to 3-cells, 2-faces, 1-edges, one discrete map from 1-edges to 2-faces, and one from 2-cells to 3-cells. □

6.2 Arrangement of space

Spatial arrangement refers to how phenomena or objects are organized and distributed in physical space. It describes the location, pattern, and relationship between elements within a volume or area. Arrangements of lines, segments, planes, and other geometrical objects are discussed in [6]. A review of papers and algorithms concerning the construction and counting of arrangement cells may be found in [7]. Space arrangements in the context of geometric computing and solid modeling have been introduced in [10] and shown in relation to solid algebras in [11, 9].

6.2.1 Homology of spaces

Homology, in mathematics, is a basic notion of algebraic topology. Two curves in a two-dimensional surface are homologous if they bind a region together and allow distinguishing between an inside and an outside. Similarly, two surfaces within a three-dimensional space are homologous if they bind a three-dimensional region within the ambient space.

In general terms, some $(d-1)$-dimensional subspaces (curves, surfaces, etc.) embedded in $\mathbb{E}^d$ are said to be homologous when their oriented union forms the boundary of a d-dimensional region of the ambient space, thereby partitioning $\mathbb{E}^d$ into two parts, interior and exterior. Intuitively, when taken together, their boundaries can be joined to enclose a portion of space, separating $\mathbb{E}^d$ into an inner and an outer region.

Chain complexes are an algebraic tool for computing or defining homology. They are used here to algebraically describe a partition of the embedding space induced by a collection of geometric objects, particularly all the cycles (holes) in the dimensions of interest—specifically, from dimension 0 to dimension 3. In this book, we refer to `Plasm` methods and algorithms to compute the boundaries of spaces as homological.

Remark 6.4
For example, consider a building with multiple hollow spaces and thin separating surfaces. Architects and civil engineers have long designed the structure of interior spaces by defining the elements that separate them and the relationships among those elements. In our approach, we abstract this notion topologically: the incidences and adjacencies among elementary separating parts are represented as a sequence of spaces of chains (subsets of cells) connected by linear boundary operators, thus forming the algebraic structure of a chain complex.

Definition 6.2 (Chain complex) Given a collection $\mathcal{S}$ of $(d-1)$-dimensional geometric objects[2] in $\mathbb{E}^d$, We compute the topology of their space arrangement $\mathcal{A}(\mathcal{S})$ as a chain complex, i.e., as a short exact sequence of linear spaces C_p of (co)chains and linear boundary/coboundary maps ∂_p and $\delta_p = \partial_{p+1}^\top$ between them:

$$C_\bullet = (C_p, \partial_p) := C_3 \underset{\partial_3}{\overset{\delta_2}{\leftrightarrows}} C_2 \underset{\partial_2}{\overset{\delta_1}{\leftrightarrows}} C_1 \underset{\partial_1}{\overset{\delta_0}{\leftrightarrows}} C_0. \tag{6.1}$$

The chain complex $C_\bullet$ fully characterizes the topology of the space partition (arrangement) induced within the ambient Euclidean space by a collection of geometric objects embedded in it. Two simple examples of space arrangements generated by STRUCT (assembly) of geometric objects are given below. Three finite atoms are generated. The fourth is the outer-cell in $\mathbb{E}^3$.

Example 6.2.1 (Arrangement of two cubes)

```
cubes = LAR(STRUCT(CUBE(1), T(1,2,3)(.5,.5,.5), CUBE(1)))
arrangement = ARRANGE3D(cubes)
VIEWCOMPLEX(arrangement, show=["FV"], explode=[1.4,1.4,1.4])
VIEWCOMPLEX(arrangement, show=["FV", "atoms"], explode
    =[1.4,1.4,1.4])
```

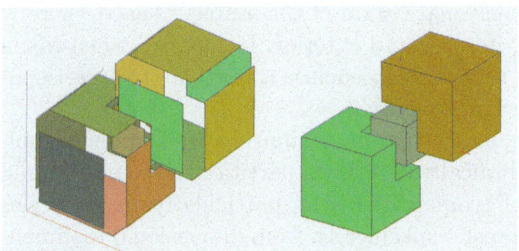

Fig. 6.2 Space decomposition carried out by two cubes: (a) exploded 2-cells; (b) exploded 3-cells of the $\mathbb{E}^3$ partition. The outer fourth atom is not shown and corresponds to the Brep of the union of the three finite cells with reversed normals.

When computing the Euler characteristic (see Definition 6.3) of arrangement value, we get, as expected:

$$\gamma = \gamma_0 - \gamma_1 + \gamma_2 - \gamma_3 = 22 - 36 + 18 - 4 = 0$$

Be careful: the faces to count (γ_2) are those in Figure 6.2a, not 6.2b! The internal ones are doubled.   □

---

[2] Examples include but are not limited to line segments, quads, triangles, polygons, meshes, pixels, voxels, volume images, B-reps, etc. In mathematical terms, a geometric object is a topological space embedded in some $\mathbb{E}^d$ [3].

Example 6.2.2 (Faces with holes) `Plasm` may utilize with the `Lar` type very general 2-cells, possibly non-convex and featuring holes. Refer to Figure 6.3b. It's important to note that the `partition` object includes all the 3D atoms of the $\mathbb{E}^3$ arrangement generated by the `Plasm` expression `ARRANGE3D (LAR(hpc))` including the holes present in the embedding space $\mathbb{E}^3$.

Conversely, `INNERS(partition)` consists only of the atoms (inners) that do not form part of the finite boundary of the exterior unbounded atoms `OUTERS(partition)`. In our scenario, this collection is connected and cannot be exploded by `Plasm` since only one hole exists in the exterior 3D space.

When dealing with a non-connected 3-chain, we would find the number of atoms in `OUTERS(partition)` equal to the first Betti number b_1. Informally, in this case, outer space has the same topology as a sphere with b_1 handles.

```
cube = CUBOID([1,1,1]);
hpc = STRUCT(cube, T(1,2,3)(.5,.5,0.75), R(2,3)(π/4), R(1,3)(π
    /4), cube);
partition = ARRANGE3D(LAR(hpc));
partition=without_outeratom(partition)
VIEWCOMPLEX(partition, show=["FV"], explode = [1.4,1.4,1.4])
VIEWCOMPLEX(partition, show=["EV"], explode = [1.4,1.4,1.4])
VIEWCOMPLEX(INNERS(partition), show=["CV"], explode =
    [1.2,1.2,2.0])
VIEWCOMPLEX(OUTERS(partition), show=["CV"], explode =
    [1.2,1.2,2.0])
```

`R(2,3)(π/4)` stands for a (multidimensional) rotation about the x axis, rotating the y and z coordinates with the usual pattern of `sin` and `cos`.

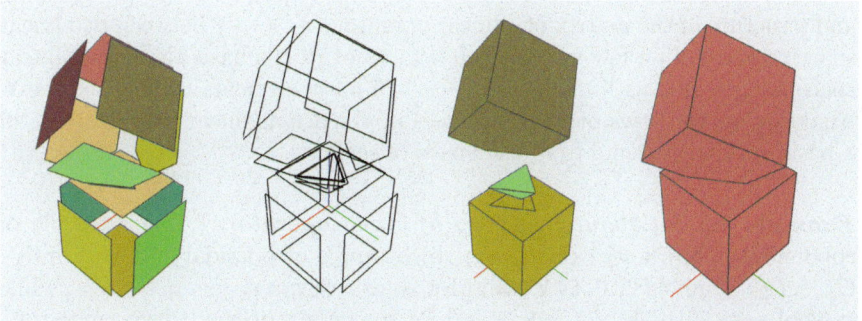

Fig. 6.3 (a) exploded elementary 2-chains (basis of C_2); (b) exploded 1-cycles boundaries of basis 2-chains; (c) exploded atoms (3-chains), standard basis $U_3 \subset C_3$ of solid algebra. (d) the connected boundary (2-cycle) of outer 3-space.

6.2.2 Introduction to arrangement pipeline

To compute a Boolean expression among instances of d-solids ($d = 2, 3$) in the input set $\mathcal{S}$, we consider the collection of their $(d-1)$-cells. Then, efficiently fragment against the others each $(d-1)$-cell $\sigma \in \mathcal{S}$, one by one—possibly in parallel, and construct its local chain complex $C_\bullet(\sigma) \subset \mathbb{E}^{d-1}$. Finally, we embed all such complexes in $\mathbb{E}^d$ and compute our goal: the yet unknown chain space C_d, and in particular, the operator $\delta_{d-1} : C_{d-1} \to C_d$.

For the embedding we use an algebraic trick, storing each $\delta_0(\sigma) \equiv \mathsf{EV}(\sigma)$ and $\delta_1(\sigma) \equiv \mathsf{FE}(\sigma)$ matrix operators as two new diagonal blocks into two larger block-matrices $\Delta_0(\mathcal{S})$ and $\Delta_1(\mathcal{S})$. This computation in dimension $d-1$, with store of $(d-1)$-complex $C_\bullet(\sigma)$ into two new blocks, is repeated (in parallel) for each $\sigma \in \mathcal{S}_{d-1}$.

Then, the block matrices must yet be reduced by topological congruence [4], identifying groups of numerically "quite" coincident vertices, giving coincident edges and faces after identification of close vertices. Eliminating duplicated cells (for $d = 0, 1, 2$), in this order, provides a partially disclosed topological partition into a chain complex C_{d-1} embedded in $\mathbb{E}^d$. The reader may think of a 2D spatial grid (or building) erected in 3D space.

Actually, we need to build the yet unknown C_d such that $\cup_{p=0}^d C_p = \mathbb{E}^d$, i.e., the major bricks of the $\mathbb{E}^d$ arrangement (partition). This main task is assigned to the Topological Gift Wrapping (`TGW`) algorithm [10]. Its input is the pair of lower-dimensional coboundary operators $\mathsf{EV} \equiv \delta_{d-3}(\mathcal{S})$ and $\mathsf{FE} \equiv \delta_{d-2}(\mathcal{S})$. The output is the operator matrix $\mathsf{CF} \equiv \delta_{d-1}(\mathcal{S}) : C_{d-1} \to C_d$. The symbols (to be read columns → rows) when $d = 3$ are given in Table 6.1.

Remark 6.5 (Meaning of linear operators) It is helpful to recall the meaning and structure of the matrix of a linear operator $\mathsf{YX} : \mathsf{X} \to \mathsf{Y}$ between two linear spaces X and Y. In a few words: each column of YX is a basis element of linear space X expressed as linear combination of basis elements of linear space Y. Analogously, the rows of YX are a basis of Y space as linear combinations of X basis with coefficients from the space of scalars.

Example 6.2.3 (About meaning of** Plasm **operators) (a) Example of coboundary (up) is $\delta_2 : C_2 \to C_3$; (b) example of boundary (down) is $\partial_3 : C_3 \to C_2$. Symbols $\mathsf{C}, \mathsf{F}, \mathsf{E}, \mathsf{V}$ stand for solid cell, faces, edges, vertices, while symbol pairs, like $\mathsf{FE:}$ $\mathsf{E} \to \mathsf{F}$, stand for sparse matrices of Plasm operators.

Therefore, the `TGW` algorithm computes one at a time, the columns of $\mathsf{FC} : \mathsf{C} \to \mathsf{F}$, i.e., for $d = 3$, the solid elements of a C_3 basis (cells of $\mathbb{E}^3$ partition) expressed as linear combination of elements of C_2 basis (boundary faces in F) with coefficients in $\{0, 1, -1\}$. Such columns are 2-cycles (closed 2-chains). In other words, in each column, we get the Brep of a solid atom of the arrangement of $\mathbb{E}^3$ induced by the dataset $\mathcal{S}$, in symbols $\mathcal{A}(\mathcal{S})$.

Table 6.1 Homological relations: (left table) topology operators in $\mathbb{E}^3$; (right table) corresponding binary relations between classes of cells.

–	C_3	C_2	C_1	C_0		–	C	F	E	V
C_3	–	∂_2	–	–		C	–	CF	–	–
C_2	δ_3	–	∂_1	–		F	FC	–	FE	–
C_1	–	δ_2	–	∂_0		E	–	EF	–	EV
C_0	–	–	δ_1	–		V	–	–	VE	–

Remark 6.6 (Notations and terminology) In Table 6.1, we illustrate the notational correspondence between chain and cochain operators and Julia sparse matrices, offering an intuitive understanding of topological operators as tools that map p-chains to $(p-1)$-chains (down) or to $(p+1)$-chains (up), respectively. These operators are typically used in Plasm to go from columns to rows via right or left matrix products.

The reader should note, in Figure 6.4b, the set of 2D atoms—displayed in exploded view and various colors for clearer distinction—of the E^2 arrangement created by randomly positioned polygons, with a random number of sides and varying radius. Each polygon is a member of the collection $\mathcal{S}$ of geometric shapes that arise from their boundary segments, the $\mathbb{E}^2$ partition. They will be called generators of a Boolean algebra (see Definition 7.6) associated with the arrangement $\mathcal{A}(\mathcal{S})$.

The generated 2-cells are, in general, neither convex nor simply connected (without holes) and constitute the set of unreducible (minimal) atoms (basis) of the C_2 linear space of 2-chains. Of course, they are also the basis of $\mathcal{A}(\mathcal{S})$ induced lattice and $\mathcal{B}(\mathcal{A})$ Boolean solid algebra.

Example 6.2.4 ($\mathbb{E}^2$ arrangement) The $\mathbb{E}^2$ space arrangement in Figure 6.4 is generated by 50 random polygons.

```
hpc = STRUCT([RandomBubble() for I in 1:50])
arrangement = ARRANGE2D(LAR(hpc))
VIEWCOMPLEX(arrangement, explode=[1.5,1.5,1.5])
VIEWCOMPLEX(arrangement)
```

Example 6.2.5 ($\mathbb{E}^3$ arrangement) In this example, a 3D space arrangement is constructed by a cube centered on the origin and by three mutually orthogonal PL cylinders.

```
CYLINDER(r,h,sides) = CIRCLE(r)([sides,1]) * INTERVALS(h)(1);
cyl = T(3)(-2)(CYLINDER(0.5,4,16));
assembly = STRUCT( T(1,2,3)(-1,-1,-1)(CUBE(2)), cyl, R(2,3)(π/2)
    , cyl, R(1,3)(π/2), cyl );
arrangement = ARRANGE3D(LAR(assembly));
VIEWCOMPLEX(INNERS(arrangement), explode=[3,3,3], show=["FV"])
VIEWCOMPLEX(OUTERS(arrangement))
VIEWCOMPLEX(INNERS(arrangement), explode=[3,3,3], show=["CV"])
```

Fig. 6.4 $\mathbb{E}^2$ arrangement with random polygons: (a) basis 2-chain $U_2 \subset C_2$; (b) Exploded cell complex. LAR cells are neither convex nor contractible.

Fig. 6.5 Isometric projection of exploded cell complex: (a) basis 2-chain $U_2 \subset C_2$; (b) 2-cycle of **OUTERS**(arrangement); (c) exploded atoms of $\mathbb{E}^3$ arrangement. Their union with reversed orientation corresponds to the finite boundary (2-cycle, i.e., closed chain) of the unbound 3-outercell, i.e., $\mathbb{E}^3$ minus the model cell-complex.

Remark 6.7 With the `arrangement` value of `Lar` type generated by the Example 6.2.5 we have for the cardinalities of chain subcomplexes:

```
julia> size(arrangement.V,2)      #=>  395      =#
julia> LEN(arrangement.C[:CF])    #=>  530      =#
julia> LEN(arrangement.C[:FV])    #=>  1458     =#
julia> LEN(arrangement.C[:EV])    #=>  1322     =#
julia> LEN(arrangement.C[:FE])    #=>  1458     =#
```

Therefore, by computing the Euler characteristic by alternating the sizes of 0-, 1-, 2-, and 3-chain bases, we have, as expected:

$$\gamma = \gamma_0 - \gamma_1 + \gamma_2 - \gamma_3 = 395 - 1322 + 1458 - (530 + 1) = 0$$

where $+1$ is the cardinality of OUTERS chains.

Remark 6.8 (Indicator of computational accuracy) This result, concerning arrangements of Euler characteristics of Euclidean spaces, guarantees the accuracy of our homological computations. If the result in 3D had been a small integer number instead of zero [10], it would indicate some local numerical inexactness and the TGW algorithm would probably not finish or conclude in error. To eliminate such possibilities, we evaluated a few numerical operations using quadruple precision numbers (Julia Float128).

6.3 Homological pipeline

In this section, we refer systematically to dimension $d = 3$ for concreteness.

The objective of Plasm homological pipeline is to disentangle the partially unknown chain complex mathematical operators $\delta_2, \delta_1, \delta_0$, called CF, FE, and EV, respectively, in Plasm (see Table 6.1). The goal is to acquire complete knowledge of the topology and geometry of the $\mathbb{E}^3$ arrangement generated by any input collection of geometric shapes (called generators) like the boundary representations (Breps) of solid objects that are combined into more complex objects of a 3D scene.

In particular, our main goal is to construct the unknown operator matrix:

$$\mathsf{FC} = \mathsf{CF}^\top \equiv [\partial_3] : C_3 \to C_2, \quad \text{where} \quad \partial_3 = \delta_2^\top,$$

and where FC represents the linear transformation between 3-chains and 2-chains. Once the 2-cells of fragmented generators are ordered, this matrix will consist of a basis of 3-chains in its rows, meaning that 3-cells (irreducible solid objects, i.e., atoms of arrangement), are expressed as linear combinations of $0, 1, -1$ scalars with the corresponding basis of 2-chains (2-cells of 3D arrangement).

Remark 6.9 In particular, by basic linear algebra, the columns of the FC matrix are the coordinate representation of unit 3-chain vectors as closed irreducible combinations of faces, providing a Brep of unit solids, i.e., a closed surface (2-cycle) made of fragmented faces.

In topological wording, each column of FC is a basis 2-cycle of the subspace of cycles $Z_2 \subset C_2$— an irreducible shell of boundary faces, see Figure

3.21. Depending on the choice of $+$ operator on chain spaces, the scalars of linear combinations (and thus the matrix values) will be either in $\{0, 1\}$ or in $\{-1, 0, 1\}$. The constructive algorithm of FC, one column at a time, is called TGW (Topological Gift Wrapping) [10]. It will be discussed later in this chapter.

The topological operators are codified as sparse matrices in Plasm, since the sparsity grows more than linearly with the number of cells, because each unit chain element has a small, bounded number of incidences with other elements.

The objective of the following pages is to discuss in some detail the computational pipeline that maps our homological methods from the input collection of geometric generators to the output arrangement of the embedding Euclidean space described by a set of polyhedral atoms.

6.3.1 Starting from a comic tale

In Figure 6.6 the homological pipeline executed by the Plasm primitive functions ARRANGE2D and ARRANGE3D is displayed step by step:

1. two Breps of generator solids in $\mathcal{S}$;
2. the exploded input collection $\mathcal{S}_2$ in $\mathbb{E}^3$;
3. 2-cell σ (red) and the set $\Sigma(\sigma)$ (blue) of possible intersection cells;
4. $\sigma \cup \Sigma$ affinely mapped on $z = 0$;
5. reduction to a set of 1D segments in $\mathbb{E}^2$ via intersection with $z = 0$;
6. pairwise segment intersections in 2D;
7. *local* 2D arrangement two columns of *local* $[\partial_2(\sigma)] : C_2(\sigma) \to C_1(\sigma)$, and
8. *global* U_3 basis of C_3. Three columns of $[\partial_3] : C_3 \to C_2$ sparse matrix, through TGW algorithm in 2D / 3D, respectively (see Section 6.3.5).

6.3.2 Generators assembly

Let $\mathcal{S}_2 \subseteq \mathcal{S}$ be the set of 2-cells of generator complexes[3], embedded in $\mathbb{E}^3$, corresponding to the set of primitive objects we want to combine. Such objects will be called generators[4] Note that $\mathcal{S}_2$ is not required to be a cellular complex, since 2-cells may mutually intersect outside of their boundaries (see Figures 6.6-1,2,3,4). It is only required that each 2-cell σ be connected and 2-manifold.

---

[3] Cell complex embedded in Euclidean space via association of position vectors to 0-cells.
[4] because they generate the space arrangement $\mathcal{A}(\mathcal{S})$, and the associated Boolean algebra $\mathcal{B}(\mathcal{S})$.

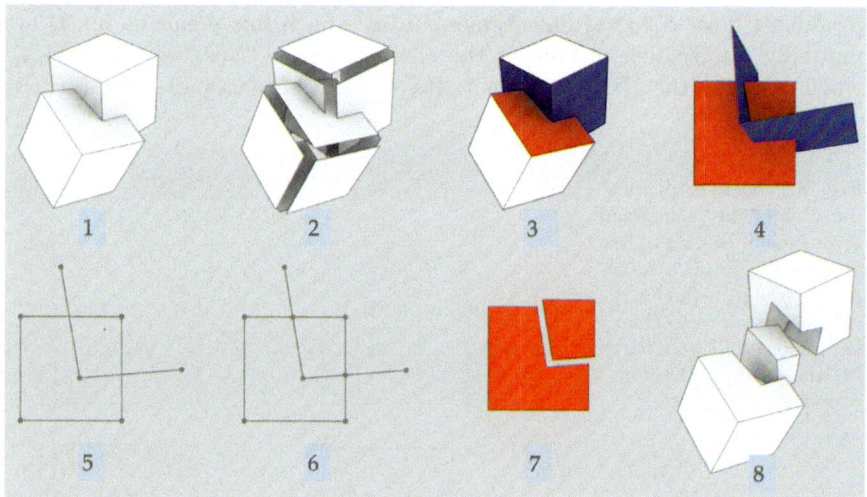

Fig. 6.6 Display of the computational pipeline. The operations are performed in parallel for each 2-face $\sigma \in \mathcal{S}_2$.

Here, we assume more general conditions. In particular, our input is a **STRUCT** or a **BOOL** alias for a solid Boolean (very general) expression that contains Breps of single solid objects and/or other **STRUCT** assemblies. This assumption is sound because this kind of input can easily accommodate symbols for Boolean operations as aliases, i.e., it can be transformed into a CSG tree of any structure and depth. Of course, a highly structured expression will also contain some or many coordinate transformations since the standard hypothesis in any assembly states that objects have coordinates referred to their own local Cartesian coordinate frame.

Hence, the first operation is to flatten the input expression and generate a **STRUCT** expression of depth one without transformations of coordinates while maintaining each object in its original position and orientation in 3D space. The second operation requires appending each other the component data structures to get just one dataset of three types **V**, **FV**, and **EV**, converted into a Julia object of **Lar** type, well known to the readers of this book.

6.3.3 Two-dimensional splitting tasks

This computational process named 2D splitting and described in Figure 6.7 consists of a few geometric steps.

Each $\sigma \in S_2$ of the input collection is mapped to subspace $z = 0$ by an affine transformation $\mathbf{Q}_\sigma$, together with the set $\mathcal{I}(\sigma) \subset S_2$ of cells potentially intersecting it. The set $\Sigma = \sigma \cup \mathcal{I}(\sigma)$ is intersected with the $z = 0$ subspace, producing a set $\mathcal{S}_1(\sigma)$ of line segments in $\mathbb{E}^2$. Such line segments in 2D are mutually intersected, producing the chain complex $C_\bullet(\sigma) = (C_{2,1,0}, \partial_{2,1})$, which provides the 2D arrangement $\mathcal{A}(\mathcal{S}_1)$. See Figure 6.7), described below:

Fig. 6.7 Basic case: computation of the regularized arrangement of a set of line segments in $\mathbb{E}^2$: (1) the input $C_1(\partial_2(\sigma) \cup (\sigma \cap \Sigma))$; (2) pairwise intersection; (3) removal of $C_1(out(\sigma))$; (4) the output 2D chain complex $C_\bullet(\sigma)$.

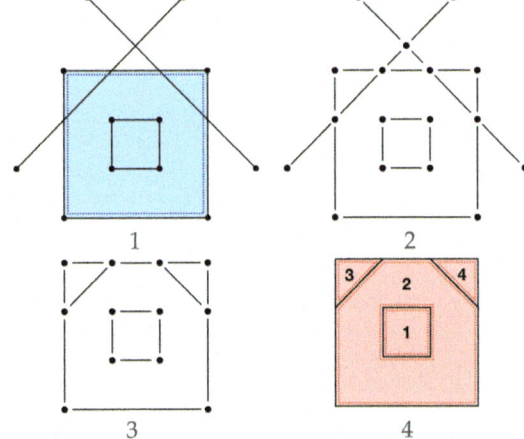

1. The input, i.e., the 2-cell σ (signed blue) and the line segment intersections of $\Sigma(\sigma)$ with $z = 0$;
2. All pairwise intersections of 1-cells;
3. Removal of the 1-subcomplex external to $\partial\sigma$ (using graph algorithms);
4. 2-chain $\sigma_1 + \sigma_2 + \sigma_3 + \sigma_4$ generated by $\sigma \cup \Sigma$ via TGW in 2D (see 6.3.5).

Care must be taken to identify the 1-cycles around the holes within the partitioned 2-cell. This will allow us to remove their outer boundary cycles by removing their columns and the outer cycle from the operator matrix. Identification is easy: each hole produces two opposite columns summing to 0. Finally, the geometric 2-complex X_σ is transformed back into $\mathbb{E}^3$ by $\mathbf{Q}_\sigma^{-1}$. In summary, the 2D splitting algorithm performs the one-to-one mapping $\sigma \mapsto X_\sigma$, by independently computing the maps $\sigma \mapsto C_\bullet(\sigma)$. The output is a collection $\mathbb{C} := \{C_\bullet(\sigma), \sigma \in \mathcal{S}_2\}$ of chain complexes, pairwise satisfying the boundary compatibility conditions, assuming the algorithm is numerically stable.

6.3.4 From two to three dimensions

Small sparse matrices of signed operators $\partial_2(\sigma) : C_2(\sigma) \to C_1(\sigma)$ have already been assembled independently in 2D for each fragmented σ, i.e., for each geometric 2-complex X_2^g, as detailed in the previous Section 6.3.3. The output of this pipeline stage is a collection $\mathbb{C} := \{C_\bullet(\sigma), \sigma \in \mathcal{S}_2\}$ of small chain complexes, one for each input 2-cell, embedded in $\mathbb{E}^3$. They were built by repeatedly applying the `TGW` algorithm in 2D (see Section 6.3.5) and then mapping the results back to 3D.

Algorithm 6.1 (*Chain Complex* Merge) Input: set of *local* chain 2-complexes; output: single *global* chain 2-complex. $\qquad\qquad\Box$

We have discussed in [4] the block diagonal marshaling matrices $[\Delta_0]$ and $[\Delta_1]$ of local coboundary matrices generated by the face splitting algorithm. The target of the Chain Complex Merge algorithm is to merge the local chains by using the equivalence relations of ϵ-congruence between 0-, 1-, and 2-cells (elementary chains). To understand all the mathematical details, the interested reader should consult the paper [4]. In particular, we reduce the block-diagonal coboundary matrices $[\Delta_0]$ and $[\Delta_1]$, used as matrix accumulators of the local coboundary chains, to the global matrices $[\delta_0]$ and $[\delta_1]$, which are representative of congruence topology, i.e., of congruence quotients between all 0-, 1-, and 2-cells, via elementary algebraic operations on columns and rows of accumulator matrices.

6.3.5 Topological Gift Wrapping

The topological gift wrapping (TGW) method discussed here from [10] is reminiscent of the "gift-wrapping" algorithm [2, 8] for computing the convex hulls of 2D and 3D discrete sets of points. We added "topological" because it alternates the application of the δ_{d-2} coboundary and the ∂_{d-1} boundary operators.

The TGW algorithm takes a couple of corresponding sparse matrices $[\partial_{d-1}]$ and $[\delta_{d-2}]$ as input and produces as output the sparse matrix $[\partial_d^+]$, augmented with the outer chain(s). A geometric embedding is used to compute the angular ordering, around some $(d-2)$-cells, of incident $(d-1)$-basis elements in the boundary's coboundary while wrapping up a $(d-1)$-cycle, as illustrated in Figures 6.8 and 1.9 .

The (minimal) $(d-1)$-cycles built from the algorithm are set as columns of $[\partial_d]$, in the incremental construction of the C_d basis.

Note also that, once the ordered sets of basis elements are fixed, columns contain the coordinate representation of $(d-1)$-cycles, built from group coefficients $(\{-1, 0, 1\}, +) \simeq \mathbf{Z}/3\mathbf{Z} = \mathbf{Z}_3$. Analogously, boundaries and coboundaries of chains are calculated by multiplying operator matrices by the proper coordinate vectors with such coefficients.

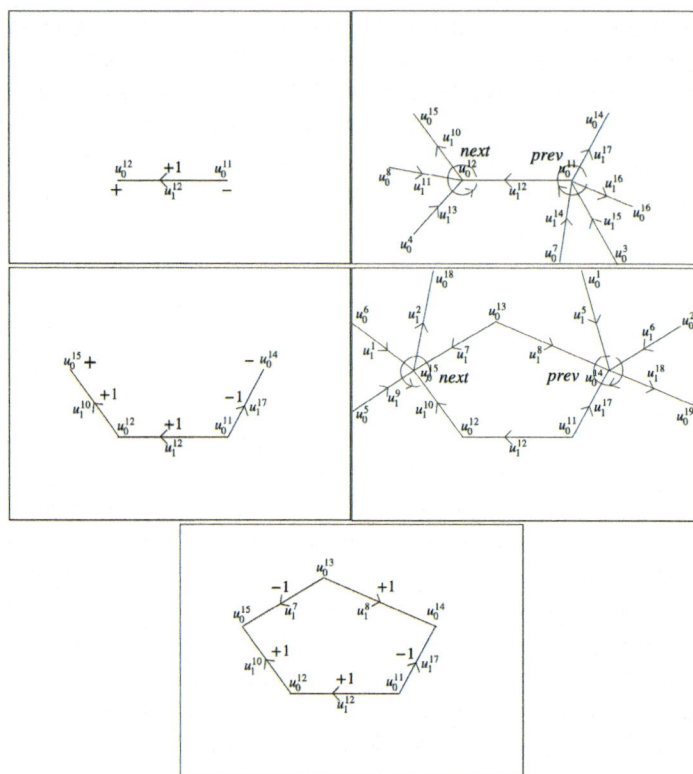

Fig. 6.8 Extraction of a minimal 1-cycle from $\mathcal{A}(X_1)$: (a) the initial value for $c \in C_1$ and the signs of its oriented boundary; (b) cyclic subgroups on $\delta\partial c$; (c) new (coherently oriented) value of c and ∂c; (d) cyclic subgroups on $\delta\partial c$; (e) final value of c, with $\partial c = \emptyset$.

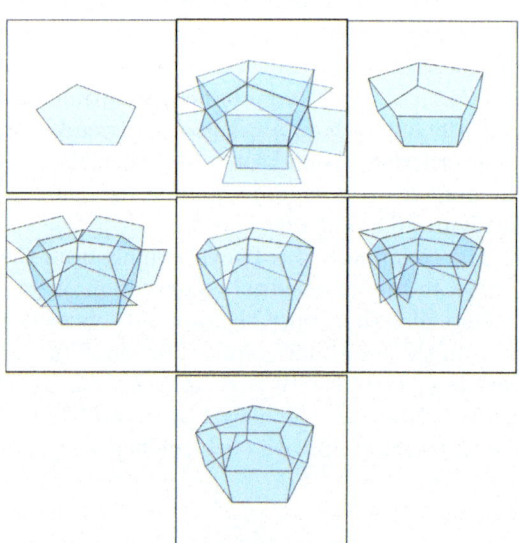

Fig. 6.9 Extraction of a minimal 2-cycle from $\mathcal{A}(X_2)$: (a) initial (0-th) value for $c \in C_2$; (b) cyclic subgroups on $\delta\partial c$; (c) first value of c; (d) cyclic subgroups on $\delta\partial c$; (e) second value of c; (f) cyclic subgroups on $\delta\partial c$; (g) third value of c, such that $\partial c = 0$, hence stop.

6.4 Commented examples

Definition 6.3 (Euler characteristic of arrangement) The Euler characteristic χ of a chain d-complex, i.e., the alternating sum of cardinalities of bases $U_k \subset C_k$ of chain subspaces of a topological space X homeomorphic to the d sphere, is either 0 or 2, depending on whether d is even or odd [10].

$$\sum_{k=0}^{d}(-1)^k \chi_k, \qquad \chi_k = \#U_k, \;\; U_k \subset C_k.$$

In particular, χ does not depend on how X is partitioned into cells. Consequently, one can speak, e.g., of the Euler characteristic of an arbitrary compact polyhedron, meaning the χ of any of its triangulations.

We are interested in this book in computing the Euler characteristic of the arrangement of spaces $\mathbb{E}^2$ and $\mathbb{E}^3$ generated by our models. We use this topological invariant to check the model reconstruction consistency. We know that it does not depend on the partition; therefore, it depends on the topology of the model to be checked.

2D/3D arrangement

Example 6.4.1 (2D arrangement examples) Many 2D unit tests have been conducted during the development and implementation of the 2D and 3D algorithmic pipeline in   Plasm. In Figure 6.10, we show two examples with fifteen squares, each having a random rotation center (the bottom-left vertex) and angles.   □

Fig. 6.10 2D arrangements of two random sets of squares : (a) arrangement atoms in different colors; (b) atoms exploded about the origin.

3D arrangement examples

Similar test examples were developed for implementing the `ARRANGE3D` function, which was later joined with the companion `ARRANGE2D` function into a single `ARRANGE` operator used within the computational `Plasm` environment `BOOL` to evaluate Boolean solid formulas in 2D or 3D.

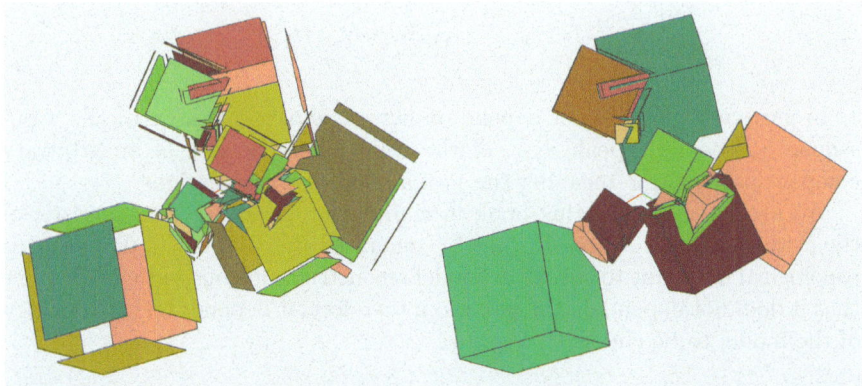

Fig. 6.11 3D arrangements of ten random cubes: (a) 2-cells of the arrangement atoms; (b) atoms in different colors exploded about the origin.

Example 6.4.2 (3D arrangement) A unit test for computation of the 3D arrangement generated by 10 cubes is shown in Figure 6.11.

Example 6.4.3 (2D/3D arrangements) In this case, we show the first examples of 2D and 3D arrangements produced by the experimental Julia package `LinearAlgebraicRepresentation.jl` in 2018 before submitting for publication (2020) the paper [10].

In Figure 6.4a, we show the 2D arrangement X_2 of $\mathbb{E}^2$ generated by a set of random line segments. The Euler characteristic is $\chi = \chi_0 - \chi_1 + \chi_2 = 11361 - 20813 + 9454 = 2$.

The arrangement of $\mathbb{E}^3$ is obtained in Figure 6.4b by merging two 3-complexes for 2×10^3 3-cells. The Euler characteristic (of the resulting 3-complex is $\chi = \chi_0 - \chi_1 + \chi_2 - \chi_3 = 8787 - 26732 + 26600 - 8655 = 0$. These counts include the outer (unbounded) 2-cell or 3-cell, respectively, that are computed by the Topological Gift Wrapping (TGW) algorithm (see Section 6.3.5).

The Euler characteristic of the two configurations is correct because the plane $\mathbb{E}^2$ arrangement is homeomorphic to the (surface) 2-sphere, whereas the 3D arrangement is homeomorphic to the 3-sphere. The Euler characteristic of the d-sphere is $\chi = 1 + (-1)^d = 2$ or 0 for even or odd space dimension d.

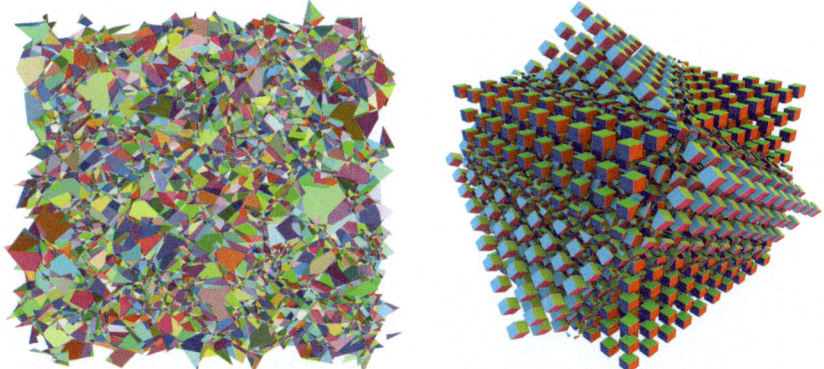

Fig. 6.12 Examples of space arrangements by merging cellular complexes: (a) 2D arrangement produced by random line segments; (b) 3D arrangement generated by two groups 10×10 of cubes.

The following example is relative to an example of $\mathbb{E}^2$ space `arrangement` generated by 50 random polygons. Note that `Plasm` also exports the `RandomBubble()` test function from the software package.

```
hpc = STRUCT([RandomBubble() for I in 1:50])
arrangement = ARRANGE2D(LAR(hpc))
VIEWCOMPLEX(arrangement)
VIEWCOMPLEX(arrangement, explode=[1.5,1.5,1.5])
```

In the above script, we have shown the typical pattern needed for this purpose. First, the set of generators, in this case, 50 random polygons, is correctly assembled within a `Hpc STRUCT` collection; then it is transformed into a `Lar` cellular complex and passed as an argument to the functions `ARRANGE2D` or `ARRANGE3D` in case of three-dimensional dataset, producing all the atoms of the `arrangement` partition.

The originated `arrangement` value is a `Lar` value describing all details of the generated set of atoms in the 2D space partition $\mathcal{A}(\mathbb{E}^2)$, including an `outer` unbounded atom, which is not explicitly drawn by `VIEWCOMPLEX` by default.

Example 6.4.4 (2D/3D arrangements)
In this section, we show the first examples of 2D and 3D arrangements produced by our experimental Julia package `LinearAlgebraicRepresentation .jl` in 2018 before submitting for publication the paper [10] (2020).

In Figure 6.14, we show the arrangement $\mathcal{A}(\mathcal{S})$ generated by the collection $\mathcal{S}$ made by the thirty-five 2-faces of 5 randomly intersecting cubes. According to [10], each 3-cell in $\mathcal{A}(\mathcal{S})$ is generated by a column of the sparse matrix of boundary map $\partial_3 : C_3 \to C_2$, with values in $\{0, 1, -1\}$.

The matrix of any linear map between two linear spaces contains by columns the basis of domain space represented in the basis of the codomain

Fig. 6.13 $\mathbb{E}^2$ arrangement with random polygons: (a) basis 2-chain $U_2 \subset C_2$; (b) Exploded cell complex. **LAR** cells are neither convex nor contractible.

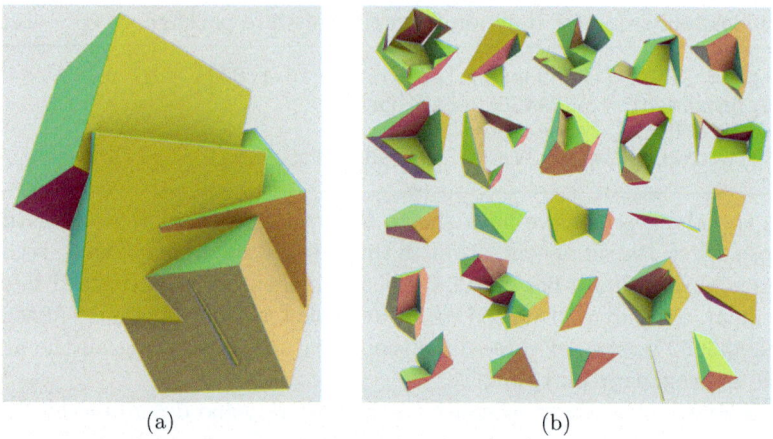

(a)   (b)

Fig. 6.14 (a) A collection $\mathcal{S}$ of five random cubes in $\mathbb{E}^3$; (b) the display of 3-cells of the generated $\mathbb{E}^3$ arrangement $\mathcal{A}(\mathcal{S})$ (not in scale, and suitably rotated to exhibit their complex shape better). The quasi-disjoint union of all atoms gives the five cubes. Note that some atoms contain holes.

space. Therefore, the columns of $[\partial_3] : C_3 \to C_2$ are 2-cycles, i.e. closed chains in C_2. In particular, elementary 3-chains are join-irreducible atoms of the CSG algebra with closed regular cells. They may be non-contractible to a point (when they contain holes) and non-convex. The outer cell is the complement of their union. Any geometric model (out of 2^{25}) from the Boolean CSG algebra generated by these five cubes is made by a subset of those 25 atoms.   □

6.4.1 Boundary

The boundary schemes, denoted Brep or brep, represent a 3-solid by representing its 2-boundary. The boundary is decomposed into 2-cells, each represented by the 1-cells of its boundary. Their two extreme 0-cells specify the 1-cells in a PL domain. The standard terminology from dimension $3 \to 0$ is: solid cells (C), faces (F), edges (E), vertices (V). Other terms of discourse include "shell" (connected component) of solid boundary and "loop" of face boundary (idem). Curved objects are usually depicted by combining various parametric (and less often algebraic) equations with topological elements. Boundary representation schemes are widely used in industry and academia and are often coupled with one or more schemes discussed below.

For a BOUNDARY operator construction and visualization with Lar datasets, readers should refer to example 7.5.3, which generates the Brep of a Boolean UNION expression. Instead, the BREP operator is used with the Hpc data structure and returns a Hpc boundary triangulation. It is mainly used to prepare a boundary model for domain integration.

6.4.2 Decompositive

In this scheme, the solid model is typically partitioned into 3-cells with the same (or similar) topology, and it usually satisfies the rules for defining cellular complexes. The decompositive scheme is mainly used for computer simulation of differential equations and is rarely used for shape design. The cells are convex and represented by their vertices and other discrete boundary points called nodes.

In the Plasm novel Boolean approach, we use a decompositive representation with general polyhedral cells, which form a partition of the embedding space $\mathbb{E}^d$, where each unit chain (cell) is represented by its boundary and called atom of the arrangement $\mathcal{A}(S) \equiv C_d$.

The 3D Brep of an atom in our approach is a column of a (sparse) matrix operator,

$$\mathsf{CF} : \mathsf{F} \to \mathsf{C} \equiv \delta_{d-1} : C_{d-1} \to C_d,$$

providing its 2-cycle, as discussed in the following.

6.4.3 Enumerative

The standard definition of this scheme resorts to some grid decomposition of the space and to listing space elements belonging to the object. Examples include hierarchical decompositions of 3-domains using octrees [1] or potrees [13] to store, transmit, visualize on the web, and reconstruct the geometry from scans of solid objects, neighborhoods, and buildings with remote sensors.

In our research with cellular complexes, we extensively use multidimensional, hierarchical, and nonhierarchical grids of cuboidal and simplicial cells (see many examples in the appendix). In particular, and not only, such geometric objects are used as the domains of manifold curves, surfaces, and varieties of higher dimension.

In the `Plasm` approach to curved geometric design, we typically `MAP` a vector function `F(u,v,w) = CONS([f(u),f(v),f(w)]) = (x,y,z)` using coordinate functions on the vertices of d-grid domains without changing the topology, and thus obtaining the PL approximated value of the curved geometric model. See the book
citePaoluzzi2003a for many curved geometry methods and examples.

Chain complex

The highly parameterized complex generator `solidMapping = THINSOLID(`
`surface)` in the parametric script `SOLIDHELICOID` below computes the local
normal by partial derivatives.

```
degree = 3
ControlPoints = [[0.,0.],[-1,2], [1,4],[2,3],[1,1],[1,2],
    [2.5,1],[2.5,3],[4,4],[5,0]]) # 10 points
knots=[0,0,0,0, 1,2,3,4,5,6, 7,7,7,7] # 14 knots
VIEW(DISPLAYNUBSPLINE(degree, knots, ControlPoints))
```

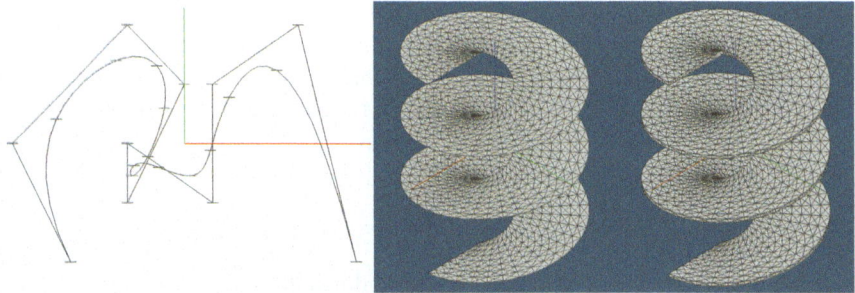

Fig. 6.15 User-defined curved 1-, 2-, and 3-manifolds: (a) Cubic non-Uniform B-spline curve and control polyline passing for the extreme control points. Look at the relation between knots, curve segments and degree: $14 = 10 + 3 + 1$; (b) highly parametric surface; (b) solid helicoid.

```
function SOLIDHELICOID(; nturns=3, R=1., r=0.5, shape=[36*nturns
    , 8], pitch=2, thickness=0.1)
    totalangle = nturns*2*pi
```

```
    grid2D = INTERVALS(36*nturns)(36*nturns) * INTERVALS(4)(8)
    Domain2D = T([2])([r])(S([1,2])([totalangle/shape[1],R-r])(
    grid2D));
    surface = p->let(u, v)=p;[v*cos(u); v*sin(u); u*(pitch/(2*pi)
    )] end
    VIEW(MAP(surface)(Domain2D))
    solidMapping = THINSOLID(surface)
    Domain3D = Domain2D * INTERVALS(thickness)(1)
    VIEW(MAP(solidMapping)(Domain3D))
end;
```

Typically, `Plasm` grids are born with integer coordinates of vertices and are scaled to proper dimensions depending on use. It is worth noting that such objects are highly non-manifold in their interiors. Interesting generation methods and use patterns can be found in the Architecture, Engineering, and Construction (AEC) design and production domains since orthogonal geometries make up most buildings.

In [9], Section 2.5, and Example 3.2.7, we demonstrate some interesting characteristics of combinatorics of product topological spaces. In particular, when these spaces have dimensions of 1 or 0, i.e., they consist of line segments or discrete points. When X, Y, Z are 1D complexes, the space X * Y * Z has a dimension of 3; if some of the product terms have dim = 0, the overall product dimension decreases accordingly and provides topological templates for intriguing geometric constructions.

6.4.4 Constructive Solid Geometry (CSG)

This scheme is highly successful in industry because its logic emulates the design process for mechanical parts, from idea generation through refinement and production. A CSG scheme causes the concept of assembly of model parts to meet together with the elementary process operations of union, intersection, and difference.

A CSG expression is often implemented as an incomplete binary tree with Boolean operations or affine transformations on non-leaf nodes and primitive objects on leaves. The original CSG schemes used solid objects (linear/non-linear equalities/inequalities) as leaves and implemented the boolops in terms of systems of polynomial inequalities. At present, a more combinable use of Brep primitives looks preferred.

The `Plasm` generalization of the CSG representation scheme will be presented and widely discussed in the next chapter, using binary Boolean algebras based on arrangements of the embedding Euclidean space.

Chain subcomplexes

`Plasm` allows for generating complexes via the topological product of others. With Julia extensions for symbol meaning, the '`*`' operator, which is naturally commutative and associative, can be used easily and correctly. Alternatively, both `*` or its `POWER` alias can be used as n-ary operators by means of `FL` combinators `INSR` (insert right) or `INSL` (insert left).

```
X = GRID([2.4,4.5,-3,4.5,2.4])
Y = GRID([7,5]); Z = GRID([3,3]);
floors = X * Y * SK(0)(Z)
VIEWCOMPLEX(LAR(floors))
envelope = LAR(STRUCT(X*SK(0)(Y)*Z, SK(0)(X)*Y*Z))
VIEWCOMPLEX(envelope)
```

Fig. 6.16 (a) 2-complex of horizontal floors; (b) 2-complex of vertical enclosures and partitions.

As we make clear in Section 7.4, `Plasm` solid Boolean algebras are more general, faster, and efficient than standard CSGs evaluation, implemented as a succession of assessments of binary operators, where partial results are used as input to a subsequent operation. In particular, they are less subject to numerical error propagation since all numerics are computed in advance in 2D, and later the algorithmic part is only topological, i.e., symbolic.

6.4.5 Primitive Instancing

This scheme typically provides users with an extendable library of geometric shapes, often described as generating functions parameterized by formal parameters (via one or more methods in Julia).

The internal representation in `Plasm` is usually Brep. It includes both default and user-defined semantic parameter values and supports 1-, 2-, or 3-variable resolution discretization (for curves, surfaces, and solid primitives), depending on an appropriate approximation of intrinsic shape dimensions.

The Julia `Plasm` package provides a fairly extensive library of predefined shapes and geometric constructions within the `fenvs.jl` module, which stands for functional environments. The many examples assist users in quickly implementing extensions.

References

1. Brunet, P., Navazo, I.: Solid representation and operation using extended octrees. ACM Trans. Graph. **9**(2), 170–197 (1990). DOI 10.1145/78956.78959. URL https://doi.org/10.1145/78956.78959 pages 237
2. Cormen, T.H., Leiserson, C.E., Rivest, R.L., Stein, C.: Introduction to Algorithms, Third Edition, 3rd edn. The MIT Press (2009) pages 231
3. Delfinado, C., Edelsbrunner, H.: An incremental algorithm for betti numbers of sinplicial complexes on the 3-sphere. Computer Aided Geometric Design **12**, 771–784 (1995) pages 96, 222
4. DelMonte, G., Onofri, E., Scorzelli, G., Paoluzzi, A.: Local congruence of chain complexes. CoRR **abs/2004.00046** (2020). URL https://arxiv.org/abs/2004.00046 pages 224, 231
5. DiCarlo, A., Paoluzzi, A., Shapiro, V.: Linear algebraic representation for topological structures. Computer-Aided Design **46**, 269–274 (2014). DOI 10.1016/j.cad.2013.08.044. URL https://doi.org/10.1016/j.cad.2013.08.044 pages 47, 137, 184, 186, 218
6. Fogel, E., Halperin, D., Kettner, L., Teillaud, M., Wein, R., Wolpert, N.: Arrangements. In: J.D. Boissonat, M. Teillaud (eds.) Effective Computational Geometry for Curves and Surfaces, Mathematics and Visualization, chap. 1, pp. 1–66. Springer (2007) pages 221
7. Goodman, J.E., O'Rourke, J., Tòth, C.D. (eds.): Handbook of Discrete and Computational Geometry – Third Edition. CRC Press, Inc., Boca Raton, FL, USA (2017) pages 221
8. Jarvis, R.A.: On the identification of the convex hull of a finite set of points in the plane. Information Processing Letters **2**(1), 18–21 (1973) pages 231
9. Paoluzzi, A., Scorzelli, G.: Computational topology, boolean algebras, and solid modeling. Computer-Aided Design **181**, 103,839 (2025). DOI https://doi.org/10.1016/j.cad.2025.103839. URL https://www.sciencedirect.com/science/article/pii/S0010448525000016 pages 37, 47, 65, 143, 218, 221, 239
10. Paoluzzi, A., Shapiro, V., DiCarlo, A., Furiani, F., Martella, G., Scorzelli, G.: Topological computing of arrangements with (co)chains. ACM Trans. Spatial Algorithms Syst. **7**(1) (2020). DOI 10.1145/3401988. URL https://doi.org/10.1145/3401988 pages 47, 94, 137, 216, 218, 221, 224, 227, 228, 231, 233, 234, 235, 251, 258, 260, 278
11. Paoluzzi, A., Shapiro, V., DiCarlo, A., Scorzelli, G., Onofri, E.: Finite algebras for solid modeling using julias sparse arrays. Computer-Aided Design **155**, 103,436 (2023). DOI https://doi.org/10.1016/j.cad.2022.103436. URL https://www.sciencedirect.com/science/article/pii/S0010448522001695 pages 47, 65, 97, 137, 184, 218, 221, 254, 260, 270, 277
12. Requicha, A.: Representations for rigid solids: Theory, methods and systems. ACM Computing Surveys **12**(4), 437–464 (1980). URL https://doi.org/10.1109/TC.1980.1675470 pages 64, 74, 138, 218, 244

13. Schutz, M., Ohrhallinger, S., Wimmer, M.: Fast out-of-core octree generation
    for massive point clouds. Computer Graphics Forum **39**(7), 13 (2020). DOI
    10.1111/cgf.14134. URL https://www.cg.tuwien.ac.at/research/publications/
    2020/SCHUETZ-2020-MPC/ pages 141, 237

Chapter 7
Boolean solid algebras

Engineers created Solid Modeling as a precise and universal language for geometry-based engineering. In this context, Plasm (Programming Language for Solid Modeling) stands out as an engineering metalanguage, offering a robust framework for representing and manipulating solid geometries. Mathematically, Plasm solid models are symbolic computer representations of expressions in algebraic systems, constructed through algorithms that mirror set operations of these algebras.

This chapter unfolds the theoretical foundations and practical applications of the Julia Plasm package in modeling Boolean algebras of solid shapes. We explore how Plasm enables the creation and manipulation of complex geometric models by leveraging homological theory and set algebra. This approach is quasi-isomorphic to Constructive Solid Geometry (CSG), expanded with n-ary operations and an explicit representation of external space.

Key topics include the algebraic relationships between solid algebras, constructive solid geometry, enumerative methods, and boundary representations of piecewise-linear (PL) polyhedra. We demonstrate that arranging a set of spatially instantiated primitive shapes corresponds to the finite Boolean algebra of regularized point sets, encompassing all possible Boolean expressions generated from a given set of primitives.

By the end of this chapter, readers will have a thorough understanding of the mathematical principles and computational techniques behind Julia Plasm's solid modeling capabilities. In particular, we show that resolving any solid algebra expression with a finite number of geometric primitives, once the spatial arrangement and specific point-set memberships are established, algebraically simplifies to Julia's set operations on bit strings, sparse matrix-vector multiplication, and shape reconstruction from algebraic topological atoms.

© The Author(s), under exclusive license to Springer Nature Switzerland AG 2026   243
A. Paoluzzi and G. Scorzelli, *BIM Geometry with Julia Plasm—Functional Language for CAD Programming*, Digital Innovations in Architecture, Engineering and Construction, https://doi.org/10.1007/978-3-031-90244-4_7

7.1 Plasm and solid representations

Woelcker and Requicha introduced a classification of the more diffused representation schemes used in solid modeling and CAD systems developed in the first twenty years in research reports and the milestone paper [38].

Remark 7.1 This classification remains relevant more than 40 years later. It offers reasoning and evaluation frameworks that have inspired much of our work with Julia Plasm. In particular, it aids in comparing today's systems and designing new ones, especially in the era of AI and generative software.

Taxonomy of representation schemes

The [38] classification is briefly summarized in the following pages to provide the user with some indication of addressing itself to Plasm scripts, with perception enhanced by knowledge of the various approaches to solid modeling.

Primitive instancing

The creation of classes of similar shapes using software primitives is a procedural method. Each object is represented by a tuple with the name of a generating procedure or function as the first element, followed by the actual values of its formal parameters. The domain and co-domain of the scheme can be divided into non-overlapping classes. There is a one-to-one correspondence between model classes and (parametric) representations of functional style. This traditional approach has been revitalized through object-oriented software development techniques and is mainly used with parametric and feature-oriented modeling systems. It requires an underlying, more abstract representation scheme where general utility methods (such as Boolean operations and integration operators) are implemented. Plasm exemplifies this paradigm by acting as a model for its users. It provides a comprehensive library of predefined Julia generator functions (also used here as coding examples) for geometries recognizable by name and user-defined parametric shapes. Notably, and quite unusually, in Plasm, the parameters of object model generators can even include parametric generators of other geometric classes as actual parameters [27].

Enumerative schemes

A solid model is described by enumerating the set of solid cells in some partitioning of the embedding space. It is possible to distinguish between schemes using sparse Boolean matrices as shape parcel enumerators and schemes based upon hierarchical space decompositions, say quadtrees and octrees, in the 2D and 3D case, respectively. In most cases, such representations are approximations of the object's space occupancy, even for linear polyhedra.

This scheme is also adopted in Plasm, e.g., by the CUBOIDGRID and SIMPLEXGRID generator functions, which accommodate basic topological cells, such as small cubes or simplices, within textured (even curved) shapes used, for example, in 3D printing and new multi-material shapes.

The hierarchical enumerative approach can be simulated in Plasm by using the operators LEFT, RIGHT, UP, and DOWN in 2D and 3D; TOP and BOTTOM in 3D, always between Hpc objects, which can easily be converted into graph structures in computer memory. In Plasm, the enumerative approach may be extended to grids that alternate similar solid shapes and empty intervals of embedding space. This method significantly speeds up the rapid development of preliminary design models, especially in architecture and civil engineering.

Decompositive schemes

The object is represented as a set of cells, usually of a given topology. Unlike enumerative schemes, the represented object is induced by the cell partition.

The BSP (Boundary Space Partition) trees belong to this type of representation [26]. As we have already seen, BSP trees are binary trees where each internal node is associated with one of the boundary hyperplanes of the object, and each leaf node represents a convex cell, either whole or empty, of the space partition induced by such a set of hyperplanes. Decompositive schemes also support more abstract and generalized representations of topology [22] and unify approaches to geometric representation and physical simulation [46].

Decompositive schemes are central in Plasm, both in the hierarchical object type Hpc, where all the leaves of the structure graph contain convex cells, and in the cell complexes of Lar type, but with the significant difference that the 3-cells are polyhedral and may be nonconvex, and even noncontractible. The homological pipeline is entirely based on the Lar scheme, except for the first step, which uses hierarchical graphs, and the last step, which may convert back to Hpc to maintain interoperability between the input and output of very complex algebraic processes.

Boundary schemes

These are the most common representation schemes used in computer graphics and in most modeling kernels, often employing specific data structures designed 'ad hoc'. Many of the earlier algorithms and procedures [53, 1, 3, 6, 7, 8, 2, 10, 13, 14, 18, 9, 20, 21, 23, 25, 30, 28, 29, 33, 35, 36, 37, 39, 51, 44, 43, 45, 48, 49, 54, 55, 57, 58, 59, 60, 61, 5, 42, 19, 16, 17] work with data structures optimized for selected classes of geometric objects, where the interior of a body is represented using its external boundary and, in particular, partitioning it into shells, faces, loops, edges, and vertices. Each type of boundary representation stores some subset of the binary adjacency relations between the various boundary entities. In particular, there is a notable difference between the boundary representations of manifolds and those of non-manifold models in the solid modeling literature.

In Plasm, the same simple Lar scheme applies to both, so it does not matter whether the number of faces incident upon an edge is two, three, four, or more. Even a single face may be incident on one edge at the boundary of a surface patch. The Plasm BREP and BOUNDARY of solid shapes are easily computed algebraically when the atoms of the homological decomposition are known (see Section 7.5) and employed by other powerful language operators, including SKELETON and OFFSET.

In particular, in Plasm, a model using the Lar data type must be visualized with the VIEWCOMPLEX primitive, which has more predefined viewing attributes than the VIEW primitive, reserved for Hpc objects. Also, Lar objects cannot be used in STRUCT or MAP expressions, but converting from Lar to Hpc type is always possible. Analogously, while all boolops apply to Hpc structures as [semantically richer] aliases of STRUCT, the BOOL results are transformers from their Hpc input to Lar datasets. Like the LAR transformer accepts only Hpc objects, BOUNDARY works only on the Lar output of BOOL.

Readers will soon find that matching the most valuable operators with the two boundary data structures they manage efficiently is easy to understand and remember.

CSG schemes

The acronym stands for Constructive Solid Geometry. In this scheme, the representations are binary trees with internal nodes representing regularized set operations or affine transformations. In contrast, leaf nodes represent primitive solids or implicit half-spaces (usually linear or quadratic). An explicit boundary representation is obtained by suitably traversing the tree primitives. In this case, we speak of boundary evaluation algorithms [40].

Remark 7.2 (CSG and Solid Boolean Algebras) In Plasm, the CSG scheme is extended to a proper Boolean Algebra of solid shapes by introducing the complement object of a cellular complex and the complement operation by inverting the normal vectors of Brep model surfaces. As someone suggested, our approach could be named GCSG for Generalized Constructive Solid Geometry. Still, we do not like the plethora of acronyms that sprouted up in Solid Modeling over the decades.

The meaning of CSG is widely recognized in all areas of computer graphics and geometric modeling, which is why we refer to our generalized approach to Boolean Solid Algebra as either CSG or Solid Algebra.

Remark 7.3 (All operations performed simultaneously) Solid operations (i.e., UNION, INTERSECTION, DIFFERENCE, and XOR) between n objects (n generators) of our CSG are performed simultaneously in Plasm, while the traditional CSG performs the solid operations one by one. This approach is implemented more efficiently and reliably and seems a good step forward in solid modeling.

Remark 7.4 (Training of a generative transformer) While evaluating a standard CSG tree yields only one solid model, our solid algebra can define the solid values for all expressions derived from the same n primitives. In theory, this allows for a low-cost evaluation of 2^m shapes within this algebra, where $m > n$ is the number of atoms in the specific spatial arrangement, yielding a large number of shape instances for training a generative transformer model. We believe this also marks a significant progress.

Some special attention should be given to the rendering of CSG models. There are typically two ways to render a constructive solid geometry model. The first way is to convert it to a Brep model and use a z-buffer. The second way is to use a CSG direct display algorithm, which is often based on a ray-casting technique: after traversing all primitives by a ray, you combine the fragmented ray intervals using the CSG expression [50, 41, 12]. This approach combines a relatively simple implementation with accuracy at the expense of high computational cost (everything needs to be repeated for every pixel of the image and for every view). A complex CSG model is displayed correctly on the screen because the rendering engine handles it. This approach is not (yet) implemented in Plasm.

Sweeping schemes

Several solids are defined in mechanical engineering, architecture, and construction applications by properly moving a base surface patch along an assigned profile curve. In this case, the representation contains both the generating surface and the curve [52]. A sweeping scheme with a parametric representation of surfaces and curves is straightforward. Approaches to this class of schemes can be reduced to a generalized Cartesian product of sets.

In particular, in Plasm, we have a family of generator functions, referred to as Minkowski operators, that operate by piecewise-linear repeated extrusion of the base cellular complex or a single cell into higher-dimensional Euclidean spaces, ultimately projecting the resulting complex back into three-dimensional space. This is particularly true for the OFFSET operator when applied to 2-complexes or 1-complexes embedded in a 3D space (see Example 3.2.1). It is important to note that the EXTRUDE function consistently maps a d-dimensional object into a $d+1$-dimensional object. The input and output of these operators are Hpc type, allowing them to be applied repeatedly to move into higher-dimensional spaces. The SWEEP operator also belongs to this class of geometric tools but is limited in the argument dimension.

Composite schemes

Some commonly used schemes use multiple types of representations. A boundary representation is typically cross-bred with another representation in a composite scheme, often a CSG or decompositive or primitive instancing [47, 40]. For example, there are some exciting composite octree/boundary representations. Geometric representations [28, 4, 34, 11], discussed in the following sections, can be considered as a composite boundary/decompositive/sweeping representation.

Import of SVG files

Plasm lacks a graphical user interface (GUI), but it can import 2D designs from svg files. SVG, which stands for Scalable Vector Graphics is the W3C standard for 2D vector graphics on the Web. The standard is based on XML and supports several primitives, including interactive exchange via mouse clicks and touch events, and animated visualization in response to user actions. The primitives can be filled with a color or a gradient. The wireframe drawing (stroke) can also be drawn with a specific width. Some features will be ported to Plasm in a future version.

```xml
<?xml version="1.0" encoding="UTF-8" standalone="no"?>
<!DOCTYPE svg PUBLIC "-//W3C//DTD SVG 1.1//EN"
"http://www.w3.org/Graphics/SVG/1.1/DTD/svg11.dtd">
<svg width="100%" height="100%" viewBox="0 0 3508 2481" version="1.1"
xmlns="http://www.w3.org/2000/svg" xmlns:xlink="http://www.w3.org/1999/xlink"
xml:space="preserve" xmlns:serif="http://www.serif.com/"
style="fill-rule:evenodd;clip-rule:evenodd;stroke-linecap:round;stroke-linejoin:
round;stroke-miterlimit:1.5;">
    <g transform="matrix(1,0,0,1,120.646,41.5716)">
        <path
d="M919.761,657.928L1111.6,1299.26L727.981,1604.36L1144.72,1827.95L1704.27,
1109L1908.3,1690.11L2510.51,1240.16L2538.6,385.068L1926.77,323.822L2095.9,882.
356L919.761,657.928Z" style="fill:none;stroke:black;stroke-width:6.25px;"/>
    </g>
    <path
d="M1930.07,1024.97L1954.12,1301.57L2380,1137.35L2337.38,933.343L1930.07,1024.
97Z" style="fill:none;stroke:black;stroke-width:6.25px;"/>
</svg>
```

Fig. 7.1 One fifth of the .svg source file generating the wire-frame part of Figure 7.2.

Currently, Plasm accepts input from external files specified by their local or remote paths and names that end with the .svg suffix. The file name and path/URL are provided as strings to the Plasm function SVG(), capitalized as the language primitives are. This function defines scalable images on the output device using a small set of graphics primitives, including lines, rectangles, circles, and splines (piecewise quadratic or cubic Bézier curves).

The SVG primitives interpreted by Plasm at the time of this writing are lines, rectangles, circles, and paths, that is, spline curves. A spline is a piecewise polynomial curve of low degree. An SVG file is usually created using an interactive graphical interface on design and animation web platforms. Figure 7.1 shows a small portion of a file of this type. Two corresponding cellular complexes generated after the SVG file parsing and Plasm interpretation are visualized in Figure 7.2.

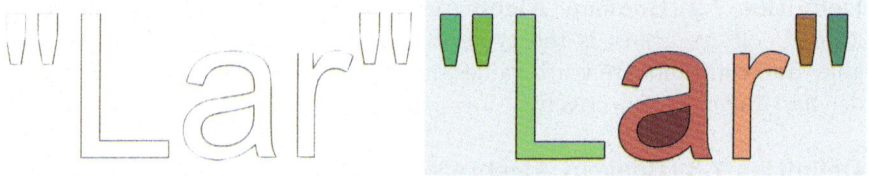

Fig. 7.2 Images generated by Plasm: (a) one-dimensional Hpc model generated by the expression SVG("filename.svg"). Note that the model of the "Lar" string includes several curved splines; (b) atoms in different colors generated by ARRANGE2D primitive. The outer (irreducible to a point) atom and cell is the ivory background.

The content of Figure 7.2 provides strong evidence of the power and importance of the homological approach to solid modeling, which the entire Plasm language is based on. In our experiment, we parsed the .svg file, extracting from the control points of its "path" splines a vector of closed 1-chains corresponding to the Plasm evaluation of low-degree Bézier curves, and converting them to closed polylines in Hpc format. This array of polylines was transformed by the LAR operator into a Lar object, then passed to ARRANGE2D to generate the corresponding arrangement of the space $\mathbb{E}^2$.

Remark 7.5 (The power of ARRANGE2D *operator)* The two drawings in Figure 7.2 were generated by CHAINCOMPLEX with show=["EV"] and show=["CV"] using the usual random colors. The first was filled out in white in the Apple Keynote application. Needless to say (no pun intended), the 2-cycle color "filling" of 2-chain atoms at the right-hand side was done by the standard visualization operator CHAINCOMPLEX of a Lar object.

Importing and parsing .svg files generated by interactive design environments and stored locally or in the cloud is a key link between the Plasm language running APIs or the REPL interpreter on a local or remote terminal and the CAD processing of file content. So far, we have only conducted a few experiments connecting the geometric language with this web tool, but it is a high-priority project for the Plasm platform.

7.2 Finite Boolean Algebras

Definition 7.1 (Finite Algebra) Let $\mathcal{A}$ be a non-empty set and the operations $\otimes_i : \mathcal{A}^{n_i} \to \mathcal{A}$ be functions of the n_i arguments. If for all i A is closed under $\otimes_i$, then the system $\langle \mathcal{A}; \otimes_1, \ldots, \otimes_k \rangle$ is called an algebra. Alternatively, the algebra $\mathcal{A}$ is a set with operations $\otimes_1, \ldots, \otimes_k$. If $\mathcal{A}$ has a finite number of elements, the algebra is said to be finite.

Definition 7.2 (Boolean Algebra) In mathematics and mathematical logic, Boolean algebra is the type of algebra in which the values of variables and constants are truth values, i.e., their value is true or false, usually denoted 1 and 0, respectively.

Definition 7.3 (Boolean Algebra's Atom) in Boolean algebra, an atom is a minimal nonzero element. Formally, an element a in a Boolean algebra $\mathcal{B}$ is called an atom if:

1. $a \neq 0$ (a is not the zero element), and
2. If $0 \leq b \leq a$, then $b = 0$ or $b = a$.

Atoms represent the smallest building blocks in the lattice structure of a Boolean algebra[1].

Definition 7.4 (Finite Boolean Algebra)
A Boolean Algebra $\mathcal{B}$ with a finite number n of atoms is a finite Boolean algebra. Every such algebra is isomorphic to the set of subsets of the first n integers $\mathcal{P}[n]$ with set operations of union and intersection, and therefore isomorphic to the Boolean algebra $\mathbf{2}^n$ of Boolean vectors of length n with bit union and intersection.

We may think of a finite Boolean algebra $\mathcal{B}$ as a set isomorphic to the power set $\mathcal{P}(X)$ of some finite set X. The power set is naturally equipped with complement, union, and intersection operations, which correspond to $-, \vee, \wedge$ operations in Boolean algebra. The complement of A is also denoted as $\overline{A}$. More precisely, we have

Property 7.1 (Boolean algebra) Any finite algebra $\mathcal{P}(X)$ is isomorphic to the Boolean algebra $\mathcal{B}(X)$, which can be represented as the set $\{0, 1\}^n = \mathbf{2}^n$ of 2^n binary strings of length n. Thus, $\mathcal{B}(X)$ corresponds one-to-one with the set $\chi_X(\mathcal{P}(X))$ of characteristic functions of the elements of $\mathcal{P}(X)$ with respect to X^2.

---

[1] The abstract structure lattice consists of a partially ordered set in which every pair of elements has a unique least upper bound or join and a unique greatest lower bound or meet.

[2] This correspondence between the Boolean algebra of subsets of a finite set and the algebra of binary strings follows from the classical algebra of Boole, and, in the general case, from Stone's representation theorem.

Vadim Shapiro noted in his doctoral dissertation [45] the isomorphism between the decompositions of space and finite algebras of sets that are uniquely and canonically represented by the union of atoms in the decomposition. Shapiro's Ph.D. thesis also introduced a hierarchy of algebras to formally define a family of Finite Set-theoretic Representations (FSR) of semi-algebraic subsets of $\mathbb{E}^d$, including representation schemes for solid and non-solid objects, such as B-reps, Constructive Solid Geometry, cellular decompositions, Selective Geometric Complexes, and others.

Definition 7.5 (Semialgebraic set) A semialgebraic set is a subset of $S \subset R^n$ defined by a finite collection of real polynomial inequalities of the form $Q(x_1, ..., x_n) \geq 0$. Finite unions, intersections, complements, and projections of semi-algebraic sets are still semialgebraic sets. Moreover, a semi-algebraic set has only finitely many connected components, and each is also semi-algebraic [24].

In this book, we represent and implement for $d = 2, 3$ the solid Boolean algebras of CSG with closed regular cells, generated by the arrangement of $\mathbb{E}^d$ induced by a collection of cellular complexes with polyhedral cells of dimension $d - 1$ [31].

Definition 7.6 (Generators) A set $\mathcal{H}$ generates the algebra $\mathcal{A}$ (under some operations) if $\mathcal{A}$ is the smallest set closed w.r.t. the operations and containing $\mathcal{H}$. The elements $h_i \in \mathcal{H}$ are called generators of the algebra $\mathcal{A}$.

The elementary shapes in Brep in a `Plasm` solid expression are the generators of its own Solid Boolean Algebra.

Definition 7.7 (Atom) An atom is an element that cannot be decomposed into two proper subsets, similar to a singleton, which cannot be expressed as a union of two strictly smaller subsets. An atom is a minimal non-zero element; a is an atom if and only if, for every b, either $b \wedge a = a$ or $b \wedge a = 0$. In the first case, we say that a belongs to the structure of b.

Definition 7.8 (Structure of algebra elements) We call structure of $b \in \mathscr{P}(X)$ the atom subset S such that b is the irreducible union of S. By extension, we also call structure of b the binary string associated with the ordered sequence of its atoms (elements of X). In other words, the structure $S(b)$ is the image of the characteristic function $\chi_X(S)$.

Relation between atoms and generators

In a Boolean algebra, atoms and generators have distinct roles and definitions. Here is a breakdown of their differences:

Table 7.1 Relation between atoms and generators

Aspect	Atoms	Generators
Definition	Minimal non-zero elements of the Boolean algebra.	Elements from which the entire Boolean algebra is derived.
Existence	Not all Boolean algebras have atoms.	All Boolean algebras have generators.
Role	Represent the "smallest" elements.	Provide a framework for constructing the algebra.
Size	Typically, many atoms in a Boolean algebra.	A generating set can vary in size, even just one element.

In summary, atoms are individual elements related to the structure and size of Boolean algebra, while generators are elements used to build or describe this algebra. When dealing with Boolean strings in finite Boolean algebras, atoms contain only one non-zero element iff the generators of the algebra do not intersect.

The fundamental property

At the core of the Boolean operations between solid rigid models implemented in our `Plasm` language, there is the following fundamental property:

Property 7.2 (Boolean atoms are unit 3-chains) There is a natural transformation [3] between d-chains defined on a spatial arrangement and the solid algebra generated by that arrangement.

Informally, the notion of a natural transformation states that a particular map between functors can be done consistently over an entire category. Natural transformations are, after categories and functors, one of the most fundamental notions of category theory [56].

In particular, unit d-chains correspond to atoms of the algebra; the $[c]$ coordinate representation (bit array) of any d-chain c generates the coordinate representation $[b]$ in boundary space: $[\partial_d][c] = [b] \in B_{d-1} \subset Z_{d-1} \subset C_{d-1}$ (see Figures 3.21, and 7.3.2). Remember, when $d = 3$, for the properties of matrices of linear operators between linear spaces, the matrix $[\partial_3] : C_3 \to C_2$ contains the domain basis (unit 3-chains) represented in the target space.

---

[3] In category theory, a branch of abstract mathematics, a natural transformation provides a way of transforming one functor into another while respecting the internal structure (i.e., the composition of morphisms) of the categories involved. Hence, a natural transformation can be considered a "morphism of functors."

7.2.1 Solid algebra expression resolution

In this section, we introduce the evaluation method for `Plasm` expressions in
solid algebra with `Hpc` objects, including variadic `UNION`, `INTERSECTION`, and
`DIFFERENCE` operators, as well as `COMPLEMENT` and `XOR`, along with parentheses
to specify the order of operations.

Remark 7.6 (Hierarchical semantics of solid forms)
Solid objects are typically defined as hierarchical assemblies of solid primitives
or more complex shapes, each represented in a local coordinate system. Most
graphics and modeling systems implement this semantics as a hierarchical
graph, where affine geometry within the nodes is defined in local systems.
Arcs are associated to affine transformations that shift the entire subgraph
rooted in the ending node onto the coordinate system of the initial node.

The `Plasm` Boolean method executes a Depth First Search (DFS) traversal
when evaluating the `Hpc` tree of the input Boolean expression, where geometry
is stored on the leaves in local coordinates, and non-leaf nodes may contain
either affine transformations or Boolean operators (see Example 7.5.1).
    We have five tasks in sequence:

Algorithm 7.1 (Boolean pipeline)

1. DFS traversal to get all solid terms (algebra generators) in root coordinates
   and store local Boolean function names as placeholders in `properties`
   fields of `Hpc` nodes (see Example 7.5.2);
2. construction of the global equivalent Boolean function made by combining elementary Boolean functions, selector functions, and parentheses (see
   Example 7.5.4);
3. execution of the embedding space arrangement, with boundary generation
   of each oriented atom cycle (`TGW` algorithm)—(see Figure 7.5.2);
4. construction of the truth table of the Boolean algebra using a point-set
   membership algorithm (see Example 7.5.2);
5. application of the Boolean function equivalent to the solid input expression
   on all minterms of atoms (see Example 7.5.6).   □

In `Plasm` implementation of Boolean operations, we make use of aliases
(`UNION, INTERSECTION, DIFFERENCE, XOR`) of the hierarchical operator `STRUCT`
[4] to assembly complex Boolean formulas (see Section 7.5.2). In particular, a
function `FLATTEN` is applied to this hierarchical assembly[5] of solid terms to
obtain a single unevaluated expression used to compute the space arrangement of atoms produced by intersecting generators in world coordinates.

---

[4] The `Plasm` operator `STRUCT` is different from Julia's `struct` user type.
[5] possibly containing affine transform tensors of subexpressions

In the `Plasm` Boolean pipeline, coordinate transformations are separated from the evaluation of the Boolean formula, as shown in Section 7.5.2, Example 7.5.2, and the following examples. The traversal algorithm runs in linear time and space, since depth-first search (DFS) has $O(n)$ complexity.

Definition 7.9 (Canonical Form) In Boolean algebra, any Boolean function can be written in Canonical Disjunctive Normal Form (CDNF), also called the sum of minterms.

Each minterm (atom) represents a combination of all input variables (generators) for which the function equals "1" or "0", i.e., represents or does not represent a solid parcel of the solution. Hence, this function is expressed as the logical sum (OR) of all such minterms, also known as the Sum of Products (SOP) form (see Example 7.5.2).

Property 7.3 (Unicity of Canonical Form) The canonical form is the sum of all the minterms for which the Boolean function evaluates to 1, and it is unique for any given function (i.e., algebraic solid expression), making it a standardized representation.

An efficient point-classification algorithm [32] is run for each atom to determine the subset of generators to which the atom belongs. The Boolean $m \times n$ truth matrix (incidence relation) AG = $Atoms \times Generators$ is therefore produced by rows, providing by columns of length m the binary representation of split generators, where m is the number of arrangement atoms. Read by rows, the matrix [AG] is a set of minterms of binary variables (each associated with one generator) for each single atom.

Atomic point-set membership tests are computed efficiently in $O(nnz(\text{AG}))$ time, using spatial acceleration structures. This truth table—whose rows correspond to atoms (minterms)—is then used to generate the solid solution, selecting the atoms that produce 1's in the Boolean solution vector, initialized to the bottom term 0^m, to obtain the sum (union) of true minterms, i.e., the set of atoms solution of the Boolean solid formula.

Remark 7.7 In other words, the AG binary tabulation is the truth table for the whole generated arrangement, isomorphic to the Boolean algebra 2^m produced by generators. The Boolean function corresponding to any solid expression will be satisfied by only a subset S of AG table rows. The characteristic function $\chi_S(2^m) \to 2$ of this subset provides the binary solution of every solid expression with the same solid terms.

After constructing the arrangement of the $\mathbb{E}^d$ space, the BOOL: Hpc $\mapsto$ Lar computation, therefore, finalizes the evaluation of the specific Boolean formula by executing the Boolean sum of products described in this page and in the following sections. In this way, it computes the specific binary solution in 2^m, i.e., by space isomorphism, the Boolean string acting as the coordinate vector of the solid algebra solution chain in C_3.

Example 7.2.1 (Space arrangement from assembly *tree)* Consider the assembly constructed in Coding 7.2.1 by putting together three instances of the unit cube, suitably rotated and translated. The Lar constructor cuboidGrid of grids of cubes with "shape" [m,n,p], returns the geometry V along with, if the optional parameter is all=true, the whole collection VV, EV, FV, and CV of p-cells, with $0 \leq p \leq 3$, represented "by-vertices". □

The variable assembly is bound to Julia's composite type generated by the function STRUCT, which provides the syntax needed to define the hierarchical geometric structures to define the dimension-independent partition of the embedding space. If you substitute STRUCT with a Boolean operator, Plasm generates the corresponding solid model. Readers should try substituting one such operator into SRUCT , and BOOL into ARRANGE3D . Then convert the resulting solution solid into Brep form, by applying the BOUNDARY operator to the result object of Lar type.

Coding 7.2.1 (Assembly of three cubes)

```julia
julia> m,n,p = 1,1,1;
    lar = CUBOIDGRID([m,n,p]);
    cube = MKPOL(lar.V, lar.C[:CV]);
julia> assembly = STRUCT(cube,
    T(1,2,3)(.3,.4,.25), R(2,3)(π/5), R(1,2)(π/12), cube,
    T(1,2,3)(-.2,.4,-.2), R(1,3)(π/5), R(1,3)(π/12), cube)
julia> ARRANGE3D(assembly)
```

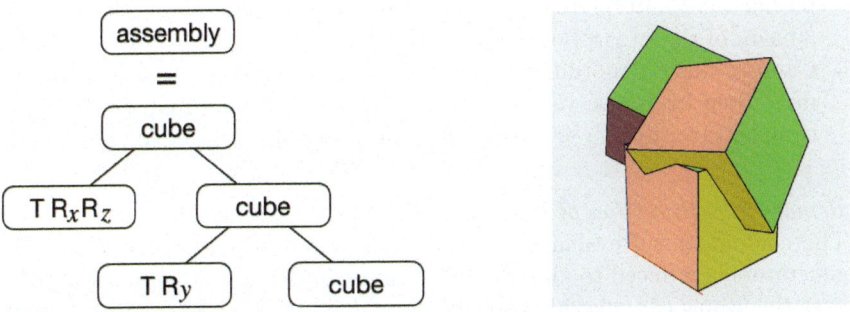

Fig. 7.3 Syntax tree of the CSG expression assembly The linearized expression STRUCT in Coding 7.2.1 corresponds to the DFS (Depth First Search) traversal of this tree.

Remark 7.8 Note that the assembly object linearizes the preorder DFS of the tree. The ARRANGE3D function applied to assembly returns in the Lar data structure the (geometry, topology) of the 3D space partition generated by it. Geometry is given by the embedding matrix W of old and new 0-cells, and topology by the sparse matrices CF, FE, EV, i.e., by $\delta_2, \delta_1, \delta_0$, of the chain complex generated by the $\mathcal{A}$(assembly) arrangement.

7.3 Subspaces of cycles and boundaries

Let us return to the mathematical (topological) ideas that generated the homological approach to Solid Modeling used by `Plasm` and discuss the algebraic method for boundary evaluation of any Boolean expression tree constructed with regularized operators of union, intersection, and difference on a number of polyhedral d-solids ($d = 2, 3$).

7.3.1 Plasm Boolean pipeline

The evaluation of a `Plasm` Boolean expression, including `boolop` operations and affine transformations, is obtained in some main steps after the reduction of all terms to global coordinates:

1. Distribute the Abstract Syntax Tree of operations in the `properties` of Boolean nodes for later evaluation. See the red `String`s in Example 7.5.2.
2. Compute an irreducible basis of closed $(d\text{-}1)$-chains (atoms) in the d-space arrangement induced by the input.
3. Map the geometry generators into the binary columns of a Boolean truth table matrix, indexed by rows with computed atoms.
4. Evaluate the Boolean function row by row (using Julia's bitwise operators), step by step, to produce a solution vector X for the input solid expression. This vector, in the canonical form of a sum of minterms (Boolean products), is interpreted as the coordinate representation of the solution 3-chain of the `Plasm` Boolean form.
5. Convert this C_3 coordinate vector of a 3-polyhedral object to some standard `Brep` by sparse vector-matrix products times the matrices of chain complex, i.e., $\partial_3 : C_3 \to C_2$, $\partial_2 : C_2 \to C_1$, and $\partial_1 : C_1 \to C_0$.

Remark 7.9 (Reduction of solid algebra to finite set algebra)
The computational evaluation of every possible solid expression with solid generators is reduced to an equivalent logical function of a finite set algebra over the atoms of a spatial arrangement and solved by Julia's native bitwise operators.

This method is implemented in `Plasm` using sparse matrices and vectors. We utilize the fact that the structure of each term in this algebra is characterized by a discrete set of points, each one computed once and for all within the interior of each atom. Set-membership classifications (SMC) concerning such single internal points of atoms compute the structure of any algebra term and, in particular, transform each solid variable (generator term) of every Boolean formula with such variables into a sparse logical array of length m, equal to the number of atoms. For an example of the structure of the

Plasm object resulting from the resolution of a Boolean expression between solid values, the reader is referred to Example 7.5.2.

7.3.2 Chain, cycle, and boundary subspaces

A *p-cycle* is defined as a p-chain in C_p without boundary; hence it is an element of the kernel Z_p of ∂_p, the red sets in Figure 7.3.2. A cycle is a chain that ∂ operator sends to zero.

A *p-boundary* is a p-chain which is the boundary of a $(p+1)$-chain. Hence, it is an element of the image B_p of ∂_{p+1}, the pink sets of Figure 7.3.2. A boundary is the image of some chain under ∂. [6]

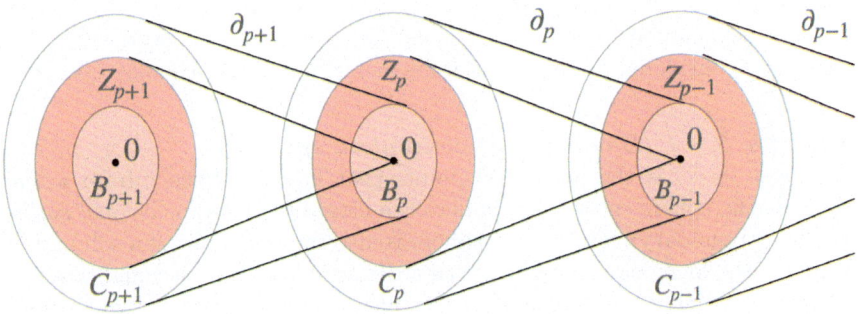

Fig. 7.4 The set of p-boundaries $B_p \subset \partial_{p+1} C_{p+1}$ is a subset of the kernel $Z_p \subset C_p$ of p-cycles, since the boundary of a boundary is empty. In other words, $\partial^2 = \partial_p \partial_{p+1} = 0$.

Property 7.4 (Columns of ∂_3 matrix are irreducible 2-cycles) Generally, the matrix of any linear operator between linear spaces, such as the boundary $\partial_p : C_p \to C_{p-1}$, gets a sense only when the bases have been fixed for both domain and codomain spaces.

In this case (since the bases already determine the ordering), each column of $[\partial_p]$ contains the coordinates of a basis vector of C_p, meaning the coefficients of its (unique) representation as a linear combination of the C_{p-1} basis. Therefore, each column of $[\partial_p]$ is a closed chain in the linear space of C_{p-1}.

It is closed (i.e., a $(p-1)$-cycle) because the constraint states that "the boundary of the boundary of each p-chain is empty." Therefore, every column of the boundary matrices, such as the k-th column $[u]_k$, is a cycle by definition.

---

[6] Standard practice with linear operators is to not use brackets for application to an argument $(\partial_{p+1} C_{p+1})$ and to remove the composition operator $(\partial^2 = \partial\partial)$.

While the matrices $[\partial_2]$ and $[\delta_1] = [\partial_2]^t$ are mathematically computable from the input data set $[\partial_1]$, the matrix $[\partial_3]$ is initially unknown and was constructed algorithmically, one column at a time, by the Topological Gift Wrapping (TGW) algorithm [31].

In Reference [31], an algorithm was presented for the transformation of the cycles basis to the boundaries basis, which is interesting from a CAD perspective. We observe that two remarks are essential for this, which concern the transformation of cycle chains to boundary chains. The first is that, by construction, the basis of 2-cycles corresponding to $[\partial_3]$ columns are (a) elementary (irreducible) and (b) non-intersecting, even in their coordinate chain representation, since they involve different rows (2-cells). The second concerns their cardinalities. It is well known that $Z_p \supseteq B_p$ or, in other words, there may be cycles that are not boundaries (see Figure 7.3.2) as discussed in the following example. So, the boundary of the outer chain in C_{p+1}, if disconnected, can be built in $B_p \subset Z_p$ by summing two or more basis cycles, whose nonzero elements are indeed in different row positions by construction. This point was considered by the Plasm implementation of the ARRANGE operators.

Example 7.3.1 (Boundary of concentric spheres)
A sphere in $\mathbb{E}^d$ is the locus of d-points that have distance r from a fixed d-point, the sphere center. The sphere is a closed $(d-1)$-dimensional surface (without boundary). The $(d-1)$-sphere is also the boundary of a solid d-ball, which is the locus of all d-points with a distance less than or equal to r from the center point. Consider the 3D space partition generated by two 2-spheres S_1 and S_2 with the same center and different radiuses $r_1 > r_2$. There are three solid cells: (a) the outer unbounded cell A, i.e., $\mathbb{E}^3$ minus the ball of radius r_1; (b) the solid shell B with thickness $r_1 - r_2$; and (c) the solid inner ball C of radius r_2. Within the two interfaces, four closed irreducible 2-cycles are generated as columns of $[\partial_3]$ in this complex, pairwise summing to zero and with opposite orientations[7]. Let us denote, from exterior to interior, as

$$[\partial_3^+] = [u_1 \ u_2 \ u_3 \ u_4] \quad \text{and} \quad [\partial_3] = [u_2 \ u_3 \ u_4], \quad \text{with} \quad u_1 = -(u_2 + u_3 + u_4).$$

If we denote the three "solid" basis elements in $U_3 \subset C_3$ (3-chain space) as A, B, C and the whole space as X, we can express them as Boolean algebra expressions:

$$X = A + B + C \tag{7.1}$$
$$A = X - B - C; \quad B = X - A - C; \quad C = X - A - B. \tag{7.2}$$

$\square$

It is easy to see that in terms of oriented boundary 2-chains.

---

[7] We denote as $[\partial_3^+]$ the matrix generated by the TGW algorithm with signed elements (in $\{-1, 0, 1\}$) and with single oriented cycles by columns, caught as OUTERS in Plasm pipeline and assembled in a single OUTER boundary.

$$\partial A = u_1; \quad \partial B = u_2 + u_3; \quad \partial C = u_4, \quad \text{with} \quad u_1 + u_2 = u_3 + u_4 = 0,$$

and hence

$$u_1 + u_2 + u_3 + u_4 = 0.$$

Finally, note that the cycles u_2 and u_3 are not boundaries:

$$u_2, u_3 \in (Z_2 - B_2) \subset C_2.$$

Coding 7.3.1 (Concentric spheres minus cube)

The Boolean implementation of Example 7.3.1 requires a slight variation to look at the interior of the inner sphere, otherwise inaccessible. Therefore, we codify a DIFFERENCE operation with three arguments: sphere, innersphere, and cube. In a variadic difference (with multiple arguments), the operational semantics "subtract from the first argument all the others." The ICOSPHERE () primitive solid may have an actual parameter to increase the boundary resolution. The innersphere is a scaled copy of the sphere, and the cube has unit sides and a corner at the origin. The hpc variable, building into an internal data structure the semantics of the Boolean expression, accepts only arguments in Hpc format and returns an Hpc. The actual computation is performed by the BOOL operator, which outputs a Lar value

```
sphere = ICOSPHERE(ICOSPHERE(ICOSPHERE()));
innersphere = S(1,2,3)(.8,.8,.8)(sphere);
cube = CUBE(1);
hpc = DIFFERENCE(sphere, innersphere, cube);

VIEWCOMPLEX(BOOL(hpc))
VIEWCOMPLEX(BOOL(hpc), explode=[1.3,1.3,1.3], show=["FV"])
VIEWCOMPLEX(BOOL(hpc), show=["CF"])
```

Fig. 7.5 Boolean DIFFERENCE sphere minus sphere minus cube visualization: (a) by faces; (b) by faces exploded; (c) by atoms. There is only one (outer) atom here.

The models are solid Breps; DIFFERENCE creates a thick solid sphere with one curved corner removed by the unit cube. □

In solid modeling, the word shell is used to denote the maximal connected closed surfaces in a Brep of a solid object (see, e.g., [30]). Here, a shell is every connected 2-cycle of the boundary of a 3-chain, or simply each of the boundary cycles of each connected solid component, including holes. Every single shell, including those inside to some unit 3-chain u_i, is obtained by matrix multiplication $[\partial_3^+][u_i]$, with $u_i \in U_3$ represented in coordinates by $\mathbf{e}_i$, which is the zero vector with only one 1 in position i. The result will be in $B_2 \subset Z_2 \subset C_2$, generating the Brep of an algebra atom (see the next section).

It is essential to remark again that the matrix $[\partial_3^+]$, as generated by TGW [31], contains a basis of $Z_3 \subset C_3$, plus one more column, sum of all the others.

A specific algorithm, combining appropriate columns of $[\partial_3^+]$, generates a basis $[\partial_3]$ of the validity subspace $B_2 \subset Z_2 \subset C_2$, where each 3-basis element is being expressed as a 2-cycle (b-rep), possibly unconnected. In other words, by construction, each irreducible unit 3-chain $u \in U_3 \subset C_3$ corresponds to a single connected 2-cycle in Z_2. Conversely, the boundary (cycle in B_2) of a general (sign-invariant) 3-chain, possibly unconnected, is given by the chain sum of irreducible 2-cycles in $B_2 \subset Z_2 \subset C_2$ (see Reference [32]).

7.3.3 Compendium of homological method

This section provides a brief summary of the Plasm homology methods to enhance user understanding and accessibility. For clarity, we illustrate the process with several 2D examples that encapsulate its core principles, ensuring it remains approachable for readers of any scientific or technical background.

To achieve this, we rely on two key elements: (a) a structural diagram (see Figure 7.6) to identify and extract information embedded in chain complexes and (b) step-by-step examples that illustrate the method in action.

We begin with two basic 2D solids, the *Ring* and the *Rectangle*, as shown in Figure 7.7a. Several BOOL solid forms and their corresponding oriented boundary representations (Breps) are compared in the following examples.

The 1-cell boundary, e_i, of the input shapes are used to generate an arrangement in $\mathbb{E}^2$, as illustrated in Figure 7.7 and described in [31]. In this figure, elementary 0-chains (v_i), 1-chains (e_h), and 2-chains (u_k) are clearly labeled to aid understanding.

From BOOL *forms to oriented Breps*

Two chain complexes with linear spaces and linear transformations can be defined in $\mathbb{E}^2$ using a binary or a ternary field of scalar coefficients:

1. the unsigned linear spaces C_p over the binary field $\langle \{0,1\}, + \bmod 2, \times \rangle$;
2. the signed linear space $C_p^{\circlearrowleft}$ over the ternary field $\langle \{0,1,-1\}, + \bmod 3, \times \rangle$.

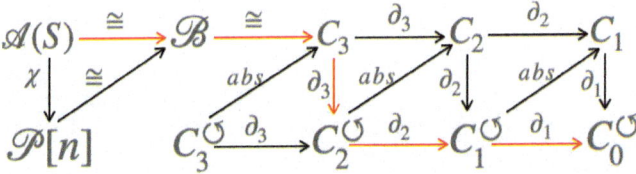

Fig. 7.6 The diagram of functors between the unsigned chain complex $C_\bullet$ over a binary field of scalars, and the signed chain complex $C_\bullet^\circlearrowleft$ over a ternary field of scalars. In red is the computational pipeline proposed in this paper.

In the diagram of Figure 7.6, $\mathscr{A}(\mathcal{S})$, $\mathscr{P}[n]$, and $\mathscr{B}$ represent, respectively: (a) the arrangement of $\mathbb{E}^d$ generated by a set $\mathcal{S}$ of primitive `Breps`, (b) the power set of natural numbers $[n] := \{1, 2, \ldots, n\}$, where n is the number of Boolean atoms in $\mathscr{A}(\mathcal{S})$, and (c) the finite Boolean algebra generated by these atoms. The following examples demonstrate some computations over curved topological polyhedra in a 2D environment for simplicity.

Example 7.3.2 *($\mathbb{E}^2$ **arrangement** $\rightarrow$ Brep *of algebraic forms*)* In Figure 7.6, the same symbol is used for arrows ∂_p between signed $C_p^\circlearrowleft$ or unsigned C_p chain subspaces. Of course, their matrix representations depend on the choice of the bases in both domain and codomain spaces. The path followed by our 3D computational pipeline is drawn in red in the diagram. The matrix of mapping $\partial_2 : C_2 \rightarrow C_1^\circlearrowleft$ is shown on the right of Figure 7.7; it contains by column the basis elements of C_2 space represented in $Z_1^\circlearrowleft \subset C_1^\circlearrowleft$ subspace of oriented 1-cycles, i.e. with coordinates in $\{0, 1, -1\}$. Some CSG forms and corresponding oriented b-reps are given in Example 7.3.3. Note in Example 7.3.9 that $\sum_{i=1}^6 \mathsf{Brep}(u_i) = 0$. □

Example 7.3.3 *(**Boundary matrix of 2-complex**)* To compute the new boundary matrix $[\partial_2] : C_2 \rightarrow B_1$, we sum the columns corresponding to the chains u_2 and u_6 in $[\partial_2^+]$. This operation effectively consolidates the contributions of these 2-cells into a linear combination that defines the subspace of 1-boundaries B_1.

Let us illustrate this process step by step. Consider the columns of $[\partial_2^+]$ associated with u_2 and u_6. Adding these columns produces a new column vector that belongs to B_1, which forms part of the basis for this subspace. This transformation is essential for refining the topological structure, as it distinguishes between oriented cycles in Z_1 and boundaries in B_1.

The resulting boundary matrix $[\partial_2]$ provides a more compact representation of the relationships between 2-cells and their corresponding 1-boundaries. This step is a critical component of computing homology groups as it prepares the data for subsequent steps in the analysis of topological features.

Let us proceed with explicit computations and verify the entries of $[\partial_2]$, ensuring consistency with the original arrangement in Figure 7.7. □

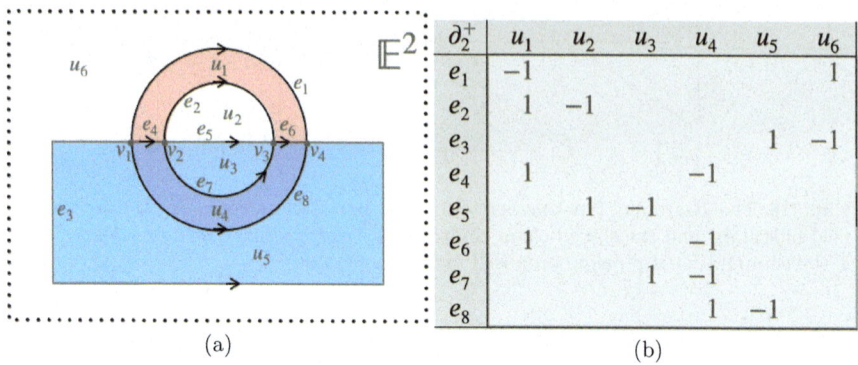

∂_2^+	u_1	u_2	u_3	u_4	u_5	u_6
e_1	-1					1
e_2	1	-1				
e_3					1	-1
e_4	1			-1		
e_5			1	-1		
e_6	1			-1		
e_7			1	-1		
e_8				1	-1	

|(a)|(b)|

Fig. 7.7 (a) The arrangement of $\mathbb{E}^2$ generated by a solid rectangle in 2D and by a solid ring in 2D. (b) The boundary mapping ∂_2^+ between the basis $U_2 \subset C_2$ of the unoriented 2-chain space and the subspace $Z_1^{\circlearrowleft} \subset C_1^{\circlearrowleft}$ of oriented 1-cycles. The sign-invariant domains of the quasi-disjoint partition of $\mathbb{E}^2$ give a finite Boolean algebra where the atoms are indicated as $u_1, \ldots, u_6$.

By performing this computation, we can explicitly verify how the 2-cells in C_2 contribute to the formation of 1-boundaries in B_1, thus providing a clear pathway to distinguish topological features such as cycles and boundaries within the arrangement. This step simplifies the representation and prepares the structure for further homological analysis.

It is worth noting that any Boolean term, such as Rectangle $= u_3 \cup u_4 \cup u_5$, also has the coordinate representation $[0, 0, 1, 1, 1, 0]$ corresponding to the chain in C_2. This representation simultaneously encodes its structure within its finite Boolean algebra.

In particular, we derive the boundary transformations illustrated below, transitioning from simple BOOL forms to oriented Brep models. The orientation of the basis elements $e_i \in C_1^{\circlearrowleft}$ is defined conventionally: an element e_i is considered positively oriented if $e_i = v_k - v_h$ with $k > h$.

Examples of symbolic conversions BOOL $\equiv$ Brep

Several examples elucidate the equivalence between representing specific subsets of 2D space as a disjoint union of atoms, denoted as BOOL, and their oriented boundary 1-chains, denoted as Brep.

Here, we have two generator domains (terms), indicating *rectangle* and *ring*. The typical pattern of all the examples is:

BOOL(Boolean form) $\mapsto$ chain $\in C_2 \mapsto$ chain $\in C_1 =$ Brep form.

Example 7.3.4 (Rectangle)

$$\text{BOOL}(rectangle) = u_3 \cup u_4 \cup u_5 = (u_3 + u_4 + u_5) \in C_2 \mapsto$$
$$\mapsto [\partial_2][0, 0, 1, 1, 1, 0]^t = [0, 0, 1, -1, -1, -1, 0, 0]^t$$
$$\equiv \text{Brep}(rectangle) = (e_3 - e_4 - e_5 - e_6) \in B_1 \subset Z_1 \subset C_1^\circlearrowleft$$

Example 7.3.5 (Ring)

$$\text{BOOL}(ring) = u_1 \cup u_4 = (u_1 + u_4) \in C_2 \mapsto$$
$$\mapsto [\partial_2][1, 0, 0, 1, 0, 0]^t = [-1, 1, 0, 0, 0, 0, -1, 1]^t$$
$$\equiv \text{Brep}(ring) = (-e_1 + e_2 - e_7 + e_8) \in B_1 \subset Z_1 \subset C_1^\circlearrowleft$$

Example 7.3.6 (Rectangle ∩ ring)

$$\text{BOOL}(rectangle \cap ring) = u_4 \in C_2 \mapsto$$
$$\mapsto [\partial_2][0, 0, 0, 1, 0, 0]^t = [0, 0, 0, -1, 0, -1, -1, 1]^t$$
$$\equiv \text{Brep}(rectangle \cap ring) = (-e_4 - e_6 - e_7 + e_8) \in B_1 \subset Z_1 \subset C_1^\circlearrowleft$$

Example 7.3.7 (Rectangle ∪ ring)

$$\text{BOOL}(rectangle \cup ring) = u_1 \cup u_3 \cup u_4 \cup u_5 = (u_1 + u_3 + u_4 + u_5) \in C_2 \mapsto$$
$$\mapsto [\partial_2][1, 0, 1, 1, 1, 0]^t = [-1, 1, 1, 0, -1, 0, 0, 0]^t$$
$$\equiv \text{Brep}(rectangle \cup ring) = (-e_1 + e_2 + e_3 - e_5) \in B_1 \subset Z_1 \subset C_1^\circlearrowleft$$

Example 7.3.8 (Rectangle − ring)

$$\text{BOOL}(rectangle - ring) = u_3 \cup u_5 = (u_3 + u_5) \in C_2 \mapsto$$
$$\mapsto [\partial_2][0, 0, 1, 0, 1, 0]^t = [0, 0, 1, 0, -1, 0, 1, -1]^t$$
$$\equiv \text{Brep}(rectangle - ring) = (e_3 - e_5 + e_7 - e_8) \in B_1 \subset Z_1 \subset C_1^\circlearrowleft$$

The linear space C_2 has dimension $6 = \#U_2$, equal to the number $m = 6$ of atoms of the solid algebra generated by the arrangement of $\mathbb{E}^2$ induced by the two primitive (generator) shapes rectangle and ring.

The 1-cell representations of unit 2-cells are given by the columns of the matrix $[\partial_2] : C_2 \to C_1$. The number m is also the length of the bit strings, giving their Boolean structure. The number of all distinct algebraic forms, i.e., the size of this solid algebra, is 2^m. □

Example 7.3.9 (Oriented boundary chain of basis elements)
Concerning Figure 7.7, the Brep of each unit (i.e. basis) 2-chain (a.k.a. atom) is computed by (sparse) matrix-product of $[\partial_2^+]$ times the unit column vectors u_k $(1 \leq k \leq 6)$ of each basis element, as shown below.

The b-reps of Boolean 2D atoms $u_k \subset \mathbb{E}^2$ and their coherently oriented boundary 1-cycles in C_1 are computed below:

$$\mathsf{Brep}(u_1) := \partial_2^+ u_1 = [\partial_2^+][u_1] \mapsto -e_1 + e_2 + e_4 + e_6$$

$$\mathsf{Brep}(u_2) := \partial_2^+ u_2 = [\partial_2^+][u_2] \mapsto -e_2 + e_5$$

$$\mathsf{Brep}(u_3) := \partial_2^+ u_3 = [\partial_2^+][u_3] \mapsto -e_5 + e_7$$

$$\mathsf{Brep}(u_4) := \partial_2^+ u_4 = [\partial_2^+][u_4] \mapsto -e_4 - e_6 - e_7 + e_8$$

$$\mathsf{Brep}(u_5) := \partial_2^+ u_5 = [\partial_2^+][u_5] \mapsto e_3 - e_8$$

$$\mathsf{Brep}(u_6) := \partial_2^+ u_6 = [\partial_2^+][u_6] \mapsto e_1 - e_3$$

Exercise 7.1 (test) Create some longer Boolean formula using rectangle and ring and then verify that the BOOL expressions of the algebra generated by the two generators and the topological sum of the boundaries of their atoms $u_1, \ldots, u_6$ coincide.

7.4 Solid and Boolean Algebras

In this section, we introduce and discuss the isomorphism between our Solid Boolean Algebra (which includes complement and variadic boolops) formed by the atoms of the spatial arrangement, and the finite Boolean algebra of binary strings in $\mathbf{2}^m$ of length m representing binary strings, where m denotes the number of atoms.

7.4.1 Solid algebraic expressions

Several user-verifiable examples have been developed within a proper testing environment to guarantee the correct implementation of the Plasm finite Boolean algebras discussed in this book. The section presents the test code and some images generated by the interactive Plasm visualization of the resulting geometric objects of Lar type.

The Random Julia package was used for this purpose. When the Random module is loaded, the default RNG is randomly seeded via Random.seed!().

This means that each time a new julia session is started, the first call to rand() produces a different result, unless `seed!(seed)` is called first.

```
using Plasm
using Random
using LinearAlgebra
Random.seed!(0)
TaskLocalRNG()
```

`TaskLocalRNG` is one of the pseudorandom number generators (PRNGs) exported by the `Random` Julia package. In particular, the `TaskLocalRNG` has a state local to its task, not its thread.

7.4.1.1 Testing framework

The `Plasm` testing framework of Boolean expressions is shown here top-down, from the main function to several small functions introducing one, two, or three instances of solid primitives. The framework also contains a growing collection of `Julia` expressions devoted to testing some simple or complex Boolean expressions in https://github.com/scrgiorgio/Plasm.jl/test/.

The `RunBooleanTest` function is structured as follows:

1. Input the test name, an array of boolean operations to execute, and an array of operation arguments; the debug mode is initialized to `false`.
2. `assembly` is initialized to the `STRUCT` of solid arguments of `Hpc` type;
3. `assembly` is converted to `Lar` type into the `lar` container of a cellular complex;
4. then the 2D or 3D embedding space is partitioned, i.e., the space arrangement $\mathcal{A}(\text{lar})$ is computed through the TGW algorithm.
5. In 3D only, `lar` is initialized to the inner atoms;
6. `debug_mode` is checked; if `true` each `atom` is visualized exploded;
7. `result` is computed by the `BOOL` function, with the actual array `lar::Lar` of inner atoms, the actual set or BoolOps to execute, the actual `input_args` for each operation, and the actual value of `debug_mode`;
8. finally, the `result` is interactively displayed together with its `title` on the related window.

```
function RunBooleanTest(name, bool_op, args; debug_mode=false)
  assembly = STRUCT(args)

  lar = LAR(assembly)
  pdim = size(lar.V, 1)

  if pdim==2
    lar = ARRANGE2D(lar)
  else
    lar = ARRANGE3D(lar)
```

```
    lar = INNERS(lar)
  end

  if debug_mode
    for atom in ATOMS(lar)
      show_debug(atom)
      VIEWCOMPLEX(atom, explode=[1.0, 1.0, 1.0], show=["V", "EV"
    ,"FV"]) end
  end    # continue ...
```

The code lines below are used to compute the `result`, show its 2-cells ("`FV`") in any case, and, in the 3D case, display the exploded atoms whose quasi-disjoint union gives the Boolean `result`.

```
result = BOOL(lar, bool_op=bool_op, input_args=[LAR(arg) for
  arg in args], debug_mode=debug_mode)

VIEWCOMPLEX(result, show=["FV"], # bot 2D and 3D result
explode=[1.2,1.2,1.2], title="$(name)$(bool_op) FV")

if pdim == 3   # if result is 3D
  VIEWCOMPLEX(result, show=["CV"],
  explode=[1.2,1.2,1.2], title="$(name)$(bool_op) CV")
end
end # function
```

7.4.1.2 Two or three basic objects

Below is a collection of small scripts for performing unit tests under different conditions. The code is self-explanatory and will not be commented on.

```
function TwoQuads()
  return [
    CUBOID([0.0, 0.0],[1.0, 1.0]),
    CUBOID([0.5, 0.5],[1.5, 1.5]) ]
end
```

Fig. 7.8 UNION, INTER-SECTION, DIFFERENCE, and XOR of two 2D unit squares.

```
function ThreeQuads()
  return [
```

```
    CUBOID([0.0, 0.0], [1.0, 1.0]),
    CUBOID([0.5, 0.5], [1.5, 1.5]),
    CUBOID([0.7, 0.7], [1.7, 1.7]) ]
end
```

Fig. 7.9 Three unit squares: (a) UNION; (b) INTERSECTION; (c) the reader should guess.

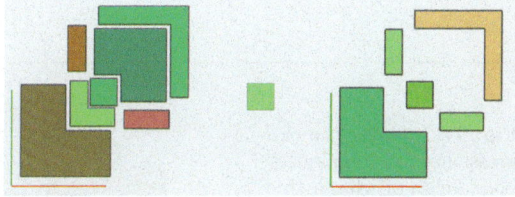

```
function TwoCubes()
  return [
    CUBOID([0.0, 0.0, 0.0],[1.0, 1.0, 1.0]),
    CUBOID([0.5, 0.5, 0.5],[1.5, 1.5, 1.5]) ]
end
```

```
function ThreeCubes()
  return [
    CUBOID([0.0, 0.0, 0.0], [1.0, 1.0, 1.0]),
    CUBOID([0.5, 0.5, 0.0], [1.5, 1.5, 1.0]),
    CUBOID([0.7, 0.7, 0.0], [1.7, 1.7, 1.0]) ]
end
```

Fig. 7.10 Three unit cubes: (a, c) exploded boundary 2-cells; (b, d) exploded 3-cells. The reader should guess the expressions.

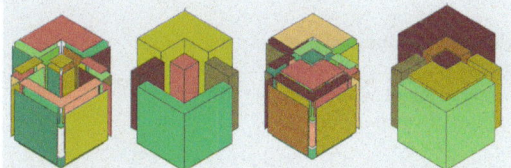

7.4.1.3 Structured methods

The following methods allow multiple use of different primitives within the same Julia programming scheme. Note that the output of the MechanicalPiece method is used by RunBooleanTest as input to the STRUCT operator.

```
function MechanicalPiece(primitive :: Hpc)
```

```
    return [
      CUBOID(
        [-1.0,-1.0,-1.0],
        [+1.0,+1.0,+1.0]),
      R(1,2)(π/2)(primitive),
      R(2,3)(π/2)(primitive),
      R(1,3)(π/2)(primitive)
    ]
  end
```

Fig. 7.11 Guess the expressions: (a,c) exploded boundaries of what?; (b,d) exploded (atoms) of what?

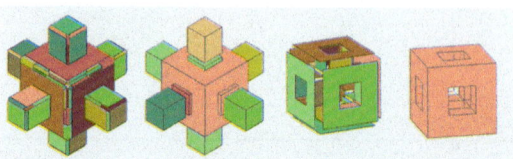

```
function MechanicalPiece(symbol::Symbol)
  if symbol==:cube
    return MechanicalPiece(CUBOID( [-0.4, -0.4, -2.0], [+0.4,
    +0.4, +2.0]))
  elseif symbol==:cylinder
    return MechanicalPiece(T(3)(-2)(CYLINDER([0.5,4])(8)))
  elseif symbol==:tube
    return MechanicalPiece(T(3)(-2)(TUBE([0.3,0.5,4.0])(4)))
  else
    @assert(false)
  end
end
```

Fig. 7.12 Guess the expressions: (a,c) exploded boundaries of what?; (b,d) exploded (atoms) of what?

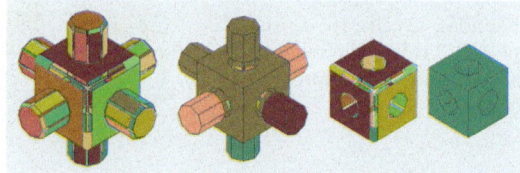

7.4.1.4 Growing Collection of Unit tests

The following small **Plasm** (**for**) loop scripts enable repeated computations with different Boolean operators and generators in 2D and 3D. The reader should find the test environment in the **Plasm/test/** directory, run the tests, and/or predict the results.

```
# 2d tests
for bool_op in [Union, Intersection, Difference, Xor]
  RunBooleanTest("boolean/TwoQuads", bool_op, TwoQuads())
end
```

```
for bool_op in [Union, Intersection, Difference, Xor]
  RunBooleanTest("boolean/ThreeQuads", bool_op, ThreeQuads())
end
```

```
# 3d tests
for bool_op in [Union, Intersection, Difference, Xor]
  RunBooleanTest("boolean/TwoCubes", bool_op, TwoCubes())
end
```

```
for bool_op in [Union, Intersection, Difference, Xor]
  RunBooleanTest("boolean/ThreeCubes", bool_op, ThreeCubes())
end
```

7.4.1.5 Examples of user-defined Boolean expression

Here, the script syntax allows for the combination of previously computed atoms. Any Boolean expression will work. Look in Plasm/test/ directory and execute the examples.

```
function UserBoolOp(v::Vector{Bool})::Bool
  return Difference([v[1],Union(v[2:end])])
end
```

```
for bool_op in [Union, Intersection, Difference, Xor]
  RunBooleanTest("boolean/MechanicalPiece/cube", bool_op,
    MechanicalPiece(:cube))
end
```

```
for bool_op in [Union, Intersection, Difference, Xor]
  RunBooleanTest("boolean/MechanicalPiece/cylinder", bool_op,
    MechanicalPiece(:cylinder))
end
```

```
# BROKEN
if false
  for bool_op in [Union, Intersection, Difference, Xor]
```

```
      RunBooleanTest("boolean/MechanicalPiece/tube", bool_op,
      MechanicalPiece(:tube))
    end
  end
```

7.4.2 Solid operations in Plasm expressions

Certainly, the main operators in Boolean algebra are the standard functions UNION, INTERSECTION, DIFFERENCE, and XOR. In Plasm, they are all defined as variadic maps[8] OP : Hpc^n → Hpc, where op ∈ {UNION, INTERSECTION, DIFFERENCE, XOR}, transforms n arguments of Hpc type into a single Hpc result. Even when $n = 2$, such functions must be prefixed to their arguments, similar to the STRUCT operator.

Remark 7.10 (Boolops as STRUCT *aliases)* Specifically, we assume that the input to Boolean expressions resembles a STRUCT that includes solid objects, affine transformations, and/or other STRUCT assemblies. This syntax can support symbols for Boolean operations as aliases for STRUCT; that is, it can be transformed into an Hpc assembly (tree) with various structures and depths.

At execution time, the input Hpc is translated into a (hierarchical) STRUCT with the field properties = Dict("bool_op" => Plasm.OP), where "bool_op" can represent any Plasm Boolean functions such as the inner functions "Union", "Intersection", "Difference", or "Xor", as shown in Example 7.5.2. During the traversal of Hpc by the BOOL operator, the following occurs:

1. Plasm BOOL applies the ARRANGE operator (2D or 3D) to the hierarchical dataset and assembles the various topological objects (V, EV, FE, FV) into a single Lar object [32]. In particular, ARRANGE computes the δ_2 operator of the induced chain complex.
2. Plasm computes the appropriate binary string for each atom (i.e., each row in the Boolean resolution matrix), using a point-set containment algorithm, thus constructing the truth table for Boolean functions based on the spatial arrangement.

Remark 7.11 (Unitary resolution) Notably, all Boolean subexpressions are resolved simultaneously from source to object code by generating the binary representation of each 3-chain atom (minterms). This method removes the need to build the Constructive Solid Geometry (CSG) tree step by step.

---

[8] Variadic map refer to a function that accepts a variable number of arguments, particularly an arbitrary quantity of arguments.

Finally, the compiler will resolve the binary minterms above, providing the 3-chain expression of the Boolean solution, which is the proper subset of atoms that `Plasm` translates into the `Lar` value of the evaluated root term. This `result` can be visualized by `VIEWCOMPLEX` as any object of `Lar` type.

7.4.3 Plasm examples

Example 7.4.1 (Boolean operations) Some solid algebra expressions are computed and visualized in Figure 7.13 and 7.14. A description of the `Plasm` syntax for Boolean solid expression is given in Section 7.4.2 and Script 7.5.1.

```
cube = T(1,2,3)(-1,-1,-1)(CUBE(2))
cyl  = T(3)(-2)(CYLINDER([0.5,4])(8))
hpc = DIFFERENCE(
        cube,
        UNION(cyl, R(2,3)(π/2), cyl, R(1,3)(π/2), cyl)
      )
result = BOOL(hpc)
VIEWCOMPLEX(result, explode=[3,3,3])
VIEWCOMPLEX(result, explode=[3,3,3], show=["CV"])
```

```
hpc = UNION(
        cube,
        UNION(cyl, R(2,3)(π/2), cyl, R(1,3)(π/2), cyl)
      )
result = BOOL(hpc)
```

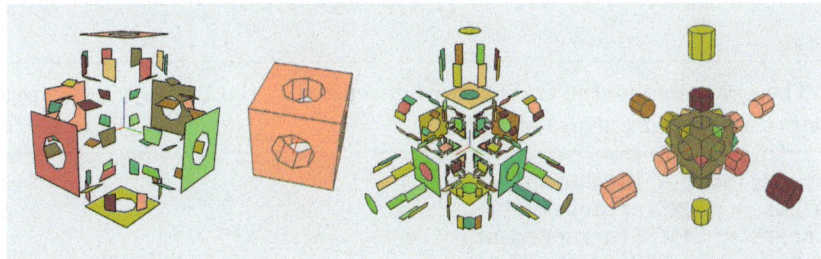

Fig. 7.13 (a) result of **DIFFERENCE** exploded; (b) single atom of **DIFFERENCE** result; (c) **UNION** exploded; (d) atoms of **UNION** result exploded.

The new user interface for solid Boolean expressions was used here. It is discussed in Section 7.4.2.

Fig. 7.14 (a) c\(x1∪x2∪x3); (b) x1∪x2∪x3; (c) c\(x2∪x3), where c, x1, x2, x3 are the algebra generators, **cube** and **cyls**, respectively.

Exercise 7.2 (UNION of random tetrahedra) In this example, we show the arrangement $\mathcal{A}(\text{lar})$ generated by the cellular complex **LAR(hpc)**, where **hpc** utilizes the Julia package to create a tetrahedralisation using **TetGen.jl** package of a set of random points. (see Plasm.jl/test/random-tetgen.jl). By exercise, the user should convert the code into the **BOUNDARY** of **UNION** of all generated tetrahedra.

```
hpc = MKPOLS([PointNd(p) for p in eachrow(pointlist)],[Cell(tet)
    for tet in eachrow(tets)])
lar = LAR(hpc)
```

First, we visualize edges, faces, and full-dimensional cells of the **lar** object, i.e., 1-, 2-, and 3-cells of it:

```
VIEWCOMPLEX( lar, show=["EV"], explode=[1,1,1] )  # show edges
VIEWCOMPLEX( lar, show=["FV"], explode=[2,2,2] )  # show faces
VIEWCOMPLEX( lar, show=["CV"], explode=[2,2,2] )  # show full-
    dim cells
```

Then, we compute the **arrangement** generated by the **lar** object, its **atoms**, **inners** and **outers** arrays of Breps.

```
arrangement = ARRANGE3D(lar)
atoms = ATOMS(arrangement)
inners = INNERS(arrangement)
outers = OUTERS(arrangement)
```

In Figure 7.15d, we show the **BOUNDARY()** shell bounding the atoms generated in the arrangement of 3D space.

```
VIEWCOMPLEX( outers, show=["CF"] )  # show boundary
```

Fig. 7.15 cellular complex of TetGen.jl example: (a) 1-cells; (b) 2-cells exploded; (c) tetrahedral 3-cells exploded; (d) the outer chain complex.

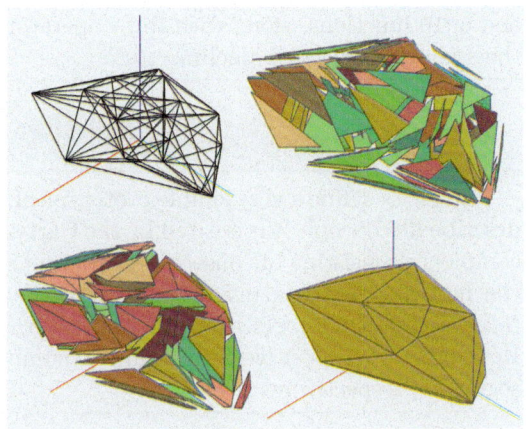

Exercise 7.3 (Arrangement of random cubes)

In this example, we demonstrate the $\mathbb{E}^3$ arrangement created by twelve cubes with random size, position, and orientation. This example was defined as a correctness test in coding the ARRANGE pipeline, a very important operator of the Plasm platform. See the file Plasm.jl/test/arrange3d.jl to view and/or use the Plasm testing platform.

Fig. 7.16 Robustness test with 12 random cubes: (a) exploded 2-cells of the arranged complex; (b) exploded 3-cells of the arranged complex.

The reader should try to extract the exterior shell of such a random complex, either as OUTER complex or as BOUNDARY(UNION(...)).

Example 7.4.2 (Development of a building frame)

An application context where a Domain Specific Language (DSL), such as the geometric generative Plasm language, may be highly valuable is in the AEC (Architecture, Engineering, Construction) technical field. In this example, we present a few lines of code used to develop a computer prototype for the structural geometry of a small building.

It is interesting to see that both the number of floors and the planar dimensions of the solid model are primarily independent of the code length.

First, let us load Plasm and define the SK alias for the SKELETON extraction function from a cellular complex. It may also be useful to note that the SK and `GRID` functions, along with the $*$ operator, combine `Hpc` datasets. Hence, they work in the same algebra.

```
using Plasm
SK = SKELETON
```

Now, let's define a very simple prototype of a small building, called `idea`, to describe its concept. It is created by the Cartesian product of three 1D cellular complexes consisting of consecutive adjacent segments of lengths specified by the numbers in the array arguments of the `GRID` primitive. This function takes a list of numbers as input and returns a sequence of line segments. Remember that positive and negative numbers represent solid and hollow segments, respectively.

```
X = GRID([2.4,4.5,-3,4.5,2.4])
Y = GRID([7,5])
Z = GRID([3,3])
idea = X * Y * Z;
VIEWCOMPLEX(LAR(idea), explode=[1.2,1.2,1.2])
```

Fig. 7.17 cellular 3-complex of a simplified building: (a) axonometric view of the complex; (b) perspective view of exploded 2-cells.

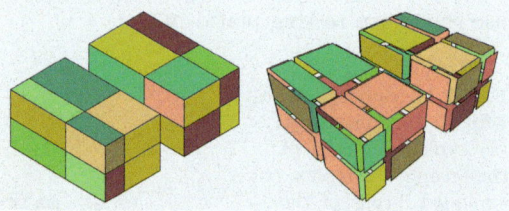

Next, we advance our example by creating specialized subcomplexes that correspond to various building subsystems, utilizing a bit of "combinatorics" of chain complexes.

```
building110 = X*Y*SK(0)(Z);    VIEWCOMPLEX(LAR(building110))
building101 = X*SK(0)(Y)*Z;    VIEWCOMPLEX(LAR(building101))
building011 = SK(0)(X)*Y*Z;    VIEWCOMPLEX(LAR(building011))
```

```
building1_110 = SK(1)(building110);
building1_101 = SK(1)(building101);
building1_011 = SK(1)(building011);
VIEWCOMPLEX(LAR(building1_110))   # perpendicular to z axis
VIEWCOMPLEX(LAR(building1_101))   # perpendicular to y axis
VIEWCOMPLEX(LAR(building1_011))   # perpendicular to x axis
```

The structural frame is now set up in the `Plasm` object `framexyz`. The `OFFSET` instances in the x and z directions create structural curbs around the floors, while the `OFFSET` longitudinal instance in y forms the beams.

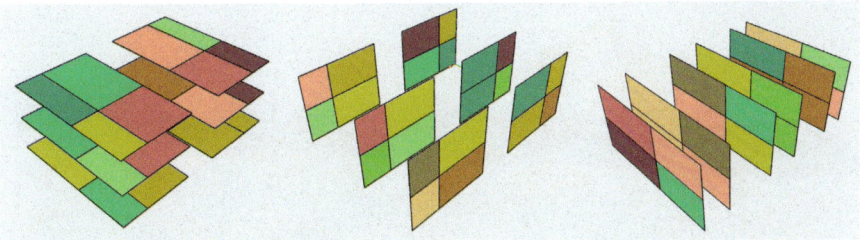

Fig. 7.18 cellular complex of simple building: (a) perspective view of horizontal z-sections (building110); (b) vertical y-sections (building101); (c) vertical x-sections (building011).

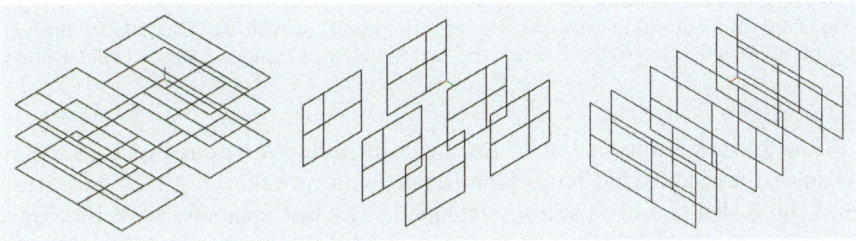

Fig. 7.19 Cellular 1D complex: (a) axonometric views of horizontal xy-frame (building1_110); (b) vertical xz-frame (building1_101); (c) yz-frame (building1_011).

```
floors  = OFFSET([ .2, .2, .2])(building110);
framey  = OFFSET([ .2, .2, .2])(building1_101);
framex  = OFFSET([ .2, .2,-.4])(building1_011);
framexyz = STRUCT(framex, framey, floors);
```

The two images in Figure 7.21 correspond to 2- and 3-cells of framexyz object.

```
VIEWCOMPLEX(ARRANGE3D(LAR(framexyz)), show = ["FV"], explode
    =[1.2,1.2,1.2] )
```

The assembly of building solid components is shown in Figure 7.4.3 using the atoms of the 3D arrangement object created by the expression ARRANGE3D (LAR(framexyz)).

Here, we select the two kinds of atoms generated by the ARRANGE3D operator.

```
atoms = ARRANGE3D(LAR(framexyz))
inners = INNERS(atoms);
outers = OUTERS(atoms);
VIEWCOMPLEX(inners, show = ["CF"] explode=[1.2,1.2,1.2] )
VIEWCOMPLEX(outers, show = ["CF"] explode=[1.2,1.2,1.2] )
```

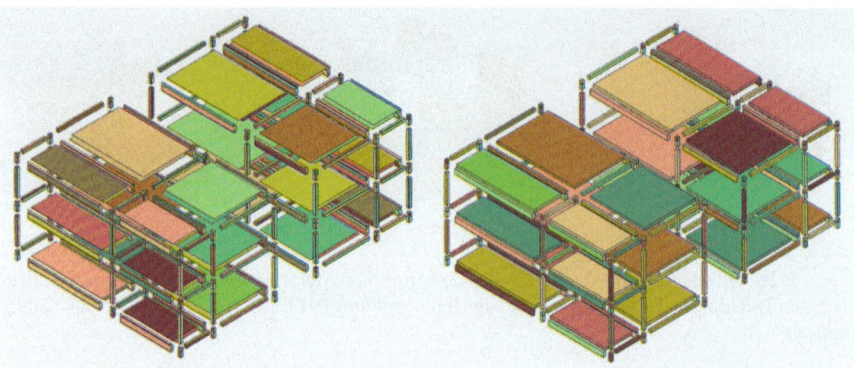

Fig. 7.20 Inner atoms in the arrangement $\mathcal{A}(\mathsf{framexyz}) = \mathsf{ARRANGE3D}(\mathsf{LAR}(\mathsf{framexyz}))$:
(a) parallel view of exploded 2-cells ; (b) parallel view of exploded 3-cells (solid atoms).

The **inners** atoms of the $\mathbb{E}^3$ arrangement induced by our input geometry **framexyz** are shown in Figure 1.20. In particular, we show both the 2-skeleton and the 3-skeleton of **framexyz** exploded. The last coincides with the chain complex stored within the object **inners** of Julia type **Lar**. In particular, the array **inners.C[:CF]** contains 242 3D 3-cell atoms, stored as lists of 2-cell faces.

Fig. 7.21 Two atoms in **OUTERS** (the bounded boundary of 3D space $\mathbb{E}^3$): (a) first connected component of Brep; (b) second connected component of Brep.

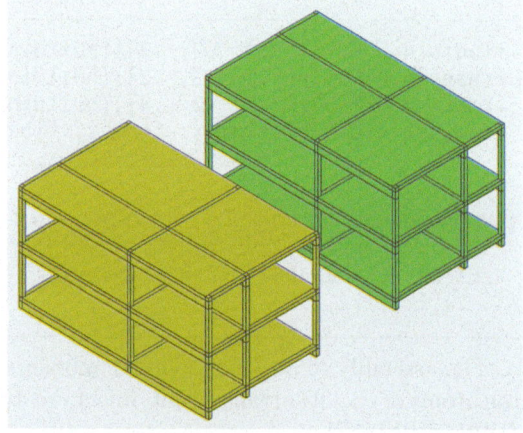

Of course, we already know that the surfaces of the outer arrangement atoms coincide with the Brep of the union of all the 242 inner atoms, possibly with reversed normals, depending on the choices made in implementing the platform visualizer.

Remark 7.12 (CSG is not a Boolean algebra) The CSG scheme is not a Boolean algebra of solid shapes. In particular, it lacks the universal set, and

Fig. 7.22 Views of atoms in OUTERS, i.e., the bounded boundary of 3D space $\mathbb{E}^3$): (a) perspective view at extrados level of the slab of the floor ; (b) perspective under that level, with Brep details.

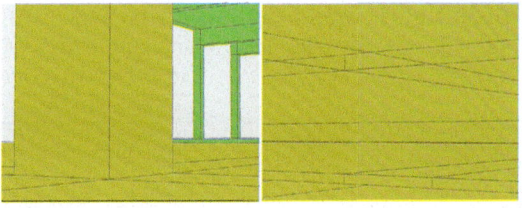

therefore the complement operation. Conversely, the Julia `Plasm` approach creates the specific Boolean algebras induced by any finite collection of, say n, solid objects and can efficiently encode, compute, and visualize any one of 2^m variadic Boolean expressions between them, where m is the number of atoms of $\mathbb{E}^d$ arrangement induced by the n primitives.

7.5 Atoms × generators = truth table

An atom of a Boolean algebra is an element that has no non-trivial proper subelements. According to the standard usage, atoms and generators represent the fundamental elements of algebras [15] and the basic building blocks of spatial arrangements, respectively, with a generator being a subset of atoms, which may be a singleton.

Definition 7.10 (Atoms in arrangement and algebra) The set of atoms of a Boolean algebra $\mathcal{B}(\mathcal{S})$ is one-to-one with the basis elements of the d-chain arrangement $\mathcal{A}(\mathcal{S})$ generated by the input collection $\mathcal{S}$.

Property 7.5 (Structure of generators in a solid Boolean algebra)
The structure of each generator shape $\mathsf{S} \subset \mathcal{B}$ is a union "$\cup$" of arrangement atoms, precisely corresponding to the true values of its column in the truth table `AG`; as a chain in C_d, it is a sum "$+$" of basis d-chains.

7.5.1 Julia and Boolean operations

Our research work [32] pointed to providing a general, well-grounded, and modern methodology to evaluate any solid algebraic expression, producing a value of the solid algebra defined by a language expression in the context of geometric design. So, we have extended the CSG approach with variadic operators and the universe set, identified with the ambient Euclidean spaces $\mathbb{E}^3$ and $\mathbb{E}^2$ by now, but extendible to higher dimensions.

Table 7.2 The *atoms* × *generators* Boolean matrix AG, relative to Figures 6.2 and 6.3 with 4 atoms, outer included, where both $Cube_1$ and $Cube_2$ are split in two parts.

AG	$Cube_1$	$Cube_2$
$atom_0$	0	0
$atom_1$	1	0
$atom_2$	1	1
$atom_3$	0	1

With this intention in mind, we have implemented in `Plasm` the isomorphism between the arrangements of geometric spaces and the set algebras of their atoms, coinciding with the basis of the linear chain space C_d.

The `TGW` algorithm [31] computes the ∂_d matrix one row at a time, generating the coordinate vector in $\mathbf{Z}_3 \simeq \{-1, 0, 1\}$ of each unknown unit vector $u_d \in C_d$ (solid atom) as linear combination of vectors in $U_{d-1} \subset C_{d-1}$. In this way, `Plasm` discovers the coordinates of facets of all basis u_d chains, and hence their boundary representation.

This sparse matrix `FC` is used to build the breps of all atoms in the arrangement of $\mathbb{E}^\ulcorner$. With all atoms' `breps` available, we can proceed to characterize each atom by only one internal point. Then, we compute the structure of algebra generators by containment testing whether each such point is contained within polyhedral primitives represented by their breps, that is, by their $(d-1)$-cells, prior to the decomposition performed by the arrangement.

The final step is the simplest computationally, where we leverage the isomorphism between our solid algebra —Boolean formulas written in `Plasm` using solid primitives, even very complex ones —and the set algebra of atoms. The reader should remember that for any finite set $\mathcal{S}$, its power set $\mathcal{P}(\mathcal{S})$ — the set of all subsets of S — is closed under union, intersection, difference, and complement operations. Importantly, this set $\mathcal{P}(S))$ is isomorphic to the set of binary strings of a specific length, i.e., the size m of the atom set.

Suppose m is the number of atoms `A` in the $\mathbb{E}^d$ arrangement generated by any collection of n solid generators `G`. The last part of the `Plasm` Boolean pipeline produces a Julia sparse $m \times n$ incidence matrix `AG ≡ boolmat::Matrix{Bool}`; see Table 7.2 for a simple example.

Any solid algebra generator is represented in `AG` as a binary column. Therefore, for any possible Boolean expression of solid algebra, it is sufficient to compute the required combination of Boolean terms by evaluating each row (either to 1 or to 0) with the Boolean function, producing the binary m-vector of the evaluated expression, that is, the final result. Any expression between solid models is thus reducible to the selection of true minterms among binary strings of length n (the number of generators).

The binary m-`Vector{Bool}` generated from evaluating the truth table row by row also serves as the sparse coordinate vector of the binary solution, which, multiplied by the boundary matrix, yields the `Brep` of the resulting solid. Finally, evaluate the implementation status of our `Plasm` Solid Algebra, which has not yet achieved v1.0.0. While the topological component of `ARRANGE3D` is stable and robust, the numerical component requires further testing.

7.5.2 Stepwise Boolean example computation

This section discusses the `Plasm` interpretation of a straightforward Boolean expression. The language directs the Julia compiler to accurately compute the `UNION` of two equal objects translated reciprocally in x, y and derived as the `DIFFERENCE` of two cubes. Each of the two (unnamed) objects is created by two (named) unit `cube` instances, where the second `cube` instance is translated in x, y, z relative to the first.

Coding 7.5.1 (Boolean expression) The `hpc` object created here is structured as a composition of `Plasm` primitives with hierarchical semantics. It is similar to the `STRUCT` operator, which transforms its argument list into the embedding coordinate frame. The Julia expressions of `cube`, `hpc`, and `result` objects are unnecessarily type-coded, but types are included for the reader's benefit.

```
using Plasm
cube = CUBE(1)::Hpc

hpc = UNION(
   DIFFERENCE( cube, T(1,2,3)(.5,.5,.5), cube),
   T([1,2])([0.75,.25]),
   DIFFERENCE( cube, T(1,2,3)(.5,.5,.5), cube))::Hpc
result = BOOL(hpc)::Lar
brep = BOUNDARY(result)

VIEWCOMPLEX(result, show=["FV"], explode=[1.2,1.2,1.2])
VIEWCOMPLEX(result, show=["CV"], explode=[1.5,1.5,1.5])
```

Coding 7.5.2 (Boolean formula interpretation) In the following script, we present a readable listing of the dynamic storage utilized by the `Plasm` primitive function `UNION` during the traversal of its expression in Coding 7.5.1. Notice how the first row of the `UNION` in the previous script transforms into the `"bool_op"` => `Plasm.Union` expression in the next script, where `Union` serves as a logical function, similar to the value in the `"bool_op"` => `Plasm` `.Difference` dictionary elements of `Hpc` node `properties`:

```
Hpc( T=MatrixNd(4) properties=Dict{String, DataType}("bool_op"
    => Union)
  Hpc( T=MatrixNd(4) properties=Dict("bool_op" => Plasm.
    Difference)
```

Fig. 7.23 The resulting
exploded model from the
UNION of two DIFFER-
ENCE of two CUBE: (a)
visualization of 2-chains;
(b) visualization of 3-
chains.

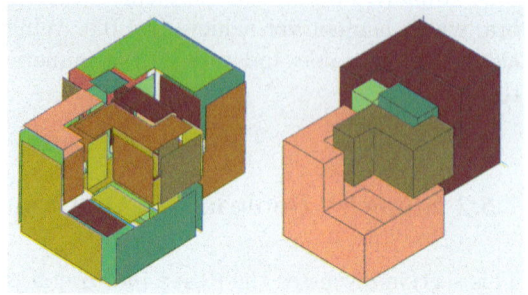

```
Hpc( T=MatrixNd(4) properties=Dict{Any, Any}()
  Geometry([[0.0, 0.0, 0.0], [1.0, 0.0, 0.0], [0.0, 1.0,
0.0], [1.0, 1.0, 0.0], [0.0, 0.0, 1.0], [1.0, 0.0, 1.0],
[0.0, 1.0, 1.0], [1.0, 1.0, 1.0]], hulls=[[1, 2, 3, 4, 5, 6,
 7, 8]])
  )
Hpc( T=MatrixNd([[1.0, 0.0, 0.0, 0.0], [0.5, 1.0, 0.0, 0.0],
  [0.5, 0.0, 1.0, 0.0], [0.5, 0.0, 0.0, 1.0]]) properties=
Dict{Any, Any}()
  Geometry([[0.0, 0.0, 0.0], [1.0, 0.0, 0.0], [0.0, 1.0,
0.0], [1.0, 1.0, 0.0], [0.0, 0.0, 1.0], [1.0, 0.0, 1.0],
[0.0, 1.0, 1.0], [1.0, 1.0, 1.0]], hulls=[[1, 2, 3, 4, 5, 6,
 7, 8]])
  )
)
Hpc( T=MatrixNd(4) properties=Dict("bool_op" => Plasm.
  Difference)
  Hpc( T=MatrixNd([[1.0, 0.0, 0.0, 0.0], [0.75, 1.0, 0.0,
0.0], [0.25, 0.0, 1.0, 0.0], [0.0, 0.0, 0.0, 1.0]])
  properties=Dict{Any, Any}()
    Geometry([[0.0, 0.0, 0.0], [1.0, 0.0, 0.0], [0.0, 1.0,
0.0], [1.0, 1.0, 0.0], [0.0, 0.0, 1.0], [1.0, 0.0, 1.0],
[0.0, 1.0, 1.0], [1.0, 1.0, 1.0]], hulls=[[1, 2, 3, 4, 5, 6,
 7, 8]])
    )
  Hpc( T=MatrixNd([[1.0, 0.0, 0.0, 0.0], [1.25, 1.0, 0.0,
0.0], [0.75, 0.0, 1.0, 0.0], [0.5, 0.0, 0.0, 1.0]])
  properties=Dict{Any, Any}()
    Geometry([[0.0, 0.0, 0.0], [1.0, 0.0, 0.0], [0.0, 1.0,
0.0], [1.0, 1.0, 0.0], [0.0, 0.0, 1.0], [1.0, 0.0, 1.0],
[0.0, 1.0, 1.0], [1.0, 1.0, 1.0]], hulls=[[1, 2, 3, 4, 5, 6,
 7, 8]])
    )
  )
)
```

Example 7.5.1 (Boolean formula visualization) In this example, we
show four pairs of images corresponding to the expression in Script 7.5.1

with the root substituted by UNION, INTERSECTION, DIFFERENCE, and XOR, respectively. It is worth noting that all models refer to the same 3D arrangement, while for each boolop (Plasm Boolean operations) a different set of atoms is selected. See Example 7.5.6 in this section. □

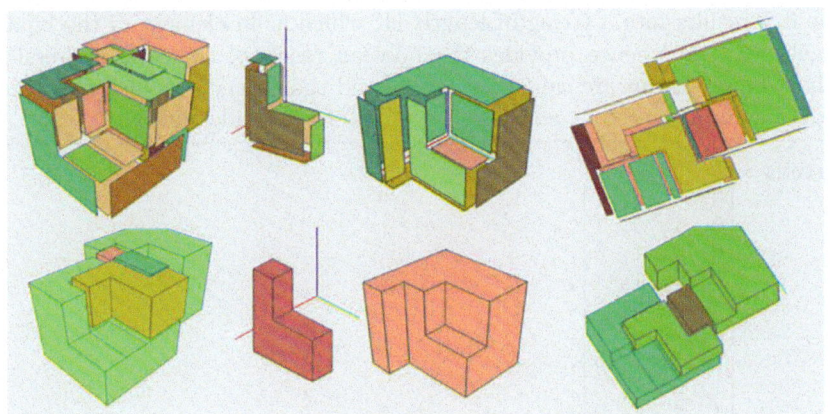

Fig. 7.24 Every column shows: (a) the exploded 2-cells; (b) the exploded 3-cells; for each of the following OP root: (a) UNION; (b) INTERSECTION; (c) DIFFERENCE; (d) XOR.

Let us note that the top image in each pair shows an exploded subset of U_2 basis complex of 2-cells (singleton 2-chains). In contrast, the lower images show the exploded U_3 subset of 3-cell atoms (singleton 3-chains), distinguishable by unitary colors, and assembled to construct the computed values of algebraic formulas. The expressions with UNION, INTERSECTION, DIFFERENCE, and XOR input Plasm statements correspond to the Lar value visualized by the last VIEWCOMPLEX statement of the source input in Script 7.5.1, containing the string show=["CV"].

Example 7.5.2 (Incidence between atoms and generators) The application ARRANGE3D(LAR(hpc)), where hpc is the value produced by the UNION statement of Coding 7.5.1, generates all 12 atoms of the spatial arrangement, creating the ordered sequence of the Boolean representation as the product (minterm) of generators for all atoms in the 3D arrangement.

Conversely, BOOL(hpc) evaluates its ::Hpc input (see Coding 7.5.2) and selects only the five atoms where results true the Boolean function that correspond to the Plasm script, which is the Union of two Difference operations of two pairs of cube objects in our example. The resulting binary string within the Boolean algebra 2^n, where n represents the number of atoms and $2 = \{0, 1\}$, is called in "sum of products" or "sum of minterms". □

We show in the following, printed by rows, the atoms::{Vector{Vector{Bool}}} with 11 rows representing the incidence relation between the inner atoms (rows) and the four generators (columns) that produce the space arrangement.

To extract from this list of terms the only atoms that form a specific solid Boolean expression within the Boolean algebra generated by the four generator cubes, a corresponding logical function $\mathbf{f} : \mathbf{2}^m \to \mathbf{2}$, where $m = 4$ (the number of generators) must be applied to each of them, accumulating the $0/1$ results into a string of length 11, which is an element of the binary algebra $\mathbf{2}^n$, and hence provides the boolean result of the solid expression. The 4-binary terms are called minterms, and the binary sum of the function values sum of minterms, and also structure of the solid solution.

```
atoms = [
   Bool[0, 0, 0, 1],
   Bool[0, 1, 0, 1],
   Bool[0, 0, 1, 0],
   Bool[0, 1, 1, 1],
   Bool[0, 0, 1, 1],
   Bool[0, 1, 0, 0],
   Bool[0, 1, 1, 0],
   Bool[1, 0, 1, 0],
   Bool[1, 1, 1, 0],
   Bool[1, 1, 0, 0],
   Bool[1, 0, 0, 0]]
```

The above are the INNERS atom terms (strings) generated by the four intersecting input cube generators. To see all the atoms generated by the 3D space arrangement, it is sufficient to apply the ARRANGE3D to the Boolean expression, or better to its (implicit STRUCT) value:

```
atoms = ARRANGE3D(LAR(hpc))
VIEWCOMPLEX(INNERS(atoms), explode=(4,4,4), show=["CV"])
```

Note the explicit INNERS(atoms) because the current length(atoms)=12, including the outer object.

Fig. 7.25 The eleven INNERS atoms generated by the input arrangement of generators, from the four instances of cube terms. The result of every solid expression in this algebra corresponds to a different subset of atoms.

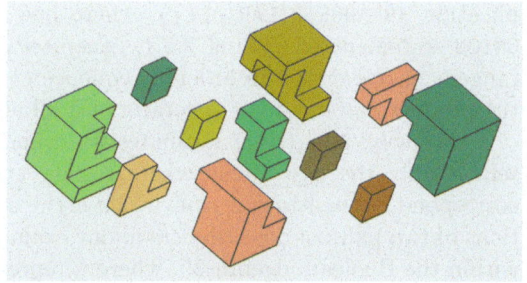

Such INNERS(atoms), to distinguish from OUTERS boundary of $\mathbb{E}^3$, are in number of 11, like the rows of *Atoms* × *Generators* relation matrix. Since the 3D arrangement induced by the generators is connected, there is just one item in array OUTERS(atoms), with atoms :: Lar=BOOL(hpc).

Example 7.5.3 (Space arrangement) The TWG algorithm of the internally instanced ARRANGE3D function produces twelve 3D atoms, inner and outer, of the 3D space (see Figure 7.5.2). Finally, the Boolean formula is applied to the rows of atoms :: Vector{Vector{Bool}} in Example 7.5.2. □

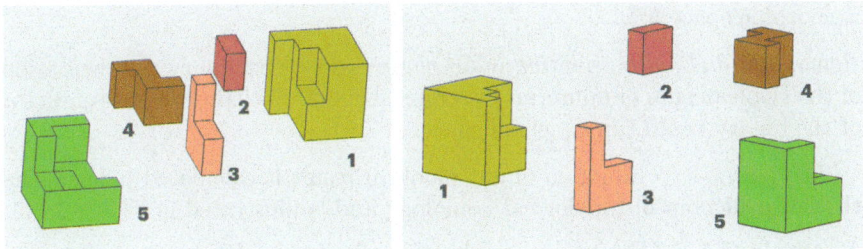

Fig. 7.26 The exploded 5 atoms selected by the **UNION** formula: (a) front view; (b) back view. The atom groups $1, 3$ and $2, 4, 5$ correspond to the two cube differences.

The atoms with a binary sequence (minterm) satisfying the formula, i.e., returning **true** when applying the corresponding logical function, enter with 1 into the result :: Hpc sequence of the original solid model formula. The Boolean solution is, therefore, in the canonical form of sum of minterms.

Example 7.5.4 (Boolean function equivalent to CSG expression) The subset of atoms to enter the canonical solution of the Boolean solid formula is selected by applying the below f Boolean function without arguments (i.e., written in pure FL style) to all minterms of atoms :: Vector{Vector{Bool}} object.

```
f = Union ∘ CONS([
      Difference ∘ CONS([S1,S2]),
      Difference ∘ CONS([S3,S4])
    ])
```

The Boolean function f, when applied to an object of Vector{Bool} type with 4 variables (the minterm of the problem), produces a Bool value (1 or 0, namely, true or false) when evaluated. We have already observed the minterms of atoms, Boolean arrays of length 4, representing the capacity of affecting the arrangement of algebra generators (the 4 cube instances in the Plasm formula that are stored in the hpc variable).

Example 7.5.5 (Boolean solution in C_3 chain space) By applying the f function to the atoms vector[9], we extract the actual structure of the binary

---

[9] Remember, from Chapter 2, that AA represents 'apply to all' in Plasm

solution from the $2^{11} = 2048$ possible configurations of bits of the Boolean algebra generated by the source expression in Example 7.5.2. This operation yields the solution in the canonical form of sum of minterms:

```
structure = AA(f)(atoms)'
  0  0  1  0  0  0  1  1  1  0  1
```

We have so generated the solution to our Solid Algebra expression as the canonical sum of minterms for the corresponding Boolean function in the isomorphic space $\mathbf{2}^m$.   □

Remark 7.13 (Boolean solution in C_3 chain space) The binary representation of the Boolean sum of minterms also represents the 3-chain coordinate vector of the binary `result` in C_3 chain space.

The boundary evaluation of any Boolean result is calculated below using the linear algebra of topological homology and is illustrated in Figure 7.5.2.

Example 7.5.6 (Boundary evaluation from result structure)

The actual final step of evaluating the `Plasm` Boolean expression of Example 7.5.2 is the generation of the geometric Brep value at the end of the Boolean processing pipeline.

Let us note that the above binary structure of a solid model as the value of the characteristic function of a solid Boolean value with respect to `Plasm` Solid Algebra is also the coordinate vector of its 3-chain, i.e., the unique vector that multiplied by the $[\partial_3]$ matrix produces the 2-chain (actually its coordinates) listing all the boundary faces.

Consider the `result::Lar` value of the `Plasm` source Boolean expression in Example 7.5.2. We know from basic algebraic topology that the boundary of a 3-chain (`structure`) is easily generated by the vector-matrix product of our vector of bits (actually a 1×11 matrix) multiplied by the 0/1 matrix [CF], viewed as the operator C → F (i.e., cells to faces) and represented below as object of type `Vector{Vector{Int64}}`.

```
CF = ARRANGE3D(LAR(hpc)).C[:CF]
12-element Vector{Vector{Int64}}:
 [1, 2, 5, 6, 7, 10, 14, 15, 16, 17  …  45, 46, 51, 52, 53, 55,
   59, 61, 62, 63]
 [8, 13, 20, 27, 31, 32, 36, 41, 45, 46, 51, 52, 56, 64]
 [4, 11, 19, 23, 35, 44]
 [26, 29, 39, 47, 50, 53, 55, 59, 61, 62, 63]
 [17, 26, 29, 34, 37, 49, 54, 58]
 [9, 12, 20, 27, 32, 38, 43, 48, 57, 60]
 [1, 5, 6, 7, 10, 14, 19, 23, 35, 40, 42]
 [25, 31, 37, 49, 57, 60]
 [3, 4, 11, 16, 18, 22, 24, 28, 30, 38, 43, 48, 54, 58]
 [25, 39, 47, 50, 56, 64]
 [2, 8, 13, 15, 21, 33, 40, 42]
 [3, 9, 12, 21, 33, 44]
```

Notice that there are $12 = 11 + 1$ atoms in the $\mathcal{A}(\mathbb{E}^3)$ arrangement, including the (unique) outer 3-chains; eleven INNERS() 3-chains plus one in OUTERS().

Recall that CF is the name used for the linear transformation from 2-chains (F) to 3-chains (C), represented in Lar as Vector{Vector{Int}}, where Int are indices of faces (F).

Currently, we have two alternatives to extract the Brep of the union solution of two differences of unit cubes, i.e., to compute its boundary faces. The first alternative is to directly use and visualize the OUTER(...) shell (2-chain); the second is to get the $\partial_3 = $ lar2cop(CF) operator and to compute the left matrix multiplication[10]:

```
∂₃ = lar2cop(CF);

@show brep_2cycle = Bool[0 structure] * ∂₃ .% 2;     #=
brep_2cycle = (Bool[0 structure] * ∂₃) .% 2 = [0 0 0 1 0 0 0 0 1
        0 1 1 0 0 0 1 0 1 0 0 1 1 0 1 0 1 0 1 1 1 1 0 1 0 0 0 1 1 0
        0 0 0 1 1 0 0 0 1 1 0 0 0 1 1 1 1 1 1 1 1 1 1 1]     =#
```

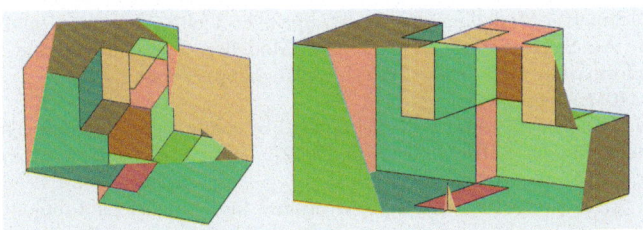

Fig. 7.27 Interior views of the Brep generated by the Plasm example in Coding 7.5.1 with UNION expression: (a) perspective view from below; (b) parallel view from above.

Coding 7.5.3 (Final Brep computation) The final passage of our stepwise computed Script 7.5.1 involves creating with Plasm the BOUNDARY of the result object. The figure above displays the Brep interior of the initial UNION expression. One image is from below; the other is from above.

```
result = BOOL(hpc)
brep = BOUNDARY(result)
VIEWCOMPLEX(brep)
```

In Plasm, the 2-boundary brep object (a 2-chain without boundary, and possibly unconnected) immediately produces a visualization in an interactive window on the computer screen through the VIEWCOMPLEX(brep) expression.

---

[10] Left multiplication functions with the same principles as right multiplication, but since the vector is on the other side of A, the column space is switched to the row space.

The generated object can also be exported as a `Lar` value to a file in certain graphics formats. Currently, as of `Plasm` v1.0.0, the only export format available is `PLY` . Other formats are in the works pipeline.

References

1. Ala, S.R.: Performance anomalies in boundary data structures. IEEE Comput. Graph. Appl. **12**(2), 49–58 (1992). DOI 10.1109/38.124288. URL `http://dx.doi.org/10.1109/38.124288` pages 245
2. Bajaj, C., Paoluzzi, A., Scorzelli, G.: Progressive conversion from b-rep to bsp for streaming geometric modeling. Computer-Aided Design and Applications **3**(5-6) (2006) pages 245
3. Baumgart, B.G.: Winged edge polyhedron representation. Tech. Rep. Stan-CS-320, Stanford University, Stanford, CA, USA (1972) pages 245
4. Bernardini, F., Ferrucci, V., Paoluzzi, A., Pascucci, V.: Product operator on cell complexes. In: SMA '93: Proceedings on the second ACM symposium on Solid modeling and applications, pp. 43–52. ACM Press, New York, NY, USA (1993). DOI `http://doi.acm.org/10.1145/164360.164378` pages 75, 90, 248
5. Bieri, H.: Nef polyhedra: A brief introduction. In: H. Hagen, G. Farin, H. Noltemeier (eds.) Geometric Modelling, pp. 43–60. Springer Vienna, Vienna (1995) pages 245
6. Bowyer, A.: SvLis Set-theoretic Kernel Modeller: Introduction and User Manual. Information Geometers (1995). URL `http://books.google.it/books?id=hYqwAAAACAAJ` pages 245
7. Braid, I.C.: The synthesis of solids bounded by many faces. Commun. ACM **18**(4), 209–216 (1975). DOI 10.1145/360715.360727. URL `http://doi.acm.org/10.1145/360715.360727` pages 245
8. Brisson, E.: Representing geometric structures in d dimensions: topology and order. In: Proc. of the 5-th Annual Symposium on Computational Geometry, SCG '89, pp. 218–227. Acm, New York, NY, USA (1989). DOI 10.1145/73833.73858. URL `http://doi.acm.org/10.1145/73833.73858` pages 245
9. DiCarlo, A., Milicchio, F., Paoluzzi, A., Shapiro, V.: Chain-based representations for solid and physical modeling. IEEE Transactions on Automation Science and Engineering **6**(3), 454–467 (2009). URL `https://ieeexplore.ieee.org/document/5071139` pages 47, 245
10. Dobkin, D.P., Laszlo, M.J.: Primitives for the manipulation of three-dimensional subdivisions. In: Proceedings of the third annual symposium on Computational geometry, SCG '87, pp. 86–99. Acm, New York, NY, USA (1987). DOI 10.1145/41958.41967. URL `http://doi.acm.org/10.1145/41958.41967` pages 245
11. Ferrucci, V.: Generalised extrusion of polyhedra. In: Proceedings on the Second ACM Symposium on Solid Modeling and Applications, SMA '93, p. 3542. Association for Computing Machinery, New York, NY, USA (1993). DOI 10.1145/164360.164376. URL `https://doi.org/10.1145/164360.164376` pages 86, 248
12. Goldfeather, J., Monar, S., Turk, G., Fuchs, H.: Near real-time CSG rendering using tree normalization and geometric pruning. IEEE Computer Graphics and Applications **9**(3), 20–28 (1989). URL `https://doi.org/10.1109/38.28107` pages 247
13. Gomes, A., Middleditch, A., Reade, C.: A mathematical model for boundary representations of n-dimensional geometric objects. In: Proceedings of the fifth ACM symposium on Solid modeling and applications, SMA '99, pp. 270–277. Acm, New York, NY, USA (1999). DOI 10.1145/304012.304039. URL `http://doi.acm.org/10.1145/304012.304039` pages 245

14. Guibas, L., Stolfi, J.: Primitives for the manipulation of general subdivisions and the computation of voronoi. ACM Trans. Graph. **4**(2), 74–123 (1985). DOI 10.1145/282918.282923. URL http://doi.acm.org/10.1145/282918.282923 pages 245

15. Halmos, P., Givant, S.: Introduction to Boolean algebra. Springer (2009). URL https://www.amazon.com/Introduction-Boolean-Algebras-Undergraduate-Mathematics-ebook/dp/B00HWUXYSG pages 277

16. Hoffmann, C.M.: Geometric and Solid Modeling: An Introduction. Morgan Kaufmann Publishers Inc., San Francisco, CA, USA (1989) pages 245

17. Hoffmann, C.M., Hopcroft, J.E., Karasick, M.S.: Robust set operations on polyhedral solids. Tech. rep., Cornell University, Ithaca, NY, USA (1987) pages 245

18. Hoffmann, C.M., Kim, K.J.: Towards valid parametric cad models. Computer-Aided Design **33**(1), 81–90 (2001). DOI http://dx.doi.org/10.1016/S0010-4485(00)00073-7 pages 245

19. Hoffmann, C.M., Vanek, G.: Fundamental techniques for geometric and solid modeling. Tech. Rep. Report Number: 91-044, Purdue University (1991) pages 245

20. Kalay, Y.E.: The hybrid edge: a topological data structure for vertically integrated geometric modelling. Comput. Aided Des. **21**(3), 130–140 (1989). DOI 10.1016/0010-4485(89)90067-5. URL http://dx.doi.org/10.1016/0010-4485(89)90067-5 pages 245

21. Lee, S.H., Lee, K.: Partial entity structure: a compact non-manifold boundary representation based on partial topological entities. In: Proceedings of the sixth ACM symposium on Solid modeling and applications, SMA '01, pp. 159–170. Acm, New York, NY, USA (2001). DOI 10.1145/376957.376976. URL http://doi.acm.org/10.1145/376957.376976 pages 245

22. Lienhardt, P.: Subdivisions of n-dimensional spaces and n-dimensional generalized maps. In: Proceedings of the Fifth Annual Symposium on Computational Geometry, SCG '89, p. 228–236. Association for Computing Machinery, New York, NY, USA (1989). DOI 10.1145/73833.73859. URL https://doi.org/10.1145/73833.73859 pages 245

23. Lienhardt, P.: Topological models for boundary representation: a comparison with n-dimensional generalized maps. Comput. Aided Des. **23**(1), 59–82 (1991). DOI 10.1016/0010-4485(91)90082-8. URL http://dx.doi.org/10.1016/0010-4485(91)90082-8 pages 245

24. Loi, T.L.: An introduction to semi-algebraic sets. Kôkyûroku Bessatsu, 1764 pp. 48–58 (2011) pages 251

25. Mantyla, M.: Introduction to Solid Modeling. W. H. Freeman Co., New York, NY, USA (1988) pages 245

26. Naylor, B., Amanatides, J., Thibault, W.: Merging bsp trees yields polyhedral set operations. In: Proceedings of the 17th Annual Conference on Computer Graphics and Interactive Techniques, SIGGRAPH '90, p. 115–124. Association for Computing Machinery, New York, NY, USA (1990). DOI 10.1145/97879.97892. URL https://doi.org/10.1145/97879.97892 pages 245

27. Paoluzzi, A.: Geometric Programming for Computer Aided Design. John Wiley Sons, Chichester, UK (2003). URL https://onlinelibrary.wiley.com/doi/book/10.1002/0470013885 pages 37, 38, 48, 73, 127, 130, 201, 206, 209, 244

28. Paoluzzi, A., Bernardini, F., Cattani, C., Ferrucci, V.: Dimension-independent modeling with simplicial complexes. ACM Trans. Graph. **12**(1), 56–102 (1993). DOI 10.1145/169728.169719. URL http://doi.acm.org/10.1145/169728.169719 pages 245, 248

29. Paoluzzi, A., Pascucci, V., Vicentino, M.: Geometric programming: a programming approach to geometric design. ACM Trans. Graph. **14**(3), 266–306 (1995). DOI 10.1145/212332.212349. URL http://doi.acm.org/10.1145/212332.212349 pages 37, 67, 245

30. Paoluzzi, A., Ramella, M., Santarelli, A.: Boolean algebra over linear polyhedra. Comput. Aided Des. **21**(10), 474–484 (1989). URL `http://dl.acm.org/citation.cfm?id=70248.70249` pages 245, 260

31. Paoluzzi, A., Shapiro, V., DiCarlo, A., Furiani, F., Martella, G., Scorzelli, G.: Topological computing of arrangements with (co)chains. ACM Trans. Spatial Algorithms Syst. **7**(1) (2020). DOI 10.1145/3401988. URL `https://doi.org/10.1145/3401988` pages 47, 94, 137, 216, 218, 221, 224, 227, 228, 231, 233, 234, 235, 251, 258, 260, 278

32. Paoluzzi, A., Shapiro, V., DiCarlo, A., Scorzelli, G., Onofri, E.: Finite algebras for solid modeling using julias sparse arrays. Computer-Aided Design **155**, 103,436 (2023). DOI https://doi.org/10.1016/j.cad.2022.103436. URL `https://www.sciencedirect.com/science/article/pii/S0010448522001695` pages 47, 65, 97, 137, 184, 218, 221, 254, 260, 270, 277

33. Pascucci, V., Ferrucci, V., Paoluzzi, A.: Dimension-independent convex-cell based Lar: representation scheme and implementation issues. In: Proceedings of the third ACM Symposium on Solid Modeling and Applications, SMA '95, pp. 163–174. Acm, New York, NY, USA (1995). DOI 10.1145/218013.218055. URL `http://doi.acm.org/10.1145/218013.218055` pages 245

34. Pascucci, V., Ferrucci, V., Paoluzzi, A.: Dimension-independent convex-cell based lar: Representation scheme and implementation issues. In: C. Hoffmann, J. Rossignac (eds.) 3rd ACM/IEEE Symposium on Solid Modeling and Applications, pp. 17–19. ACM Press, New York, NY (1995). SMA 95 pages 48, 248

35. Pratt, M.J., Anderson, B.D.: A shape modelling api for the STEP standard. In: in Fourteenth International Conference on Atomic Physics, pp. 1–7 (1994) pages 245

36. Raghothama, S., Shapiro, V.: Consistent updates in dual representation systems. In: Proceedings of the fifth ACM symposium on Solid modeling and applications, SMA '99, pp. 65–75. Acm, New York, NY, USA (1999). DOI 10.1145/304012.304019. URL `http://doi.acm.org/10.1145/304012.304019` pages 245

37. Rappoport, A.: The generic geometric complex (ggc): a modeling scheme for families of decomposed pointsets. In: Proceedings of the fourth ACM symposium on Solid modeling and applications, SMA '97, pp. 19–30. Acm, New York, NY, USA (1997). DOI 10.1145/267734.267749. URL `http://doi.acm.org/10.1145/267734.267749` pages 245

38. Requicha, A.: Representations for rigid solids: Theory, methods and systems. ACM Computing Surveys **12**(4), 437–464 (1980). URL `https://doi.org/10.1109/TC.1980.1675470` pages 64, 74, 138, 218, 244

39. Requicha, A.G.: Representations for rigid solids: Theory, methods, and systems. ACM Comput. Surv. **12**(4), 437–464 (1980). DOI 10.1145/356827.356833. URL `http://doi.acm.org/10.1145/356827.356833` pages 245

40. Rossignac, J.: IBNC: Integrated Boundary and Natural CSG for Polyhedra (Review, Simplifications, and Integration of Prior Art). Computer-Aided Design **150**, 103,296 (2022). URL `https://doi.org/10.1016/j.cad.2022.103296` pages 246, 248

41. Rossignac, J., Requicha, A.A.G.: Depth-buffering display techniques for constructive solid geometry. IEEE Computer Graphics and Applications **6**, 29–39 (1986). URL `https://api.semanticscholar.org/CorpusID:15696719` pages 247

42. Rossignac, J.R., O'Connor, M.A.: A dimension-independent model for pointsets with internal structures and incomplete boundaries. Tech. Rep. Research Report RC 14340, IBM Research Division, Yorktown Heights, N.Y. 10598 (1989) pages 245

43. Rossignac, J.R., O'Connor, M.A.: SGC: a dimension-independent model for pointsets with internal structures and incomplete boundaries. In: Geometric modeling for product engineering. North-Holland (1990) pages 245

44. Rossignac, J.R., Requicha, A.A.G.: Constructive non-regularized geometry. Comput. Aided Des. **23**(1), 21–32 (1991). DOI 10.1016/0010-4485(91)90078-B. URL `http://dx.doi.org/10.1016/0010-4485(91)90078-B` pages 245

45. Shapiro, V.: Representations of semi-algebraic sets in finite algebras generated by space decompositions. Ph.D. thesis, Cornell University, Ithaca, NY, USA (1991) pages 245, 251

46. Shapiro, V.: Solid modeling. In: G. Farin, J. Hoschekand, M. Kim (eds.) Handbook of Computer Aided Geometric Design. Elsevier Science Publishers (2001) pages 245

47. Shapiro, V., Vossler, D.L.: Separation for boundary to CSG conversion. ACM Trans. Graph. **12**(1), 3555 (1993). DOI 10.1145/169728.169723. URL https://doi.org/10.1145/169728.169723 pages 248

48. Shapiro, V., Vossler, D.L.: What is a parametric family of solids? In: Proceedings of the third ACM symposium on Solid modeling and applications, SMA '95, pp. 43–54. ACM (1995). DOI 10.1145/218013.218029. URL http://doi.acm.org/10.1145/218013.218029 pages 245

49. Silva, C.E.: Alternative definitions of faces in boundary representations of solid objects. Tech. Rep. TM-36, Production Automation Project, Univ. of Rochester (1981) pages 245

50. Tilove: Set membership classification: A unified approach to geometric intersection problems. IEEE Transactions on Computers **C-29**(10), 874–883 (1980). URL https://doi.org/10.1109/TC.1980.1675470 pages 247

51. Voelcker, H.B., Requicha, A.A.G.: Constructive solid geometry. Tech. Rep. TM-25, Production Automation Project, Univ. of Rochester (1977) pages 245

52. Vossler, D.L.: Sweep-to-CSG conversion using pattern recognition techniques. IEEE Computer Graphics and Applications **5**(8), 61–68 (1985). URL https://doi.org/10.1109/MCG.1985.276215 pages 247

53. Weiler, K.: Edge-based data structures for solid modeling in curved-surface environments. Computer Graphics and Applications, IEEE **5**(1), 21 –40 (1985). DOI 10.1109/MCG.1985.276271 pages 245

54. Weiler, K.J.: Topological structures for geometric modeling. Ph.D. thesis, Rensselaer Polytechnic Institute (1986) pages 245

55. Weiler, K.J.: The radial edge structure: A topological representation for non-manifold geometric modelling. In: M. Wozny, H. McLaughlin, J. Encarnacao (eds.) Geometric modelling for CAD applications, pp. 3–12. North-Holland, Amsterdam (1988) pages 245

56. Wikipedia: Natural transformation — Wikipedia, the free encyclopedia (2023). URL https://en.wikipedia.org/wiki/Natural_transformation. [Online; accessed 3-january-2025] pages 252

57. Woo, T.: A combinatorial analysis of boundary data structure schemata. Computer Graphics Applications, IEEE **5**(3), 19–27 (1985) pages 245

58. Wozny, M., Turner, J., Preiss, K.: Geometric modeling for product engineering: IFIP WG 5.2/NSF Working Conf. on Geometric Modeling, Rensselaerville, U.S.A., 18-22 September, 1988. North-Holland (1990). URL http://books.google.it/books?id=-BkfAQAAIAAJ pages 245

59. Yamaguchi, F., Tokieda, T.: Bridge edge and triangulation approach in solid modeling. In: T. Kunii (ed.) Frontiers in Computer Graphics. Springer Verlag, Berlin (1985) pages 245

60. Yamaguchi, Y., Kimura, F.: Nonmanifold topology based on coupling entities. Computer Graphics and Applications, IEEE **15**(1), 42–50 (1995) pages 245

61. Zhou, Q., Grinspun, E., Zorin, D., Jacobson, A.: Mesh arrangements for solid geometry. ACM Trans. Graph. **35**(4), 39:1–39:15 (2016). DOI 10.1145/2897824.2925901. URL http://doi.acm.org/10.1145/2897824.2925901 pages 245

Index

Affine space
  affine subspaces
    may not have common elements, 123
  affinely independent
    four points non coplanar, 123
    three points non collinear, 123
    two points non coincident, 123
  associate each point
    tuple of coordinates, 123
  operation
    affine action, 122
    difference, 122
    displacement, 122
  Point space
    difference of points, 123
    displacement, 123
    external operation, 123
    point plus vector, 123
Affine transformation
  Coordinates
    Homogeneous, 127
    one added, 127
  equation
    linear, 132
    Matrices, 132
  function
    bijective, 126
  invertible
    functions, 126
  Matrix
    Scaling, 132
  Plasm
    HOMO, 128
    multidimensional, 127
    multidimensional matrix, 126
    Scaling, 132
    user interface, 126

square matrices
  invertible, 126
Tensor
  applied to models, 128
Translation
  in homogeneous coordinates, 136
Algebra
  Basic building block
    atoms, 277
    generators, 277
  Boolean
    formula, 253
    sum of minterms, 284
  CSG
    formula, 253
    traversal, 253
  Definition
    atom, 250
    binary string, 251
    Boolean, 250
    elementary solid , 251
    generator, 250
    Semialgebraic set, 251
    Structure, 251
  expression
    logical function, 283
    solid formula, 283
  finite
    algebra, 250
  formula
    logical function, 255
  generators
    Structure, 277
  isomorphism
    arrangement, 252
    power set, 251
  mathematical logic

binary terms, 250
operator
  DIFFERENCE, 254
  INTERSECTION, 254
  UNION, 254
  XOR, 254
Solid
  Boolean, 251
STRUCT
  alias, 255
Arrangement
  chain complex
    linear spaces and maps, 224
  isomorphism
    Boolean algebra, 215
  pipeline
    d-polyhedra, 215
    2-chains, 225
    atoms, 225
    chains and cochains, 215
    coboundary, 224
    computation, 224
    FC,FE,EV, 224
    geometric, 215
    matrices, 224
    piecewise-linear geometry, 215
    space partition, 226
    subdivion, 224
    TGW, 224
    Topological Gift Wrapping, 224
  Space
    partition, 221
Array
  alias
    Matrix, 15
    Vector, 15
  values
    isbits, 15
    isboxed, 15
Atoms
  Brep
    enumeration, 217

Backus'
  Turing Award
    1977, 36
    FP, 36
Bicubic
  Bezier
    surface, 208
  Hermite
    surface, 208
Boolean
  binary

structure, 284
expression
  boundary evaluation, 284
logic
  minterms, 282
structure
  of solution, 284
Boundary
  3D printing
    technology, 218
  Brep
    solid objects, 217
  models
    3D printing, 218
    Breps, 217
    library, 217
    solid objects, 217

cellular complexes
  linear spaces
    bases, 91
    unique coordinates, 91
Chain complex
  coordinate
    representation, 93
  definition
    p-chain, 93, 95
    basis, 93
    Chain bases, 93, 95
    Chain coordinates, 95
    direct sum of vector subspaces, 94
    graded map, 92, 95
    graded vector space, 92, 94
    Linear p-space, 95
    linear space, 93
    linear(vector) space, 93
    sequence of linear maps, 96
    sequence of vector spaces, 96
    space partition, 92
    subcomplex, 95
    Unit p-chain, 95
    unit p-chain, 93
  graded map
    of degree k, 95
  graded vector space
    direct sum of subspaces, 95
    indexed by integers, 95
CuArrays
  facilitate
    GPU computing, 31
CUDA
  CuArrays
    high level, 31

DArray
  distributed
    set of workers, 28
Dictionary
  creation
    by generators, 13
    by passing key=>value, 13
    by passing pairs, 13
  reading
    like arrays, 13
  Standard
    keys, 13
    values, 13
Distributed
  DArray
    distributed parallel processing, 27
  Distributed arrays, 27
  multiple processes
    different abstractions, 27
    separate memory spaces, 27
Distributed programming
  primitives
    remote calls, 27
    remote references, 27
  remote references
    Future, 27
    RemoteChannel, 27
DistributedArrays
  common arrays
    zeros, ones, rand, etc, 28
  DArray
    global interface, 27
  dist
    how many chunks, 28
  init
    initialize indices for memory chunks,
      28
    tuple of index ranges, 28
  vector
    dist, 28

FL
  AA
    ApplyToAll, 42
  Conditional combinator
    if, 44
  CONS
    vector function, 42
  expression
    value, 40
  expressions
    function-valued, 41
  function
    combinator, 40

function-level
  ew functions by composition, 41
functional programming
  at Function Level, 36
functions
  application to data, 40
INSR,INSL
  combinators, 44
Julia
  application, 38
  geometry, 38
  programsequivfunctions, 38
  sequences, 38
  unary, 38
model generation
  explicit parameterization, 40
  implicit parameters, 40
parametric
  geometric model, 40
Plasm
  geometry, 40
  subset, 40
primitive objects
  characters, 41
  Hpc, Lar types (Plasm), 41
  numbers, 41
  truth values, 41
semantics of FP
  algebra of programs, 36
syntax
  Julia, 37, 40
  minimal, 38
  Plasm geometry, 37
  straightforward, 38
Folds
  interface
    distributed, 25
    sequential, 25
    threaded, 25
Functional programming style
  function
    assigned to variable name, 9
  higher order
    ex, 10
    filter, 10
    foldl, 11
    map, 10
    native injulia, 10
    reduce, 10
    reductions, 10
    standard syntax in Plasm, 10
  higher-order
    partial applications, 10
  lambda functions

curried application, 10
  higher-order, 10
lambda syntax
  assigned to named variable, 9
  stabby lambd, 9
  tuple of arguments, 9
  value-generating expression, 9
last value returned
  nested functions, 9
named functions
  may return a function, 9
  return value, 9
nested functions
  curried, 10
Functions
  arguments
    actual, 7
    formal, 7
  function call
    argument list, 8
    arguments, 8
  keyword-optional arguments
    after ;, 8
    any number, 8
  maps
    arguments, 7
    impure, 7
    non mathematical, 7
  return
    multiple values, 7
    tuple, 7
    value, 7
  Statement functions
    no function keyword, 7
    no return keyword, 7
  String
    interpolation, 8

Geometric space
  affine
    coords sum to one, 124, 125
  convex
    both constraints, 124, 125
  positive
    coordinates, 124, 125
    point, 124, 125
Geometry
  algebra
    minterm, 66
    sum of minterms, 66
  Cell
    connected, 63
    linear, 63
    possibly non contractible, 63

  possibly nonconvex, 63
cell
  boundary, 64
  closure, 64
  interior, 64
Cell complex
  finite, 64
  regular, 64
decompositive
  object partition, 65
enumerative
  space partition, 65
finite
  algebra, 66
graded set
  cells,chains, 63
operator
  GRID, 65
  PROD, 65
representation
  boundary, 64
  decompositive, 64
  enumerative, 64
Skeleton
  regular, 64
  subcomplex, 64
solid
  objects, 66
solid object
  Characteristic function, 66
GPU
  CuArrays macros
    convenient programming, 32
  sync
    sync, 31
  focus
    when performance matters, 29
  focus:hardware-specific kernels, 29
GPUArrays
  AbstractArray
    interface, 29
  interface
    for all GPU platforms, 29

Homology
  chain complex
    global, 224
    local, 224
  computation
    arrays of arrays, 227
    atoms boundaries, 225
    cycles reconstruction, 224
    expression, 224
    space maps, 224

types of voids, 224
spaces
  boundaries, 221
  cycles, 221

Integrals
  boundary
    triangulation, 106
  finite integration
    using volume elements, 103
  Mechanical properties
    structure products, 107
  Plasm
    monomial integrals, 104
  Plasm function
    M, 107
  product of structure
    surface->curve, 103
    volume->surface, 103
  solid structures
    center of gravity, 103
    moments of inertia, 103
    volume, 103
  surface
    triangulation, 104
  test
    affine transform, 106
    cube, 106
    rotation, 107
    tetrahedron, 106

Julia
  Arrays
    mutable, 6
  collections
    Tuples, 6
  combinator
    cat,hcat,vcat, 45
  concurrency
    Asynchronous tasks, 22
    GPU, 22
    Multi-threading, 22
    schedule, 23
    task x), 23
    Task, 23
    task machinery, 23
    spawn, 23
  Coroutines
    resume, 22
    suspend, 22
  data parallelism
    map, 24
    reduce, 24
  Dictionaries

key is unique, 6
  maps to a specific value, 6
  store and retrieve data efficiently, 6
Distributed
  pmap, 25
ecosystem
  packages, 33
functional programming
  combine styles, 1
Functions
  building block, 6
  collections, 6, 14
  data analysis, 6
  first-class, 6
  machine learning, 6
  passed as arguments, 6
  returned by functions, 6
GPU
  compiler, 30
  computing, 28
  CuArrays macros, 32
  data processing, 22
  first-class support, 30
  machine learning, 22
  Nvidia, AMD, Intel, etc, 28
  performance, 30
  Programming, 28
  programming, 30
  scientific simulations, 22
  vendor-neutral, 28
I/O
  coroutines, 24
iterable types
  AbstractString, 3
  Array, 3
  Dict, 3
  Range, 3
  Set, 3
iterator comprehension
  filtering, 25
  flattening, 25
  mapping, 25
map
  AA in Plasm, 42
multiprocessing
  message passing, 27
  separate memory chunks, 27
Multithreading
  ccall, 26
  composable, 26
  External libraries, 26
  limitations, 26
  loop parallelism, 26
  multi-core, 26

multiple threads, 26
parallel loops, 26
sCPU core, 26
sharing memory, 26
speed up, 26
threadid(), 26
Threads.nthreads, 26
warnings, 26
multithreading
available cores, 22
coroutines, 22
I/O-bound tasks, 22
new types
properties, 4
records, 4
structs, 4
operator
splat, 56
package
CuArrays, 22, 31
CUDA.jl, 30
Distributed, 27
DistributedArrays, 27
OpenCL.jl, 30
parallel
computed concurrently, 21
decomposition, 21
models, 22
multitasking, 21
optimal, 21
task granularity, 21
parallelism
concurrency, 21
efficient, 22
preemptive thread-based multitasking, 21
tasks, 21
thread count, 21
performance
packages, 2
Plasm
advanced geometry, 35
Booleans, 277
CNR, 35
Euler number, 46
functional programming, 36
geometric programming, 35
geometric types, 50
geometry to FL, 36
module, 32
package, 40
programming, 53
syntax, 36
primitive type

Bool, 5
Char32, 5
Float64, 5
Int32, 5
Int64, 5
productivity
ecosystem, 2
programming
CUDA.jl, 30
Programming:Julia native, 28
Programming:JuliaGPU, 28
Sets
set operations, 6
union, intersection, and difference, 6
unique elements, 6
unordered collection, 6
Sparse Arrays
CSC, 20
supertype
Any, 5
Tasks
switches in any order, 24
switching, 24
threads, 24
Tuples
single block of memory, 6
unmutable, 6
Type
Abstract, 4
final, 4
Hierarchical, 4
isbits, 4
parameterized, 4
subtype, 4
value, 4
Types
documentation, 5
method dispatch, 5
optimizations, 5
variables, 5
worker pool
f, 25
map, 25
pmap, 25
JuliaGPU
CUDA
Nvidia platform, 29
GPUArray
speed-up, 29
OpenCL
Portability, 29
vendor-specific
AMDGPU, 29
CUDA, 29

oneAPI, 29

Linear Algebra
  adjoint
    transpose, 18
  inner product
    dot, 18
  statistic
    rms, 18
Linear space
  Basis
    change, 119
    minimum set of generators, 119
    ordered set of vectors, 119
  Bernstein polynomials
    function graphs, 120
    sampling, 121
    various degree, 120, 121
  Change of basis
    from old basis, 122
    to new basis, 122
  Linear independence
    linear combination, 116
    non-zero solutions, 117
  Matrices
    Addition, 118
    Multiplication, 118
    Subtraction, 118
    vector spaces, 117
  operation
    addition, 116
    product by a scalar, 116
  Span
    generated by a subset of vectors, 118
    Generators, 118
    smallest subset, 118
  Subspace
    Bases, 118
    Coordinates, 118
    dimension, 118
    Generators, 118
    intersection of subspaces, 118
    Span, 118
    vectors, 118
  vectors
    scalars, 116
Linear topology
  piecewise
    partition, 215

Matrix
  arithmetics
    addition, subtraction, multiplication,
      and division, 15

block matrices
  consistent size, 17
expressions
  MatLab, 16
function
  size, 17
isbits
  Complex numbers, 14
  numeric types, 14
linear algebra
  determinant, 17
  eigenvalues, 17
  eigenvectors, 17
  operations, 17
  trace, 17
operations, 16
  syntax, 17
parameterized
  Number, 15
scientific computing, linear algebra, 15
Type
  complex, 17
  definition, 16
values
  range, 16
Vector
  stacked, 16
multiparadigm
  expressiveness
    expressiveness, 1
Multiprocessing
  distributed-memory
    parallel computing, 27

Pipeline
  atoms
    computation, 228
  Chain Complex
    Merge, 231
  congruence topology
    reduction, 231
  flatten
    hierarchical input, 229
  incremental construction
    C_d basis, 231
  marshaling matrices
    block diagonal, 231
  Splitting
    1-chains, 230
    boundary compatibility conditions,
      230
    pairwise 1D intersections , 230
  STRUCT
    convert to global coords, 229

input, 229
TGW
  sparse matrices, 231
three-dimensional
  chain complex, 227
topological gift wrapping
  algorithm, 231
PLASM
  combinator
    CAT, 45
PLaSM
  project
    CNR, 37
Plasm
  Affine
    General Rotation, 131
    rotation, 130
  Algebra
    Generalized CSG, 49
    sum of minterms, 49
  arrangement
    generators, 278
    pipeline, 215
  Arrangements
    spaces, 215
  assembly
    graph, 147
  atoms
    decomposition, 217
  BOOL
    Breps, 261
  Boolean
    BOOL, 283
    INNERS, 282
    methodology, 277
    OUTERS, 283
    semantics, 279
    structure, 283
  Boolean function
    FL style, 283
  boundary
    extraction, 85
    operator matrix, 285
  Brep
    extraction, 285
  building
    framework, 275
  cell complex
    combinatorics, 239
  Cellular
    boundary, 216
  Cellular complex
    non manifold, 144
  cellular models

Chain complex, 216
  topology, 216
Chains
  (co)boundary matrix, 261
  atoms, 256
  basis matrices, 257
  basis of cycles, 260
  basis of subspace, 257
  cochains, 257
  coordinate matrices, 264
  coordinate representation, 262
  coordinates, 261
  cycles vs boundaries, 261
  linearity, 261
  operations, 256
  operator matrices, 261
  orientation, 264
  oriented boundaries, 264
  sparse matrices, 256
  subspace, 256, 257
combinators
  DISTR,DISTL, 46
Conditional combinator
  IF, 44
constant function
  K, 43
CSG
  tree, 239
curve
  manifolds, 246
curved manifolds
  curves, 159
  manifolds, 159
  surfaces, 159
Curved objects
  polynomials, 237
  transfinite, 237
curves
  NURBS, 199
  Quartic, 197
  RATIONALBEZIER, 199
dataset
  DAG, 216
  Directed Acyclic Graph, 216
design
  1-complex, 175
  ABOVE, 180
  aggregation, 179
  ALIGN, 181
  alignment, 181
  assemblies, 178
  CONVEXHULL, 175
  CUBOID, 175
  DOWN, 180

floor-slab, 178
ICOSPHERE, 176
LEFT, 180
MAX, 181
MIN, 181
pillars, 178
QUOTE, 175
RIGHT, 180
SIZE, 175
skeleton, 179
SPHERE, 176
STRUCT, 179
structure, 178
sub-assemblies, 178
tools, 174
TOP, 180
trusses, 178
UP, 180
expressions
  Boolean, 49
  geometric, 49
  parametric, 49
fenvs.jl
  many tools, 48
first coordinate
  homogenous, 127
flat scheme
  chain complex, 216
Function
  order, 174
function
  applied, 38
  higher-level, 38
generators
  CIRCLE, 165, 167
  CIRCUMFERENCE, 167
  CONE, 165
  CROSSPOLYTOPE, 163
  CUBE, 133
  CUBOID, 163, 175
  CYLINDER, 169
  FINITECONE, 165
  HELICOID, 172
  Higher order, 171
  ICOSAHEDRON(), 176
  library, 174
  NGON, 167
  OCTAHEDRON, 133, 163
  parametric, 163
  Platonic solid, 176
  profile, 174
  solid, 163
  SOLIDHELICOID, 172
  solids, 163

SQUARE, 133
structure, 278
TORUS, 171
TRUNCONE, 165
TURBOPUMP, 174
unit circle, 168
Geometric models
  BIM, 115
  CAD, 115
  geometric types, 115
  parametric functions, 115
Geometry
  flat, 216
graphics
  SVG, 248
Hierarchical
  representation scheme, 216
hierarchical cellular schemes
  convex cell, 216
Hierarchical Polyhedral Complex
  Hpc, 216
higher order
  INSL, 11
  INSR, 11
homogeneous coordinate
  Position, 216
Homogeneous coordinates
  linear, 128
  normalized, 128
Homology
  algorithm, 260
  boolop, 256
  boundary, 256, 257, 261
  chain, 257
  cycle, 257
  diagram, 261
  equivalence, 256
  implementation, 256
  representation, 257
  shell, 260
Hpc
  complex, 86
Identity function
  ID, 43
IF
  higher-order function, 44
incidence
  atoms and generators, 278
input
  SVG, 248
Integration
  finite formulas, 101
  structure products, 101
Julia

application, 38
  mostly unary, 38
  programs*equiv*functions, 38
  symbolic modeling, 159
Lar
  subfolder, 49
Linear Algebraic Representation
  Lar, 216
mapping
  coordinate functions, 160
modules
  fenvs, 48
  geometry kernel, 47
  main, 48
  Plasm.jl, 48
  viever.jl, 50
  visualizers, 47
operator
  ABOVE, 180
  ARRANGE, 270, 272
  ARRANGE2D, 228
  ARRANGE3D, 281
  BERNSTEINBASIS, 207
  BEZIERSURFACE, 208
  BILINEARSURFACE, 207
  BOOL, 283
  BOUNDARY, 273
  Boundary, 285
  CONVEXHULL, 166
  CUBOID, 175
  CUBOIDGRID, 213, 245
  DIFFERENCE, 281
  DOWN, 180
  EXTRUDE, 212
  GR, 132
  GRID, 213, 273
  HERMITESURFACE, 208
  HOMO, 128, 132
  INTERSECTION, 281
  INTERVAL, 190
  INTERVALS, 191
  INTSTO, 130
  JOIN, 77, 165
  LEFT, 180
  MAP, 57, 161, 169, 196
  MAX, 181
  MIN, 181
  MINKOWSKI, 212
  MKPOL, 55, 166
  OFFSET, 212, 247
  OUTER, 273, 285
  OUTERS, 283
  Q, 175
  QUOTE, 175

RATIONALBEZIER, 199
RIGHT, 180
SEL, 192
SIMPLEXGRID, 213, 245
SK, 273
SKELETON, 273
SOLIDHELICOID, 239
STRUCT, 148, 277
SWEEP, 212
Tensor product, 206
TENSORPRODSURFACE, 207
THINSOLID, 172, 205, 239
TOP, 180
UKPOL, 166
UNION, 273, 281, 285
UP, 180
VIEWCOMPLEX, 57, 93, 170
XOR, 281
output
  SVG, 248
Parametric modeling
  mapping of complexes, 159
  parametric functions, 159
  topological transformations, 159
Pipeline
  three-dimensional, 227
polygon
  cellular complex, 211
polyline
  cellular complex, 211
porting
  Julia, 37
  Python, 37
  Scheme+C++, 37
ports
  native Common Lisp, 47
  Python and C++, 47
  Scheme and C++, 47
primitive
  INTERVALS, 166
primitives
  flexibility, 160
  geoid, 160
  hollow sphere, 162
  sphere, 160
  spherical coordinates, 160
  tensors, 160
product
  *, 56
  PROD, 56
Properties
  attach properties to geometry, 156
  dictionary, 156
Representation

affine geometry, 218
  algebraic topology, 218
  computer programs, 218
  definition, 218
  functional languages, 218
  scheme, 218–220
  types, 219
  values, 219
representation
  manifold, 88
  nonmanifold, 88
representation scheme
  hybrid, 216
scheme
  Boolean, 237
  Brep, 237
  buildings, 239
  CSG, 239
  decompositive, 237
  mapping grids, 238
  non-hierarchical, 238
  potree, 238
  product, 240
schemes
  Boundary, 245
  Composite, 248
  Decompositive, 245
  Enumerative, 244
  enumerative, 245
  hierarchical, 245
  manifolds, 246
  Minkowsky, 247
  partitioning, 245
  Primitive instancing, 244
  primitive instancing, 248
  Procedural, 244
  set operations, 247
  solid, 244
SIMPLEX
  primitive, 79
solid
  BOUNDARY, 211
  Hpc, 212
  manifolds, 246
  SOLIDIFY, 211
solid algebra
  finite algebra, 256
Solid Boolean Algebra
  atoms, 246
STRUCT
  Hpc, 148
  linearized in traversal, 150
  nested hierarchically in storage, 150
surface

  manifolds, 246
symbolic modeling
  applications, 159
  generators, 159
  geometric models, 159
  primitives, 159
  transformers, 159
Tensor
  Elementary rotations, 129
  equation, 131
  General Rotation, 131
  linear, 128
  matrix, 128
  notation, 130
  planar rotation, 128
  rotation, 130
  rotation angle, 128
Tensors
  dimension-independent, 136
TGW
  algorithm, 278
Topology
  foundation, 62
  integration, 101
triangulation
  coherently oriented, 88
Type
  Geometry, 216
Types
  Geometry, 50
  Hpc, 50
  Lar, 50
Product of complexes
  cell complex
    $dim = n + m$, 90
  operation
    associative, 91
    commutative, 91
    distributive, 91
    no parentheses, 91
  topology
    algebraic, 91
    combinatorial, 91
Program composition and application, 38
programming paradigms
  Julia
    multiparadigm, 1

regularized arrangement
  2D
    computation, 230
Representation
  scheme
    orthogonal geometries, 239

topology
  combinatorics, 239
Rpresentation
  schemes
    Requicha, 218

Set
  collection
    no duplicates, 14
    unordered, 14
  Datatype
    support union, intersection, and
      difference, 14
  mutable containers
    fast membership, 14
    insertion testing, 14
    unique values, 14
Simplex
  cochain
    volume, 84
  extraction
    faces, 84
    facets, 84
  extrusion
    algorithm, 86
    from d, 86
    to $d + 1$, 86
  facets
    extraction, 84
  Lar
    complex, 85
Simplicial complex
  boundary
    facets, 84
  boundary complex
    generation, 85
  Cube
    tetrahedrization, 86
  Delaunay
    triangulation, 86, 88
  dimension
    intrinsic, 79
  grids
    via extrusion, 88
  higher-dimensional
    boundary, 81
  Hpc
    convex hulls, 83
  Lar
    dataset, 80
  MKPOL(W,FW)
    VIEW, 88
  multidimensional
    facets, 84

operator
  extrusion, 87
order
  maximum simplex, 80
orientable
  coherently oriented, 83
orientations
  relative interior-exterior, 83
oriented facets
  remove one vertex, 84
Plasm
  coding, 85
  SIMPLEX, 80
  SIMPLEXFACETS, 89
simplex
  coding, 77
  multidimensional, 77
triangulation
  rules, 78
  well-formed, 78
Simplicial models
  affinely independent
    pointset, 77
  dimension
    embedding, 79
  faces
    combinatorics, 80
  Lar
    subcomplexes, 83
  Simplex
    convex set, 77
  triangulation
    coherently oriented, 88
    Hpc, 80
Solid
  building
    dwelling, 277
  export
    Brep, 286
    OBJ, 286
    PLY, 286
  expressions
    assembly tests, 267
    Boolean, 269
    Boolean result, 266
    combination, 269
    multiple boolops, 268
    quasi-disjoint union, 266
    solid, 264
    syntax, 269
    testing framework, 265
    unit cubes, 267
    unit tests, 266, 269
    user-defined, 269

isomorphism
  Boolean algebra, 264
object
  BOOL, 285
  Brep, 285
operations
  3-chain solution, 270
  no CSG-styls, 270
  unitary resolution, 270
Solid Modeling
  language
    for engineers, 243
  practical applications
    computational techniques, 243
  theoretical foundations
    constructive solid geometry, 243
    mathematical principles, 243
Solid models
  boundary
    Euler Characteristics , 138
  Chain
    subcomplexes, 139
  CSG
    binary tree, 145
  Decompositive scheme
    cellular complexes, 141
  hierarchical
    multidimensional, 141
  Instancing scheme
    Plasm library, 146
  Parametric scheme
    instancing, 146
  Plasm
    Boolean algebras, 146
    cellular complex, 139
    currying, 147
    decompositive scheme, 139
    grids, 143
    multidimensional, 141
    parametric, 143
    primitive functions, 146
    primitive solids, 146
    topological spaces, 143
  polyhedral complex
    Hpc, 139
    Lar, 139
  Representation Schemes
    Ari Requicha, 137
Space
  arrangement
    algebraic, 215
    Boolean algebra, 215
    pipeline, 215
Sparse Arrays

compressed sparse column
  CSC, 19
coordinate format
  non-zero element, 19
  storage by triples, 19
function
  I,J,V arrays, 21
  sparse(), 21
implementation
  CSC, 20
inverse sparse
  findnz, 21
inverse sparsevec
  findnz, 21
SparseMatrixCSC
  AbstractSparseMatrixCSCTv,Ti, 20
SparseMatrixCSCTv, Ti
  column pointers, 20
  row indices, 20
  stored values, 20
  storage by triples
  three input arrays, 19
Sparse Matrix
  large datasets
    improve efficiency, 19
    reduce memory usage, 19
Spline
  curves
    approximating, 198
    interpolating, 198
STRUCT
  contain
    affine transformations, 150
    Hpc objects, 150
    polyhedral complexes, 150
  input/output
    Hpc objects, 150

TGW algorithm
  input
    output, 231
Topology
  adjacency
    chains, 187
    sparse matrices, 186
  algebraic approach
    incidence, 186
  basic concepts
    ideas, 62
    methods, 62
  boundary
    external space, 62
    representation, 62
  cell

decomposition, 62
Chain complex
    boundary, 183
    coboundary, 183
chain complex
    space partitions, 61
Cochain
    integration, 102
Cochain complex
    graded vector space, 101
Cochains
    measure or evaluate chains, 103
cochains
    domain integral, 101
    dual space, 101
computing
    Boolean expression , 62
    solid shape, 62
duality
    chains, 102
    cochains, 102
incidence
    computation, 185
    CSC, 184
    data structures, 186
meshes
    graphics, 61
methods
    BIM, 61
    CAD, 61
operator
    boundary, 101
    coboundary, 101
Plasm
    Boolean method, 63

chain complexes, 63
space
    arrangement, 62
Transfinite
    surfaces
        Bezier, 203
        Coons patch, 203
        cubic, 203
Tuples
    immutable
        distinct values, 12
        memory block, 12
        type, 12
    tuple
        function, 12
        returns a tuple, 12

Variables
    LATEX names, 2
    Greek letters, 2
    initialized
        declaration, 2
    names
        undeclared, 2
Volume integration
    complexity
        O(#E), 104
    Plasm
        closed formulae, 104
        polyhedral volumes, 104
    properties
        centroid, 103
        inertia, 103
        volume, 103

The manufacturer's authorised representative in the EU is Springer
Nature Customer Service Centre GmbH, Europaplatz 3, 69115 Heidelberg,
Germany. If you have any concerns regarding our products, please
contact ProductSafety@springernature.com

Printed and bound by CPI Group (UK) Ltd, Croydon, CR0 4YY

23/04/2026

02095586-0003